S0-CAA-428

Solder Paste Technology

Principles and Applications

Solder Paste Technology

Principles and Applications

Colin C. Johnson
Joseph Kevra, Ph.D.

Notices

Microtrac®	Leeds and Northrup
Fluorinert®	3M
Multifluor®	Air Products and Chemicals, Inc.
Galden®	Montedison
Tacky Tester™	Austin American Technology
LaserGuide™	Creative Automation Company
Accu-Coat™ and **Vidalign**™	Aremco Products, Inc.
Avisas®	Affiliated Manufacturers, Inc.
Contaminometer®	Protonique, S.A.
Freon®	E.I. du Pont de Nemours & Co.

FIRST EDITION/FIRST PRINTING

Library of Congress Cataloging-in-Publication Data

Johnson, Colin C.
Solder paste technology : principles and applications / by Colin C. Johnson and Joseph Kevra.
p. cm.
Includes index.
ISBN 0-8206-3203-4
1. Solder pastes. I. Kevra, Joseph. II. Title.
TT267.J68 1989
671.5′6—dc20 89-31855
CIP

TAB BOOKS Inc. offers software for sale. For information and a catalog, please contact TAB Software Department, Blue Ridge Summit, PA 17294-0850.

Questions regarding the content of this book should be addressed to:

Reader Inquiry Branch
TAB BOOKS Inc.
Blue Ridge Summit, PA 17294-0850

Acquisitions Editor: Larry Hager
Technical Editor: Lisa A. Doyle
Production: Katherine Brown
Book Design: Jaclyn B. Saunders

To Val and Sue

For all the lost weekends

Contents

Foreword

The source of the very first solder paste or solder "paint" has been lost in history. However, plumbers and builders had been using these materials for some time before the electronics industry even existed. So what's so new about this product to warrant a scientific book all its own?

In conjunction with today's fast-growing surface-mount technology, solder pastes offer some unique advantages. Pastes are the preferred method of pre-placing flux and solder into tiny electronic joints because of their ease of dispensing, favorable tackiness, fast solderability, and other beneficial properties.

My personal involvement with solder paste began with IBM's exploration in hybrid circuit processes late in the 1950s. These experiments yielded the earliest true application of the surface-mount concept. Later developments produced the first electronic-grade thick-film solder at Alpha Metals (the company with which the authors and myself have been or are presently affiliated). Solder pastes were slow to find wide application in the rest of the electronic industry. However, today, solder paste has reached maturity along with the advances of surface-mount technology for the printed circuit industry.

Solder paste formulation and use requires a thorough knowledge of chemistry (fluxes), metallurgy (solder alloys), physical chemistry (rheology), and automated electronic manufacturing techniques. The authors of this book have all these qualifications and more. My association with Colin Johnson goes back more than 15 years when we worked together in Europe. At that time, he was already an established expert in metal-joining techniques associated with the semi-conductor and other electronics industries—techniques such as component construction, hybrid circuit assembly, and printed circuit soldering. His knowledge of solder preforms and other understanding of the industry are equally admirable.

Joseph Kevra provides the scientific touch to this book. His Ph.D. in physical chemistry makes him uniquely qualified to oversee the research and development of paste, a position he has held for 10 years at Alpha Metals. Yet Joe is not just a theoretical chemist—he has years of experience in supervising the production of chemical materials. Thus, he is intimately involved in the development of processes and quality systems.

Howard H. Manko
International Consultant on
Solder and Cleaning Technology

Preface

Solder pastes, sometimes referred to as solder creams, have been used in the electronic industry for more than 25 years. This fact might come as a surprise to manufacturing and other engineers who are now wrestling with the problems associated with entry into the new surface-mount technology presently used in printed wiring board assembly. Many of these engineers are wondering why their employers hired design staff to change the old soldering process from one that was almost routine to an alternative method that often involves radical re-education of production personnel.

It has been less than ten years since the publication of the second edition of the first comprehensive handbook on soldering by Manko (*Solders and Soldering*, McGraw-Hill Book Co., 1979). However, the subject of solder paste occupies little more than a half page in that book. The intention of our work is to provide what we believe is sadly lacking in the electronics industry today: a handbook on solder pastes. We had originally intended to concentrate only on the basic product itself, but when we began to define subjects for the different sections of the book, we soon realized that to achieve our ultimate goal, we had to cover every conceivable aspect of solder paste, both as a material and all its applications.

Our goal is to provide a reference manual that can be used by all those involved in solder paste and its usage, be it research chemists, production and design engineers, sales personnel, or buyers. For too long, solder paste has been shrouded in a fog of mystery.

We would like to thank all our colleagues at Alpha Metals for their support—in particular to Denis Mohoric, Bob Amabile, and Ron Bulwith. We would also like to thank Ed Belino for preparing samples and specimens, Paul Lotosky for his valuable word-processing advice, and Joan Callander for her patience and help.

Introduction

There are many ways to create a sound metallurgical bond in electronics applications by use of solder. In mass-production, the techniques of wave, drag, or dip soldering are widely employed.

When quality joints must be created that require accurate placement of a predetermined quantity of metal, then the number of application methods available is much more limited. Those that find most favor at the present time involve the use of solder preforms and solder pastes. Hand soldering, which requires a hand-held soldering iron and a wire solder alloy with one or more inner cores of flux, is not regarded by many as a reliable method of soldering for modern electronics. The subject appears in this publication only in the later chapters when discussing touch-up and repair work.

Solder preforms are basically stampings from foil or ribbon of different alloys with varying shapes, sizes, and thicknesses, often available with a flux core. Discs, washers, rectangles, and squares are the most widely used shapes, and these preforms together with spheres play an important role in electronic soldering today, particularly in the semiconductor industry.

Although some of the larger users of preforms have devised means of automatically dispensing or loading these onto, for example, a printed wiring board, the handling of such parts can create severe problems, particularly in light of the fact that some of the spheres being placed have a diameter of 0.005 inch (0.127 mm) or less.

At one time, the designations *solder paste* and *solder cream* were used to differentiate the quality of the product. ''Cream'' was associated with a high-quality product designed for electronic components, while the term ''paste'' was reserved for those products used in non-electronic applications such as soldering automobile radiators, etc. A paste or cream is a blend of a powdered solder alloy and a flux vehicle. The term "solder paste" is now generally accepted as standard terminology in the industry. In this book, only the term "paste" is used to describe the product, irrespective of its application.

The similarity between a preform and a paste ends here. Most preforms can be provided with an inner core of flux, as already described. Although other preforms are frequently coated with flux, they are still solid in form and can be handled as individual parts fairly easily. A paste, on the other hand, is a solder alloy in the form of a fine powder that is suspended in a flux-containing medium. This medium can range in consistency from a fairly thin liquid to a paste-like material similar in appearance to a heavy grease. Pastes normally contain several ingredients other than flux, as this book explains. These other components provide the final product with different characteristics to suit particular application needs. Whatever modifications to the formulation are made, paste is normally fairly wet and sticky (after blending the powder and flux) and incapable of being handled in the same way as a preform. Thus, the application techniques are more complex and the product selection is much more critical.

It has long been recognized that solder paste provides a unique method of soldering a wide variety of assemblies. Manko states that solder was probably used some 5,000 years ago; whether it was employed then in the form of paste is unknown. In modern times, it was probably the automotive industry that first adopted solder as a production material for body filling and soldering radiators and gas tank necks, as well as in the plumbing trade where it was used for joining lead, steel, and later, copper pipes. Initial deposition methods were crude, and the quality of the ingredients used in manufacturing the paste would be completely unacceptable by today's electronic industry standards.

Later, in the 1960s, solder paste began to replace dip soldering in the hybrid circuit or thick-film industry, and more diverse and sophisticated methods were devised for heating the solder paste to melt the alloy.

As the electronics industry grew, solder paste became increasingly important. Recent years have seen a exponential growth in its use with the advent of surface-mount technology. Solder paste is now a so-called "high-tech" product. Increasing demands are constantly being made on performance requirements. Solder paste is a complex substance, and the user must be knowledgeable with regard to all of the parameters and test methods usable to predict paste behavior. Paste will certainly remain a major metallurgical joining method for some time to come, and we hope that this book will be helpful to veteran users of solder paste as well as those who might be considering its use for the first time.

This book addresses all areas of solder paste. A casual observer might be quite surprised at the many diverse fields that are affected by the study of solder paste technology. The first part of the book answers the question "What is solder paste?" Chapter 1 reviews the importance of solder powder particle size and shape in addition to alloy composition. Chapter 2 considers the different flux classifications. Chapter 3 discusses formulations more comprehensively, while Chapter 4 is devoted to rheology and viscosity measurement. Methods of paste deposition, reflow, and residue cleaning are considered in Chapters 5 through 7. Chapter 8 reviews criteria for the selection of solder paste and the various processes for which it can be employed, while Chapter 9 attempts to deal with all the problems, great and small, that the user of solder paste can expect to encounter. Finally, Chapter 10 is devoted to a detailed discussion of the test methods in current use in the industry.

1

Powder Processing and Classification

There are a number of publications dealing with the subject of metal powders, but it is not the intention here to discuss the technology of these powders in great depth. A basic understanding of the methods by which powder is manufactured is, however, useful to anyone involved with solder paste. The two major processes employed frequently produce two slightly different forms of powder in terms of particle shape, in particular; and as a user of the final product, the paste, you should be aware of what your paste supplier has to do to ensure that the powder you are incorporating into your product is as good as it can be. The quality of the powder really does play a vital role in the performance of the paste, and that is the focus of this first chapter.

Many articles have been written on solder alloys, and every handbook covering electronics manufacturing processes, especially printed wiring board assembly, contains one or more sections devoted to bonding by means of solders. No work on solder paste would be complete without a discussion of the alloys that can be used, especially as there are many that can be, and already are, supplied as powder in solder pastes, which would simply not be appropriate for machine- or hand-soldering. There are even one or two alloys that have been ''discovered'' exclusively in connection with a paste application.

The portion of this chapter dedicated to alloys is also further sub-divided into sections dealing with specific elements, such as indium, bismuth, and antimony, which are often the subject of questions, as every solder paste vendor knows. All three, and others, have their specific uses and advantages and drawbacks, which are covered in some detail. The purity of solders is likewise addressed.

The physical characteristics of the powder are next considered, in particular, shape, size, and surface condition, which are all very important in their own way in affecting the performance of the paste in deposition and reflow.

Finally, some attention is paid to the handling of powder, which, depending on the alloy and particle size distribution, can be critical.

POWDER PROCESSING METHODS

Powder metallurgy is a complex subject, and several methods are employed to produce powder, which can be loosely described as either chemical or mechanical. Among the many industrial uses for metal in the form of powder are (1) the manufacture of iron and steel machine components of high strength fabricated from sintered compacted powder, (2) metal bearings sufficiently porous to enable them to be impregnated with lubricant, and (3) refractory metals, such as tungsten and molybdenum, which are used in the aerospace industry. In the solder industry (what this book is concerned with), all such processes are kept highly secret, and the occasional paste vendor might even be reluctant to reveal whether or not his or her product is being manufactured from powder produced in-house or supplied by an outside source! In any event, the actual processing method is usually not revealed.

Chemical Processing

Processing is normally undertaken by either *precipitation* or *reduction* techniques. In the former, metal solutions of the type similar to those used in electroplating can be subjected to a electric current, and metallic ions from the solution are deposited onto a cathode. The coating produced is generally of a brittle nature, which can be removed and broken up to produce powder particles.

In a reduction process, a liquid or solid agent, or a gas such as hydrogen is used in conjunction with a metallic compound in oxide form. The hydrogen reacts with the oxide to form pure metal and steam, i.e., lead oxide (PbO) + hydrogen (H_2) = lead (Pb) + water (H_20), the latter being continuously removed by replenishment of the hydrogen flow. In a liquid or solid process, the by-product of the metal oxide has to be separated and removed. The rather brittle original oxide compound can easily be mechanically milled into fairly fine particles prior to the reduction process, so the final metal particles can be reasonably closely controlled.

The major disadvantage of chemical processing is that powders cannot be produced in this way from alloys. It is important to remember that for solder pastes, which normally contain an alloy rather than a single metal component, the powders are produced from a pre-alloyed material and *not* from two or more separately cast elements. A process incorporating hydrogen would also not be practicable for any but the high-melting-point elements used in solders, as it does not have effective oxide reduction capabilities much below 300 degrees C or 572 degrees F.

Mechanical Processing

A major method for producing powder is milling, but this requires elements or alloys of sufficient brittleness (such as brass and bronze) to allow them to be worked in this fashion. Such a process would be feasible only for solders with similar characteristics, for example bismuth/tin, but this would almost certainly introduce additional, more serious problems with regard to deterioration of surface solderability.

The powder for solder alloys is invariably created by atomization of the metal. A number of different atomization methods are employed in the metal industry, including a vacuum process. With soft solders, however, air or gas atomization are the most common. In this case, the metal is broken up into particles direct from the melt (after being heated in a crucible made of a material such as graphite to its liquid state in air or, sometimes, depending upon the alloy being processed, under a blanket of inert gas, such as nitrogen or argon).

The liquid alloy is transferred from the crucible to a tundish or reservoir from which it passes as a stream through a nozzle. It is then broken up by jets of inert gas into droplets, which solidify into spheres (FIG. 1-1). In a similar concept, the molten metal falls upon a so-called metal spinning disk or cup (FIG. 1-2). The solder is again broken up into droplets as it is ejected from the edge of the disc or cup. The droplets cool in the air before falling into a collection chamber. The cooling can be accelerated by directing inert gas or even cool water or oil upon the droplets.

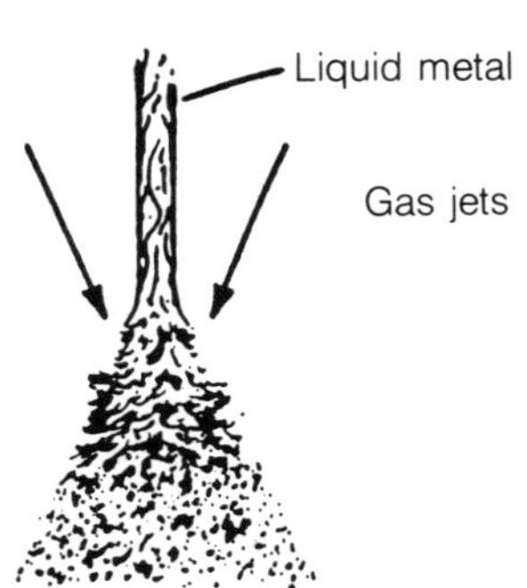

Fig. 1-1. Gas atomization. (American Society for Metals, Metals Handbook, 9th ed., vol. 7: Powder Metallurgy, *1984.)*

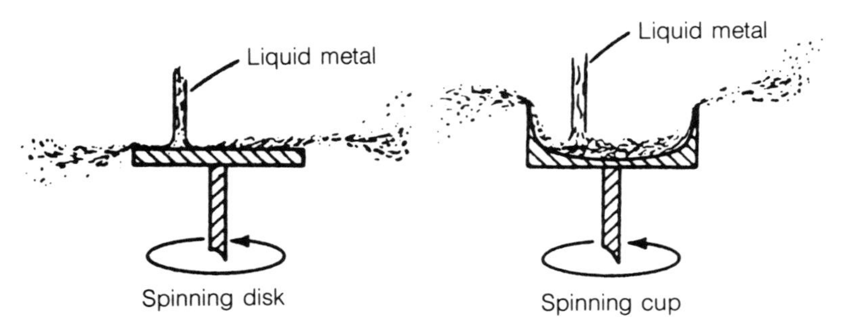

Fig. 1-2. Spinning disk and spinning cup. (American Society for Metals, Metals Handbook, 9th ed., vol. 7: Powder Metallurgy, *1984.)*

In either case, the process can be performed horizontally through a tunnel, or in the case of air atomization, vertically in a tower arrangement. The former is regarded as the most effective, because the atomization tunnel can be fairly easily extended, space permitting, to incorporate zones for in-line air classification of the particle sizes. The length or height of the atomization tank depends on whether or not solidification of the particles is to be achieved during their journey into the final deposition chamber or by more rapid means such as external cooling by air or water.

The powder can be collected by means of a purpose-made cyclone system fitted with a device for dust separation. In air atomization, the size, shape, and angle of the nozzle, the gas flow velocity and temperature, and the pouring rate for the molten solder play a significant role in determining the size and shape of the powder. In the spinning process, these characteristics are greatly influenced by the rotation speed and the shape of the edge of the disk or cup. In both cases, the process temperature is of vital importance so that premature solidification does not occur prior to formation of the particles. Equally vital is fast cooling of the particles after they have formed to prevent them from sticking together. This also has to be compatible with the need emphasized by Hirschhorn for sufficient cooling time for the creation of spherical shapes, so that natural surface tension forces have all the time they need to reduce the surface area of the particles to a minimum while they are still in a liquid, and therefore mobile, condition.[1]

Figure 1-3 is a schematic of typical water and gas atomization processes; however, the atomization of solder in water is not generally recommended because of the tendency of the metals to react with water or steam to form surface oxides.

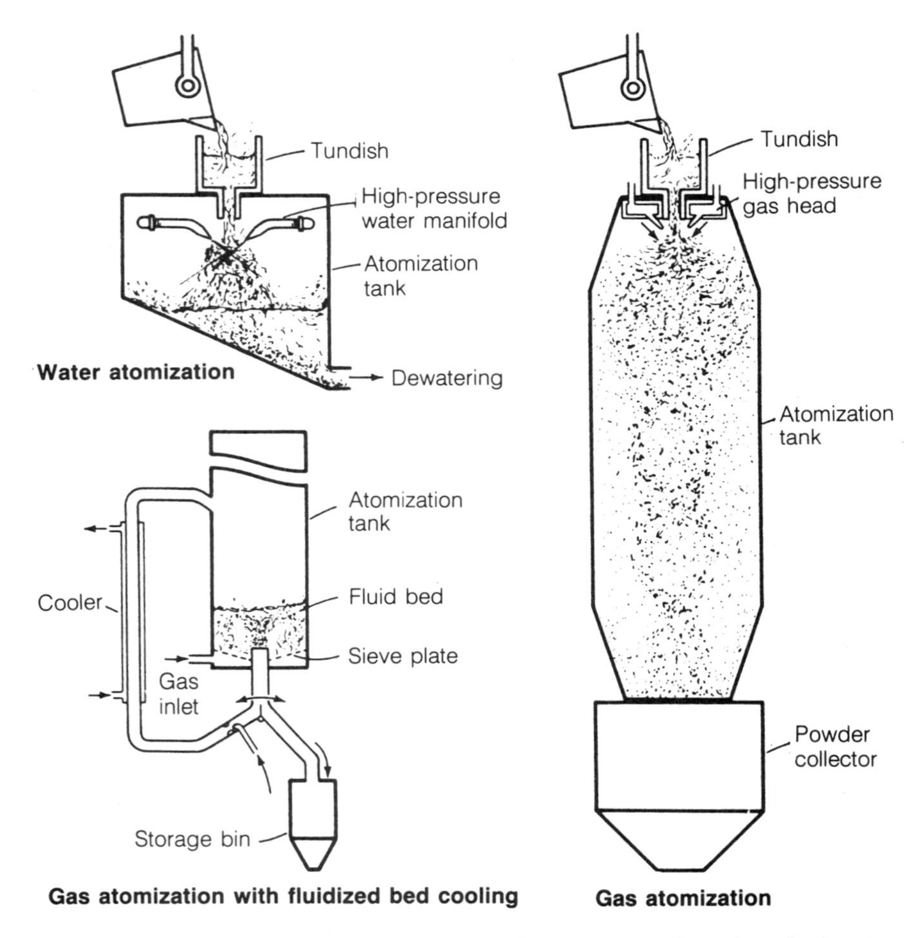

Fig. 1-3. Schematic of water and gas atomization processes. (American Society for Metals, Metals Handbook, 9th ed., vol. 7: Powder Metallurgy, *1984.)*

During the atomization process, a thin oxide film normally forms on the surface of the particles. However, this tends to have a protective effect by preventing further oxidation provided that, in the case of high-lead-containing alloys in particular (which oxidize rapidly), conditions do not exist that would promote additional oxidation. High humidity levels can cause severe reactions with the oxidized surfaces of such alloys, and powders containing higher than 95 percent lead should be stored either in dry nitrogen or kept frozen. It is by no means unusual for an alloy powder such as the 1 percent tin/97.5 percent lead/1.5 percent silver to turn completely black in a fairly short period of time in improper storage conditions.

Powder surfaces can rapidly degrade through abrasion caused either by friction of the individual particles or the effect on them of contact with a metal sieve during particle size separation or measurement. For sensitive alloys, size classification in air or atmosphere is definitely preferred. In this case, the powder is blown through a series of chambers. The larger particles of greatest weight fall first, and depending upon the number of zones available, the remaining particles are sorted by gravity into different sizes, until only very fine, unclassifiable powder remains. Both air flow rates and chamber sizes are very important parameters for successful air classification of powders.

ALLOYS

For the manufacture of solder paste used in electronic assembly, so-called "soft solder alloys" are used. These are normally defined as having a melting point below 450 degrees C (842 degrees F), but for the overwhelming majority of electronic bonding applications, the range is between 179 degrees C (354 degrees F) and 327 degrees C (620 degrees F)—the melting point of pure lead. The electrical conductivity of the most widely used alloys is only about 11 to 13 percent of that for copper, from which most circuitry in the printed wiring board industry is built. But the relatively large mass of solder used to form a reliable joint has a greatly increased current carrying capability and does not affect the electrical properties of the copper. In comparison with the higher-melting-point materials, such as those used for brazing, soft solders do have the disadvantage of relatively low mechanical strength, and *creep*, or movement of the solidified metal under stress, occurs at fairly low temperatures. On the other hand, soft solders are fairly inexpensive, and a further very important advantage is that their low melting points enable assemblies to be bonded that would not be able to tolerate the much higher brazing temperatures.

There are certain criteria to be observed when selecting an alloy for a solder paste:

- it should be easily processed into powder;
- it must form a metallurgical bond with the surfaces to be joined;
- the alloy should be of high purity;
- in its molten state, the alloy should have good flow properties and capillary action;
- in most solder paste applications, it should be able to wet the surfaces to be bonded by means of rosin-based fluxes;
- it must not soften or re-melt at the temperature at which the device will be operating;

- its melting point should be compatible with both components and substrate material;
- it should have as narrow a plastic or pasty range as possible (defined shortly);

The working temperature of the alloy needs to be approximately 25 to 60 degrees C (77 to 140 degrees F) above its melting point for good bonding. This can vary according to the alloy being employed, as well as reflow conditions. Remember that soldering relies on wetting or diffusion, not melting of the base metal for joining, which is why a solder with a melting point of 183 degrees C (361 degrees F) can be reliably used to bond to copper and nickel with melting points of 1083 degrees C (1981 degrees F) and 1455 degrees C (2651 degrees F), respectively.

None of the basic requirements listed differ from those applicable to other forms of soldering with the exception of the need for the alloy to be available in the form of powder.

The data shown in TABLES 1-1 and 1-2 is based on information from several sources. The tensile and shear strength values quoted for the different alloys can vary to some extent between reference books, technical papers, and paste manufacturers' literature, so they should be used as a guide only. Such properties greatly depend on the testing parameters used.

Table 1-1. Melting Temperatures of Tin-Lead Alloys

Alloy	Melting Range °C		Melting Range °F		Density	
	Liquidus	Solidus	Liquidus	Solidus	gm/cm^3	lb/in^3
100% Sn	232		440		7.28	0.2628
63Sn/37Pb	185	183	365	361	8.40	0.3032
60Sn/40Pb	189	183	372	361	8.50	0.3069
55Sn/45Pb	200	183	392	361	8.68	0.3134
50Sn/50Pb	216	183	421	361	8.87	0.3202
40Sn/60Pb	235	183	455	361	9.28	0.3350
30Sn/70Pb	260	183	500	361	9.72	0.3509
25Sn/75Pb	274	183	525	361	9.96	0.3596
20Sn/80Pb	280	183	536	361	10.21	0.3686
15Sn/85Pb	288	227	550	441	10.48	0.3783
10Sn/90Pb	302	268	576	514	10.75	0.3881
5Sn/90Pb	320	310	608	590	11.06	0.3993
3Sn/97Pb	322	314	612	597	11.22	0.4051
100% Pb	327		621		11.35	0.4098

Table 1-2. Mechanical Properties of Tin-Lead Alloys

Alloy	Tensile Strength lbf/in²	Tensile Strength N/mm²	Shear Strength lbf/in²	Shear Strength N/mm²
100% Sn	1800	14	2600	20
63Sn/37Pb	7700	60	5500	43
60Sn/40Pb	7600	59	5600	43
55Sn/45Pb	6400	50	5600	43
50Sn/50Pb	6000	47	5200	40
40Sn/60Pb	5400	42	4800	37
30Sn/70Pb	5000	39	4600	36
25Sn/75Pb	4800	37	4400	34
20Sn/80Pb	4800	37	4200	33
15Sn/85Pb	4600	36	4000	31
10Sn/90Pb	4400	34		
5Sn/90Pb	4000	31		
100% Pb	1800	14	1800	14

Binary Tin-Lead Alloy Systems

When discussing soft solders for electronics, one is apt to think automatically in terms of two-part binary tin (chemical symbol Sn, from the Latin *Stannum*)/lead (Pb, from *Plumbum*) alloys. As Langan rightly points out, these represent only a proportion of the solder alloys permitted by most United States and international standards covering the subject.[2] It is, however, certainly true that solder alloys used in the assembly of printed wiring boards are normally composed of tin and lead in ratios close to the eutectic point. The active constituent in most soft solders is tin, which promotes wetting of the majority of common metals; with many, the wetting is accompanied by the formation of layers of intermetallic compound at the solder/substrate interface. The presence in a metallographic section of a compound layer does not necessarily mean that good wetting of the surfaces has taken place, and the absence of a compound layer does not always indicate that wetting has not occurred. The solubilities in tin-lead solder of the common base metals such as copper and nickel are quite low, as is illustrated in FIG. 1-4. With nickel in particular, the very thin layers of intermetallic compounds formed at the interfaces during a normal soldering operation might not be visible even with the aid of an optical microscope.

Tin has a natural affinity with many other metals. Tin usually reacts with most of the metal surfaces to which it is being bonded, e.g. copper, to form intermetallic compounds that, although strong, are also brittle.

The *eutectic composition* is a distinct alloy that has a single melting or solidification point rather than a melting range. The melting point of the eutectic is always less than that of the individual parent metals. In the tin-lead phase diagram of FIG. 1-5, the eutectic point is marked as 61.9 percent tin and 38.1 percent lead, although the 63/37 is accepted in the electronic industry as the de facto eutectic.

Compositions other than the eutectic do not solidify immediately upon cooling; instead, they have a temperature range in which they are partly liquid and partly solid. This region, often called the *plastic* or *pasty* range, becomes

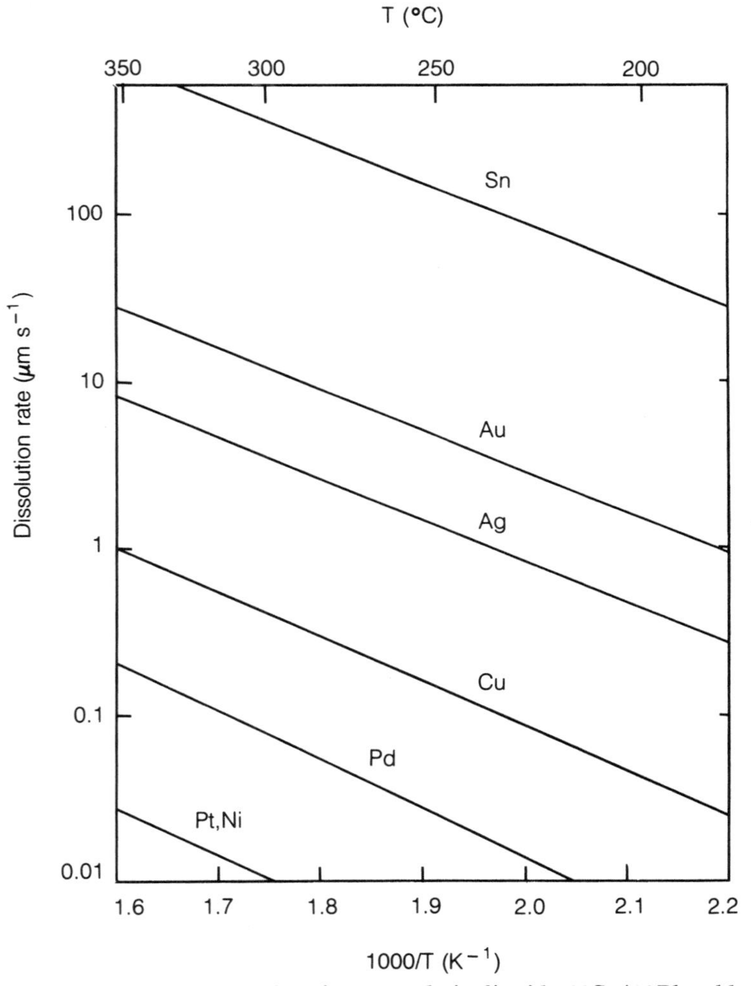

Fig. 1-4. Dissolution rates of various metals in liquid, 60Sn/40Pb solder. (R.J. Klein Wassink, Soldering in Electronics *(Ayr, Scotland: Electro-chemical Publications Limited, 1984.))*

more pronounced the farther one moves away from the eutectic. When selecting an alloy, always remember that in the plastic range, tin-lead solders have no strength, and even after solidification at up to 30 degrees C (86 degrees F) below the eutectic, they are considerably weaker than they are at room temperature. Binary tin-lead solders lose their strength as the temperature increases towards their melting point and are thus not considered suitable for stressing above 100 degrees C (212 degrees F). For these applications, tin-antimony, tin-silver, and high-lead-containing alloys incorporating small percentages of silver and tin are used. All lead-rich alloys with little or no tin, however, have markedly inferior wetting properties, and their high liquidus temperature can also present problems of oxidation of the base metal and rapid charring and decomposition of the flux during soldering.

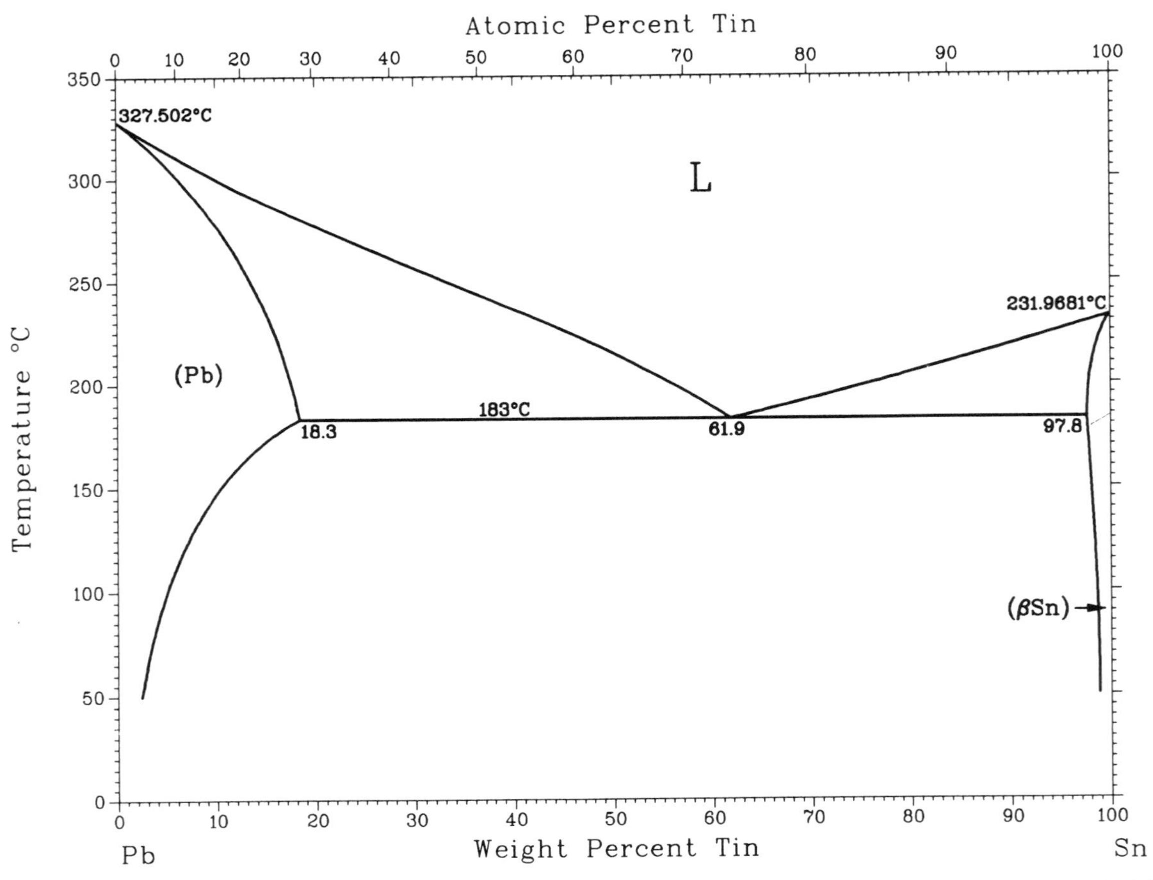

Fig. 1-5. Tin-lead phase diagram. (American Society for Metals, Metals Handbook, 8th ed., vol. 8, *1973.)*

Alloys at or near the eutectic are therefore the most desirable for soldering, including solder paste reflow. This is why the 63/37 and 60/40 percent ratios have been accepted in the electronic industry as the most reliable for the tin-lead alloy.

The 63/37 and 60/40 percent tin-lead solders, with their excellent mechanical, metallurgical, thermal, and electrical properties, meet most requirements for solder paste where it is to be used for attaching and bonding leaded and most leadless components to printed wiring boards. As well, it provides selective solder pads, or *bumps*, on the board for subsequent reflow and joining of similar parts. Both alloys can be processed into good quality powder with excellent surface solderability and good control of particle size and shape. There is little cost difference between the two alloys as far as actual metal costs are concerned, and the powder processing charge that would be included in the price for the solder paste would likewise be very similar.

However, because the 60-percent tin material is rarely used now in either surface-mount or hybrid-circuit assembly, the powder is not blown either so frequently or in such large quantities, and lead times for product made from this alloy could suffer as a result. The user is therefore recommended to adopt the eutectic 63/37 alloy rather than the 60/40 with its plastic range between solidus and liquidus points. Although short, it can still be a cause of disturbed joints, created by movement of the solder in the joints while still in the soft, pasty state. Such movement can, for example, occur due to conveyor vibration immediately after reflow.

Dunn[3], in his study, determined that the lower viscosity of the 63/37 alloy enhances its flow properties. He found the 60/40 material to be more viscous and sluggish, with shorter wicking distances, although in some instances the latter characteristic might be a benefit with regard to reducing wicking (see Chapter 9).

There is no advantage to increasing the tin content in tin-lead solder above 63 percent, because the melting point (and price) increase while strength decreases.

Other tin-lead alloys are used in solder paste for different applications. These would include the 5Sn/95Pb and 10Sn/90Pb ratios for component assembly. In this case, the high melting points (respectively 299 degrees C/570 degrees F and 312 degrees C/594 degrees F) of these lead-rich alloys would ensure they don't soften or re-melt, causing the devices to fall apart during subsequent reflow or soldering operations with the conventional 63Sn/37Pb or 60Sn/40Pb materials.

In non-electronic applications, the 40Sn/60Pb and 30Sn/70Pb alloys in paste form are employed in the assembly of automobile radiators. These compositions have adequate strength from sub-freezing to well above the temperature of boiling water.

Lead alone has no application in electronics soldering, and although tin has in the recent past been used widely, especially in parts of Europe for printed wiring board plating, the only purpose it has served in terms of solder paste was as part of a so-called "ratioed" alloy (see Chapter 9). Pure tin in powder form is frequently not very solderable, and it is often blamed to be a contributor in whiskering (see also the section "Impurities in Solder" later in this chapter.)

Additional Elements Used in Tin/Lead Solders

As already mentioned, other elements can be added to tin/lead solders to improve their mechanical properties. Often, other metals are added to the tin-lead alloy or used with tin and lead separately in different compositions for various reasons of strength and other specific physical properties (TABLES 1-3 and 1-4). These other elements include antimony, bismuth, cadmium, indium, and silver.

Antimony. Antimony (Sb, from *Stibium*) is an element that is almost never used alone. Its major use has always been as an additive to lead in automobile storage batteries to increase strength and corrosion resistance. It performs the same function in solder alloys and also serves to form low-melting-point eutectics for a number of different metal combinations.

Tin-lead alloys containing up to about 3 percent antimony have slightly greater strength than the binary alloys and can be used in less exacting applications for economic reasons (FIG. 1-6). However, brass components should not be soldered with antimonial alloys, because antimony can combine with zinc from the brass to cause brittleness in joints. Antimony wets similarly to tin, but on brass its affinity for zinc is much greater than that of tin for the copper component, so when using antimonial solders on brass, it is very easy for a joint to become brittle. A solution for this is to nickel-plate the brass.

When tin prices reached a very high level some years ago, small quantities (2.5 to 3.0 percent) of antimony were used to replace the more expensive tin, as in so-called "functional alloys," which have plastic ranges similar to those of the 63/37 and 60/40 tin-lead alloys typically employed in machine soldering.[4] As an example, one of the most popular of these alloys patented by Manko, 52Sn/45Pb/3Sb, has a melting range of 187 to 194 degrees C (368 to 382 degrees F), which compares with 183 to 190 degrees C (361 to 379 degrees F) for the 60Sn/40Pb combination.

The alloy 28Sn/70.5Pb/1.5Sb, with a solidus of 185 degrees C (365 degrees F) was for many years used as a filler to hide the joints between the welded sections of automobile bodies.

Between 0.2 and 0.5 percent, antimony has until now been required by the U.S. Federal Specification QQ-S-571E to be included in certain tin-lead solders,

Table 1-3. Melting Temperatures of Solder Paste Alloys

Alloy	Melting Range °C		Melting Range °F		Density	
	Liquidus	Solidus	Liquidus	Solidus	gm/cm^3	lb/in^3
48Sn/52In	118	118	244	244	7.30	0.2635
50Sn/50In	125	118	257	244	7.30	0.2635
52Sn/48In	131	118	268	244	7.30	0.2635
58Bi/42Sn	138	138	280	280	8.56	0.3090
58Sn/42In	145	118	293	244	7.30	0.2635
80In/15Pb/5 Ag	149	142	300	288	7.31	0.2833
100In	157	157	315	315	7.31	0.2639
43Sn/43Pb/14Pb	163	144	325	291	8.99	0.3245
70In/30Pb	174	160	345	320	8.19	0.2956
60In/40Pb	185	174	365	345	8.52	0.3075
62Sn/36Pb/2Ag	189	179	372	354	8.41	0.3036
50In/50Pb	209	180	408	356	8.86	0.3198
96.5Sn/3.5Ag	221	221	430	430	7.36	0.2657
40In/60Pb	225	195	437	383	9.30	0.3357
99Sn/1Sb	235		455		7.27	0.2624
95Sn/5Sb	240	232	464	450	7.25	0.2617
95Sn/5Ag	245	221	473	430	7.39	0.2667
19In/81Pb	280	270	536	518	10.27	0.3707
80Au/20Sn	280	280	536	536	14.51	0.5237
92.5Pb/5Sn/2.5Ag	296	287	565	549	11.02	0.3978
10Sn/88Pb/2Ag	299	268	570	514		
92.5Pb/5In/2.5Ag	300		572		11.02	0.3978
95Pb/3Sn/2Ag	306	305	583	581		
97.5Pb/1Sn/1.5Ag	309	309	588	588	11.28	0.4072

Table 1-4. Mechanical Properties of Solder Paste Alloys

Alloy	Tensile Strength lbf/in^2	Tensile Strength N/mm^2	Shear Strength lbf/in^2	Shear Strength N/mm^2
48Sn/52In	1700	13	1600	12
50Sn/50In	1700	13	1600	12
80In/15Pb/5Ag	2500	19	2100	16
100In	600	5	900	7
43Sn/43Pb/14Pb	6100	47		
60In/40Pb	4100	32		
62Sn/36Pb/2Ag	7000	54		
50In/50Pb	4700	36		
96.5Sn/3.5Ag	8900	69	4600	36
40In/60Pb	5000	39		
95Sn/5Sb	5900	46	6000	47
95Sn/5Ag	8000	62		
19In/81Pb	5600	43		
92.5Pb/5In/2.5Ag	4600	36		
97.5Pb/1Sn/1.5Ag	4400	34		

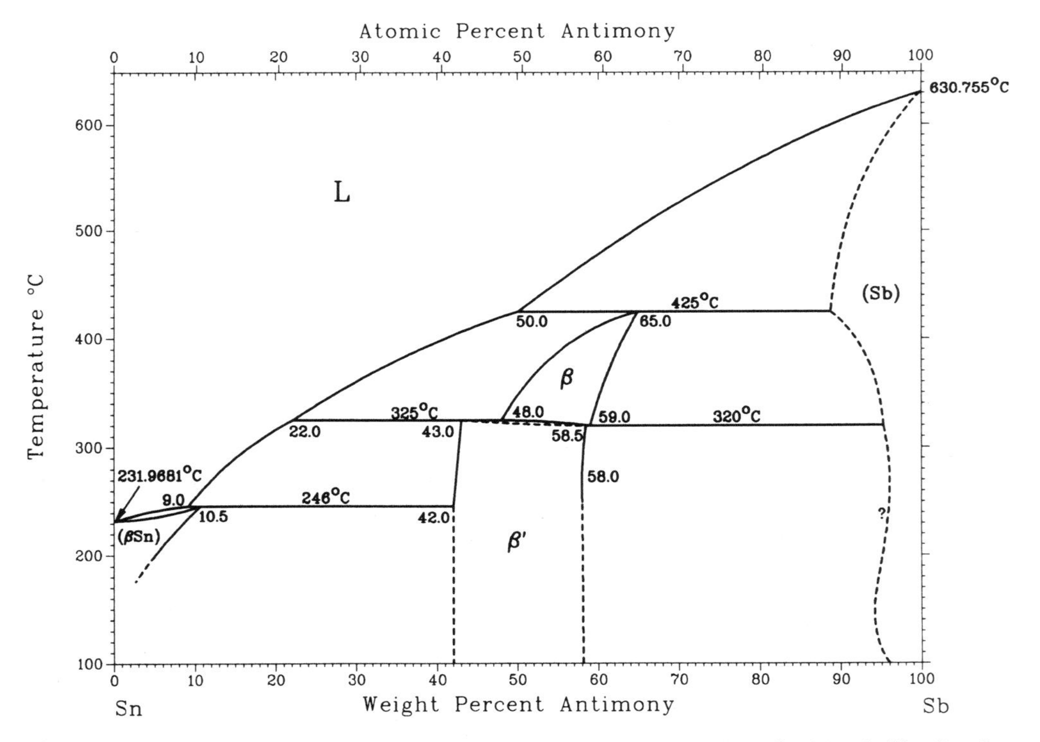

Fig. 1-6. Tin-antimony phase diagram. (American Society for Metals, Metals Handbook, 8th ed., vol. 8, *1973.)*

ostensibly to retard the formation of tin pest. *Tin pest* is a change that takes place in the crystalline form of the metal at low temperatures, producing a gray, amorphous powder. It has been claimed, however, that the risks normally associated with this "disease" are minor, because it normally occurs only in unalloyed tin or tin that has been contaminated by aluminum or zinc, and it does not affect alloys of tin containing more than 5 percent lead. In soldering terms, up to 0.5 percent antimony can be tolerated, since at this level scarcely any change in normal tin-lead soldering characteristics is detectable. It is interesting to note, however, that the latest proposed revision of this specification will in the future restrict the antimony content in to a level of 0.12 percent. At levels of 1 percent and more, deterioration in spread is noticeable, and at much higher than 5 percent, embrittlement of the solder can result.

One of the most widely used antimony-containing alloys is the 95 percent tin/5 percent antimony, which provides superior creep strength and good thermal fatigue resistance. It is used in soldering button diodes down onto rectifier bridges in the manufacture of automotive alternators, and with a melting range of 232 to 240 degrees C (450 to 464 degrees F), it has occasionally been selected to replace the more expensive 95 percent tin/5 percent silver composition. As an alloy, it is probably even better known in the plumbing sector, where new regulations forced the replacement of lead by antimony in the repair and joining of metal pipes carrying water for human consumption.

The combination 99 percent tin/1 percent antimony is very familiar to the semiconductor industry, where it has for years been used for die-attach applications, particularly for diodes.

Another well-known semiconductor composition is the so-called "J" alloy, consisting of 65 percent tin/25 percent silver/10 percent antimony.[5] This was developed as a compromise between the so-called hard and expensive materials (gold-tin, gold-germanium, and gold-silicon) and the softer lead-rich solders, with power-cycling values approaching those of the gold alloys. The "J" alloy is not presently known to be commercially available as a paste. This might be due to the difficulty in producing it in powder of acceptable particle shape and size, or perhaps to a lack of interest in semiconductor production of any soldering product requiring the use of a flux.

A ternary tin-lead-antimony alloy containing 10 percent antimony was tried in a paste several years ago following work between one of the authors and a leading U.S. manufacturer of axial capacitors. The intention was to improve high temperature (260 degrees C/500 degrees F) strength in the bond between the chip termination and nail-head lead while keeping the alloy melting point to a minimum to avoid reflow of the tin-lead plating on the leads and charring of the rosin flux. This works very successfully, despite the fact that it has a low nominal melting temperature range of 240 to 250 degrees C (465 to 482 degrees F). The alloy at room temperature is fairly brittle as you might expect from a composition containing such a level of antimony, but its mechanical properties are still acceptable to the industry.

Bismuth. Bismuth (Bi) is added primarily to lower the melting point of the alloy, but it reduces the mechanical strength of the joint and in high concentrations can produce poorly alloyed and brittle joints. In the consumer electronics industry in Europe, it has been used at a concentration of 2 to 3 percent to impart to the solidified solder a matt appearance to facilitate visual inspection of joints.

Bismuth is also of interest as an ingredient for alloys that do not contract upon solidification. When approximately 50 percent or more bismuth is present, the alloy will, in fact, expand during solidification in direct contrast to other metals used in soldering, with the exception of antimony, which behaves in similar fashion. The characteristic is advantageous, of course, where, for example, lead-to-hole ratios in through-hole design printed wiring boards are great enough to create voids when a conventional tin-lead solder is used, and the bismuth-containing alloy expands to fill these. Occasionally, depending upon its composition and the amount of bismuth present, the alloy will neither shrink nor expand.

The bismuth-containing alloys require special fluxes that are active at lower melting temperatures, and such solders have a high rate of oxidation. For this reason, powder produced from bismuth-loaded alloys tends to degrade very rapidly and has to be incorporated almost immediately into a suitable flux vehicle. Sizing should be undertaken by air classification rather than screens, because the latter tend to promote abrasion of the powder particles, leading to further surface degradation.

In general, few bismuth-containing alloys are suitable for incorporation in solder paste. One of the worst in this regard is the 42 percent tin/58 percent bismuth eutectic (FIG. 1-7). However, the 43 percent tin/43 percent lead/14 percent bismuth, with a melting range of 114 to 163 degrees C (291 to 325 degrees F) has been successfully used for step or sequential soldering (see Chapter 8). A 46 percent tin/46 percent lead/8 percent bismuth alloy with a reported melting point of 172 degrees C (341 degrees C) is used by a major Japanese manufacturer. This alloy is claimed to have superior wetting characteristics on copper to those nominally experienced with conventional tin-lead products.

A major application for bismuth has really nothing to do with soldering as such. Low-melting-point alloys containing this element are used as "fusible" alloys in automatic fire sprinkler systems, voltage surge protection devices, and lightning arresters, usually as a preform, which is either stamped from foil or compacted from powder. Their sole function is to respond to significant increases in temperature by melting, thus releasing pressure on whatever safety mechanism they might have been holding inoperative.

Cadmium. Cadmium (Cd) has also been used to lower the tin-lead alloy melting points, as in the celebrated ternary 50Sn/32Pb/18Cd with a liquidus temperature of 145 degrees C (284 degrees F). But the highly toxic nature of cadmium and the reluctance of most of the solder paste manufacturers and their outside suppliers to produce cadmium-containing powder has all but eliminated the availability of product in this alloy. It seems likely, in any case, that the future use of cadmium will be increasingly restricted by environmental regulations, if not entirely prohibited.

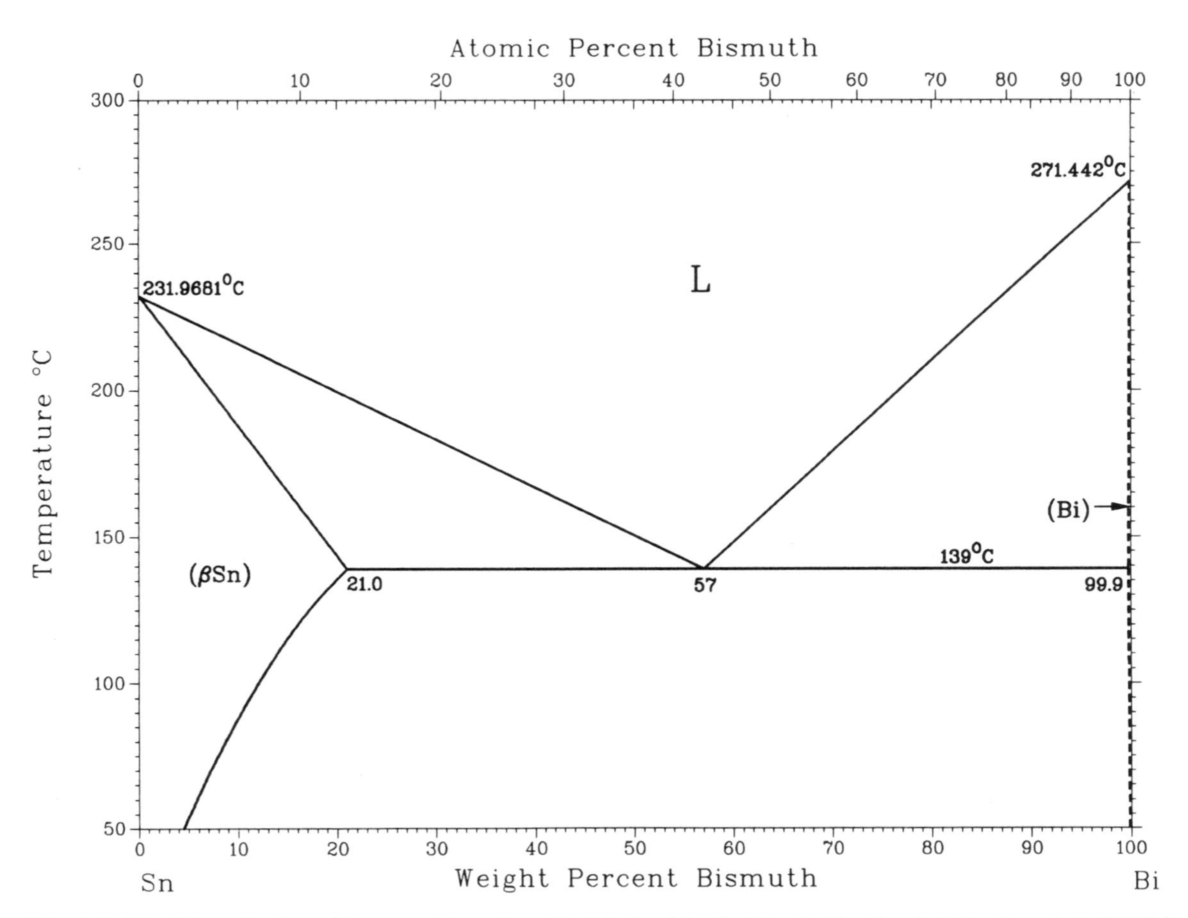

Fig. 1-7. Tin-bismuth phase diagram. (American Society for Metals, Metals Handbook, 8th ed., vol. 8, *1973.)*

Gold. Only one alloy incorporating tin and/or lead in combination with gold is normally available as a solder paste, and this is the composition 80 percent gold/20 percent tin, with a eutectic of 280 degrees C or 536 degrees F (FIG. 1-8). It is used exclusively for soldering down onto gold metallizations when conditions preclude the introduction of any other alloy.

The gold-tin eutectic has high strength and generally solders well, even with a mildly-activated rosin flux system. Its high cost, which is typically more than 100 times the normal selling price of the tin-lead eutectic, encourages very judicious use of the alloy.

Indium. Much has been published on the subject of indium (In), which has a number of important attributes. Unfortunately, these are often outweighed by the high cost of this metal and the very poor soldering characteristics of some of the more useful indium alloys.

Indium itself tends to oxidize very quickly, especially when it is supplied

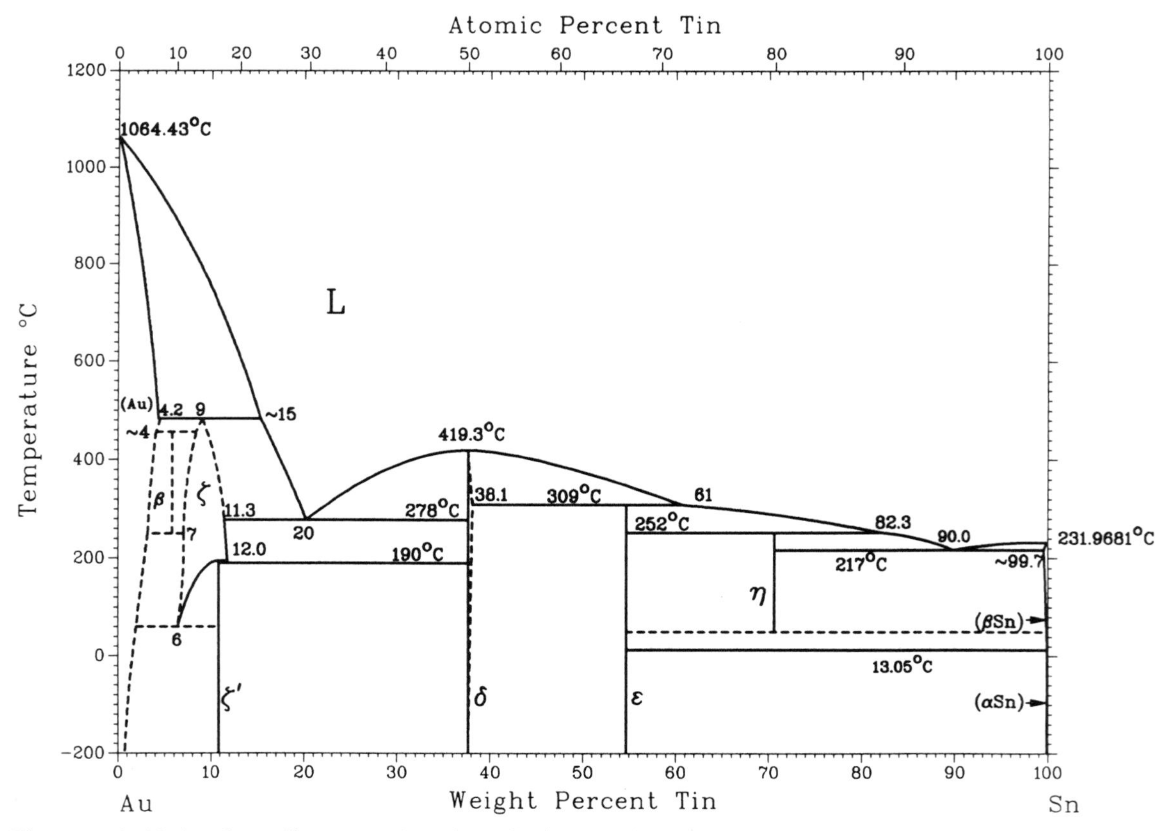

Fig. 1-8. Gold-tin phase diagram. (American Society for Metals, Metals Handbook, 8th ed., vol. 8, *1973.)*

in a form providing a large surface area, such as powder. The indium oxide formed has a high melting point and is very difficult to remove. This means that a specially activated flux is needed to promote wettability. The high-indium-containing alloys also provide low strength in comparison with tin-lead and other solders.

Its major advantage is its ductility and malleability, which enable indium-containing alloys to withstand thermal stress fatigue testing much more successfully than tin-lead solders. For this reason, certain alloys, such as the 92.5Pb/5In/2.5Ag, have been widely adopted by manufacturers of silicon power devices for die-attach; these components experience considerable temperature cycling during operation. Work by Wild has shown that both the indium-tin and indium-lead solders produce much slower crack propagation during low cycle fatigue testing, both at room temperature and at 100 degrees C (212 degrees F) than does the 63Sn/37Pb.[6]

The plasticity of indium, which to some extent is retained even down to cryogenic temperatures, also offers benefits in the bonding together of materials with mismatched thermal coefficients of expansion, such as glass and metal. This is particularly useful in the mounting of larger leadless ceramic components, such as chip carriers, to conventional printed wiring board materials, such as epoxy/fiberglass laminate. The difference in rates of expansion between these materials during heating and cooling often causes micro-cracking in solder joints of tin-lead and tin-lead-silver. The ductility of indium, normally supplied as part of an indium-lead alloy, can greatly reduce this phenomenon. It should be emphasized, however, that substantial thermal coefficient mismatches cannot be compensated merely by the selection of an indium alloy.

One of its other major applications in paste form for surface mounting is in sequential soldering, whereby an assembly is joined in a series of steps using alloys melting at reducing temperatures, so that bonds already created are not disturbed by the current soldering operation. The comparatively narrow plastic range (FIG. 1-9) of the indium-lead alloys makes this an ideal family of solders for such a process. The plastic range is normally no greater than 25 degrees C (77 degrees F), so the operating temperatures need rarely be much more than 50 degrees C (122 degrees F) above the solidus of the alloy.

Finally, indium has the important characteristic of being a very limited scavenger of gold and therefore finds application in soldering down onto gold thick- and thin-film circuitry. In this instance, indium is again used in conjunction with lead in a variety of alloys according to the melting temperature preferred. If the subsequent operating temperature of the device exceeds 125 degrees C (257 degrees F), indium-based solders are generally not recommended, as solid-state diffusion of gold may occur. In such cases, one must rely on the gold-tin (80/20 percent) eutectic solder.

Phase diagrams were developed by Karnowsky and Rosenzweig[7] for the gold-tin-lead and by Karnowsky and Jost[8] for the gold-indium-lead ternary systems. Jost noted from these diagrams that at 250 degrees C (482 degrees F), 50-percent-by-weight indium-lead solder must dissolve approximately 1-percent-by-weight gold before a solid protective layer of gold-indium intermetallic ($AuIn_2$) is able to form to prevent further scavenging.[9] Thirty-seven-percent-by-weight tin-lead solder must dissolve approximately 13-percent-by-weight gold before the solid gold-tin intermetallic ($AuSn_2$) is created. It was concluded that the eutectic tin-lead alloy dissolves approximately 13 times more gold at a temperature of 250 degrees C (482 degrees F) than the 50 percent indium/50 percent lead composition.

As the result of further work, it was determined that because approximately 1 micron (40 microinches) of gold is required to form the protective $AuIn_2$

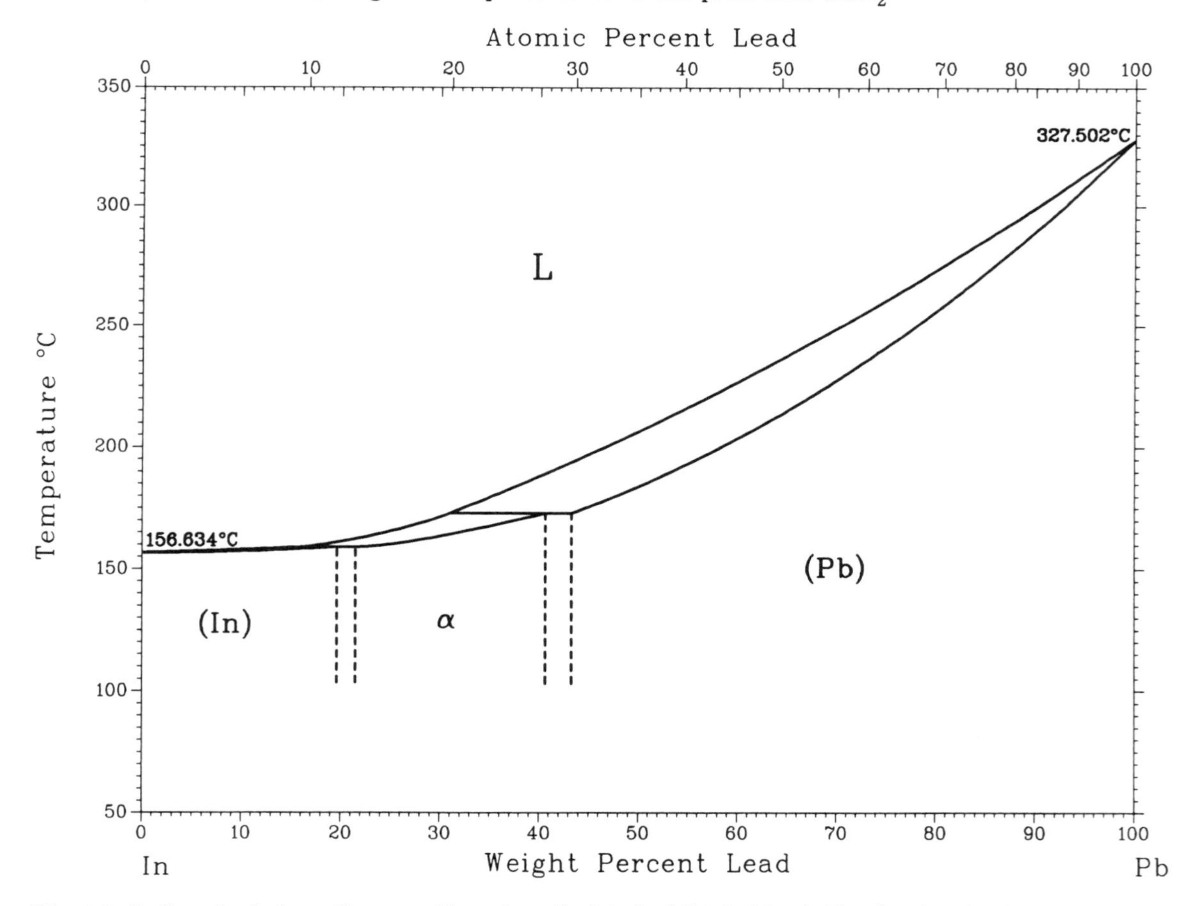

Fig. 1-9. Indium-lead phase diagram. (American Society for Metals, Metals Handbook, 8th ed., vol. 8, *1973.)*

intermetallic layer, lead-indium alloys should not be employed on gold thicknesses less than this measure. Aging tests subsequently revealed that to avoid problems resulting from the growth of other intermetallics, it is advisable not to use indium-lead solder on gold films thicker than 10 microns (400 microinches).

Silver. Silver (Ag, from *Argentum*) is used either to impart strength to the alloy or to reduce silver scavenging.[10] Silver scavenging is the attack by liquid solder upon surface metallization containing silver. As the majority of thick-film inks used on ceramic substrates in the hybrid circuit industry have a high silver content that rapidly dissolves into molten tin resulting in the destruction of the silver-containing conductor traces, sufficient silver must be present in the solder to reduce this rate of dissolution, or *leaching*, from the metallized pads and the circuitry (FIG. 1-10). The same condition also applies to the component thick-film silver terminations such as those of ceramic chip capacitors, still being used for surface-mount applications and leadless ceramic chip carriers.

For thick-film hybrids, the alloy employed for bonding is normally 62 percent tin/36 percent lead/2 percent silver, which provides joints of excellent tensile shear strength. This is basically the 63 percent tin/37 percent lead composition already mentioned, to which 2 percent silver has been added (FIG. 1-10). The addition of the silver only slightly affects the melting range—179 to 189 degrees C (354 to 372 degrees F). There is a eutectic of the alloy at 179 degrees C (354 degrees F), which is 62.5 percent tin/36.1 percent lead/1.4 percent silver, but few users actually specify this exact composition, and most seem to be under the wrong impression that the 62/36/2 ratio really represents the eutectic.

The solubility of silver in eutectic tin-lead solder is about 5 percent. A silver content of approximately 2 percent or less does not cause any deterioration in soldering results and, indeed, improves the shear strength at room temperature by 10 percent and at 100 degrees C by 20 percent. At higher silver levels, however, grittiness of the solder surface is noticeable due to the intermetallics that are formed. Alloys with up to 4 percent silver are employed to reduce silver scavenging, but except in step-soldering processes, their melting points are generally too high to permit their use for routine thick-film or surface-mount reflow operations.

An alloy commonly used for sequential soldering is the eutectic of tin and silver (FIG. 1-11)—96.5 percent tin/3.5 percent silver—particularly for thick-film conductor inks susceptible to silver scavenging or leaching. This composition is especially useful for bonding lead-frames to the edge connectors on the ceramic substrates because it will usually not re-melt at the reflow temperatures later employed for soldering components elsewhere on the same assembly with a lower melting point alloy, such as the 62 percent tin/36 percent lead/2 percent silver.

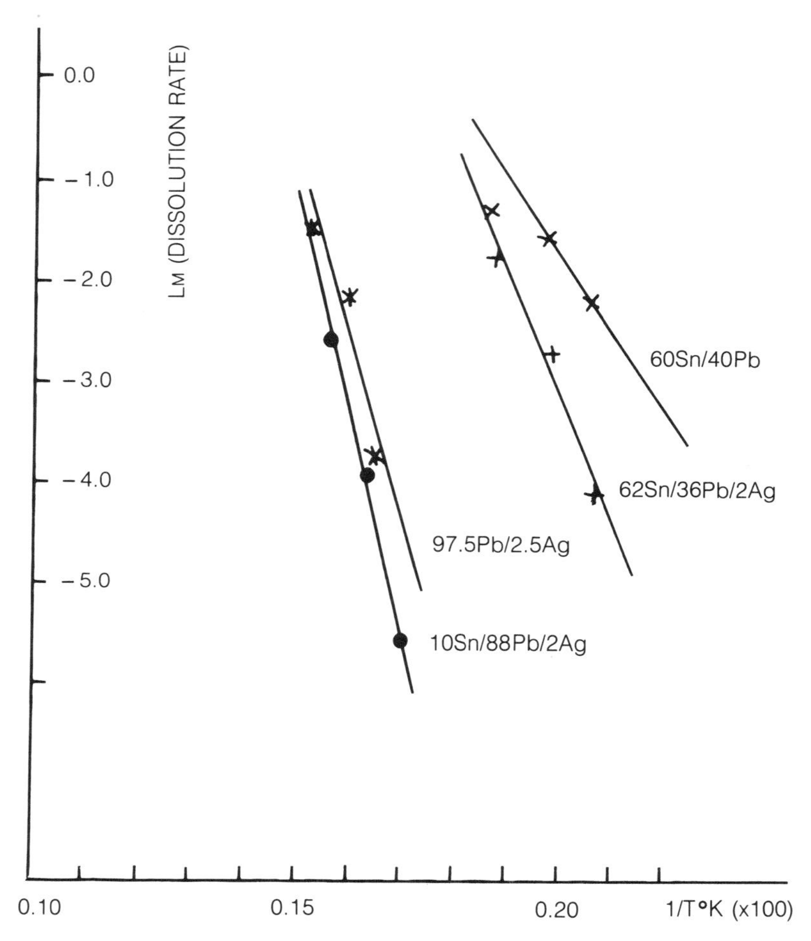

Fig. 1-10. The dissolution rate of silver in the tin/lead solder alloy is significantly reduced by the addition of small additions of silver to the solder. (R.A. Bulwith and C.A. MacKay: American Welding Society, "Silver Scavenging Inhibition of Some Silver-Loaded Solders," Welding Journal Research Supplement 64, *1985.)*

Some of the popular high-melting-point tin-lead-silver alloys are the 10/88/2, 5/92.5/2.5, and 3/95/2 alloys with melting points ranging between 290 and 306 degrees C (566 to 583 degrees F). These are normally used either in automotive "under-the-hood" applications, where operating temperatures are too severe for the otherwise stronger 96.5 percent tin/3.5 percent silver eutectic or 95 percent tin/5 percent silver composition or for the assembly of leaded components. The addition of the small amounts of silver again provide an improvement in joint strength.

If you look through tables of standard alloys published by the solder paste manufacturers, there are no tin-lead-silver compositions shown between the melting point of 95 percent tin/5 percent silver at 245 degrees C (473 degrees F) and 10 percent tin/88 percent lead/2 percent silver at 290 degrees C (554 degrees C). Two Alpha Metals metallurgists, R.A. Bulwith and C.A. Mackay, succeeded in bridging this gap with two new alloys for users requiring an intermediate melting point range of 260 to 280 degrees C (500 to 536 degrees F). Moreover, they have a desirable narrow plastic range of about 10 degrees C (18 degrees F). Both alloys exhibit excellent wetting and joint strength characteristics.

There are a number of other ways of reducing the effects of loss of adhesion due to silver scavenging. These include coating the conductor with a barrier layer of a less soluble metal, such as nickel or copper; by minimizing the process time and temperature; and by limiting the amount of solder at the interface.

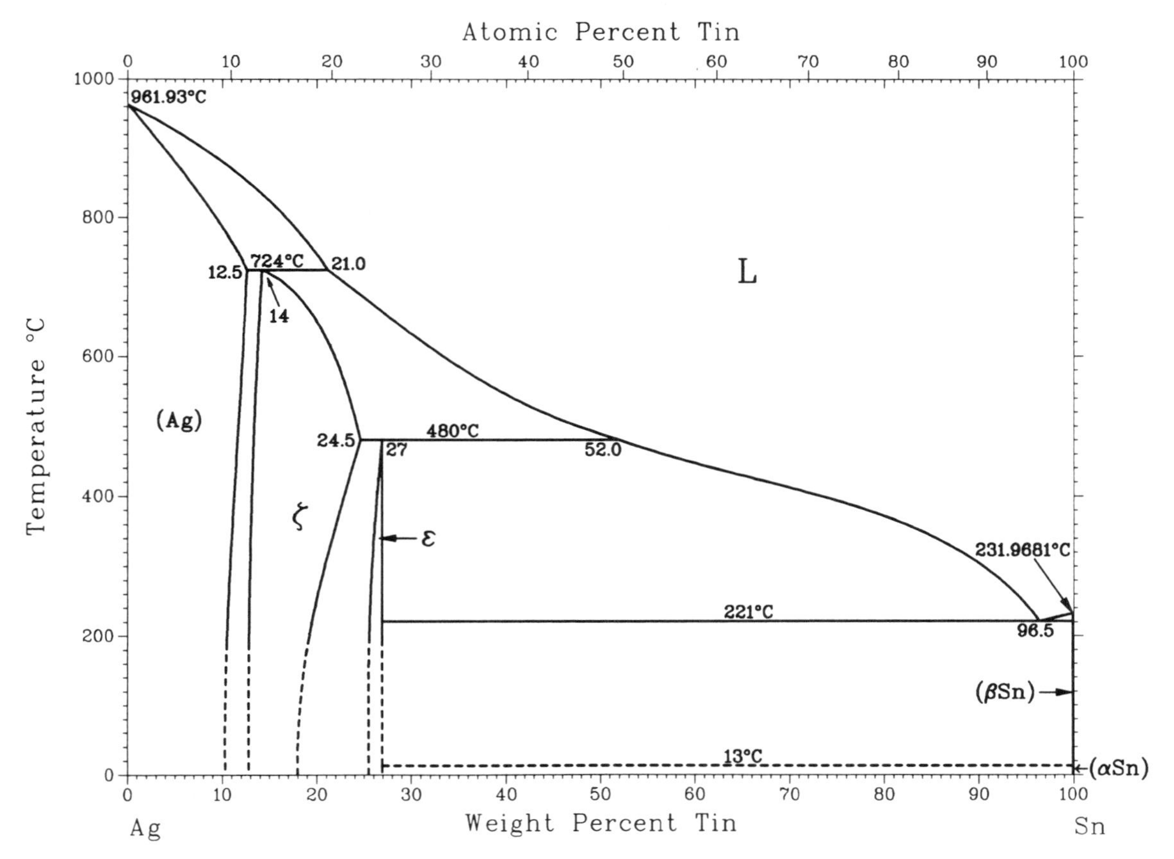

Fig. 1-11. Tin-silver phase diagram. (American Society for Metals, Metals Handbook, 8th ed., vol. 8, *1973.)*

Other Additives. Other additives include cerium (Ce) and tellurium (Te), which are claimed to refine the grain structure of tin-lead alloys. Similar effects have been reported for the addition of about 1 percent magnesium to eutectic tin-lead solder which, like tellurium, is said to provide a greater shear strength plus a marked reduction in the loss of strength and hardness values during storage, after solidification. We do not recommend the addition of any of these elements to solder paste alloys. Too little evidence has been presented as to their alleged benefits, and insufficient data is available with regard to possible deleterious effects that could occur during reflow.

Phosphorus has been added to soft solders, especially tin-lead systems, to bring about a reduction in the amount of visible oxide formed.[11]

Solderable Surfaces

It is beyond the purpose of this volume to discuss in any detail the solderability of the surfaces to be reflowed using solder paste. It would, however, be appropriate to list for the inexperienced reader the principal alloys recommended for the different metallizations and applications.

The surfaces most commonly encountered in solder paste reflow applications for epoxy/fiberglass printed wiring boards (surface-mount technology) are:

Bare Copper
Electroplated tin-lead
Infrared-fused electroplated tin-lead
Hot-air-leveled tin-lead

In most cases, both for the bare copper and tin-lead coated boards, the eutectic 63 percent tin/37 percent lead alloy will be suitable. The fusible coatings, such as 63 or 60 percent tin-balance lead, which melt below the reflow temperature of the same alloy, do not oxidize very rapidly under normal conditions and provide the base metallization on boards and substrates with a reasonable degree of solderability protection. This is especially the case for an electroplated coating that has been fused or reflowed, or a molten coating deposited by the hot air leveling process. Porous coatings, such as those electroplating frequently provides, offer rather short-term protection.

Exceptionally, if chip components such as capacitors with silver-palladium terminations are being mounted and silver scavenging or leaching is a concern, then the composition 62 percent tin/36 percent lead/2 percent silver should be used.

A majority of such terminations are now provided with a barrier of nickel and a final layer of high-melting-point alloy, such as 10 percent tin/90 percent lead, obviating the need for silver-bearing solder paste.

The 63 percent tin/37 percent lead and 62 percent tin/36 percent lead/2 percent silver are virtually the only two alloys now supplied in the form of solder paste as standard to the industry for the reflow temperature range of 215 to 250 degrees C (419 to 482 degrees F), unless a step-soldering operation is involved. This particular aspect is discussed in Chapter 8.

Thick-film hybrid circuits use the following:

Copper
Silver
Silver-palladium
Silver-platinum
Gold
Gold-palladium
Gold-platinum

Standard tin-lead systems are used in conjunction with the copper thick-film ink, while silver-bearing alloys are generally regarded as essential for the pure silver and silver-palladium compositions.

The use of thick-film copper for hybrid assemblies has been steadily increasing over the past few years despite initial, severe, solderability problems. The employment of silver-bearing solder paste is in this case not normally required or desirable unless component terminations call for their use.

The reason for the popularity of the silver-palladium alloys is that they are relatively inexpensive while being very compatible with both gold and aluminum wire bonding systems. They share with pure silver, however, the disadvantage that they tend to become rather unsolderable if stored for a prolonged period of time. Mechanical cleaning is then required for restoring solderability.

Thick-film gold conductor inks are much less utilized in the hybrid circuit industry than silver compositions, mainly for cost reasons. Some of the higher temperature tin-lead alloys with reduced tin content can be used for reflow with gold compositions containing palladium or platinum, since these elements reduce the rate of solution of the gold in tin. This can be a matter of seconds, even at low reflow temperatures of 215 degrees C (419 degrees F). Hence, for pure gold, 80 percent gold/20 percent tin is normally used.

The chief advantages of gold as a metallization are that it is ideal for eutectic die-bonding, it has high thermal conductivity, and it is generally very solderable.

There has been an increase in the use of indium-lead alloys in conjunction with gold metallizations, especially the 60 and 50 percent indium compositions with melting points of 185 and 209 degrees C (365 and 408 degrees F) respectively. As already mentioned elsewhere, the solvency of gold in the indium-lead alloys is very low. If the gold-tin eutectic is not to be used, then it is recommended that only indium-lead alloys be considered if dissolution of the the gold into the tin is likely to cause a problem. The gold-tin intermetallic formed is extremely brittle and will produce mechanical failures when subjected to stresses such as flexing or vibration. This intermetallic increases with temperature and time, and to some extent with a rise in the tin content of the alloy leads to a dramatic loss of solder joint strength.

As an indication of the natural affinity of gold for tin, approximately 15 percent by weight of the noble metal can be dissolved in liquid solder of 60 percent tin/40 percent lead at a temperature of 250 degrees C (482 degrees F).

Where tin-lead alloys must be used, reflow temperatures and times should be kept to a minimum to avoid excessive gold dissolution.

It is not practicable to make a gold-loaded tin-lead alloy to prevent leaching in the same way as the silver-bearing compositions are designed because at least 25 percent gold would be needed. The 64.5 percent tin/32.5 percent lead/3 percent gold is the eutectic of this ternary alloy.

It should be noted that the presence of silver in the solder alloy does not significantly reduce the solubility of gold. The high strength of the 80 percent gold/20 percent tin alloy is desirable in microwave applications where elevated energy levels would cause the failure of conventional soft solder joints.

Finally, it is worthwhile to consider nickel and nickel alloys, such as Kovar and Alloy 42, which are usually very difficult to solder to and might require a pre-cleaning or micro-etching process using a solution of inorganic acid. Nickel platings, except electroless nickel, in cases where phosphorous content is higher than about 5 percent, are normally fairly solderable.

In conclusion, paste alloy selection is normally made according to the requirements of the user based on the physical properties of the metals involved. Considerations include mechanical strength and electrical characteristics, the nature of the materials to be joined, and any reflow temperature—not melting temperature—constraints.

Impurities in Solder

As a general rule, solder paste manufacturers with a reputation to protect can be relied on to supply alloy of good to excellent quality, meeting either of the widely-accepted United States Specifications QQ-S-571E or ASTM B-32 (TABLE 1-5). On occasion, the manufacturers might be called upon to meet more stringent requirements, prompted either by their own perceived needs or those of their customers.

For reference purposes, what follows is a summary of the more commonly encountered, or less frequent but significantly undesirable, impurities.

Table 1-5. Solder Alloy Specifications: 63Sn/37Pb.

Major Constituent Ranges and Maximum Impurity Levels Allowed

(Percentages by weight)

Element	Symbol	QQ-S-571E	ASTM B-32
Tin	Sn	62.5–63.5	62.5–63.5
Lead	Pb	Balance	Balance
Antimony	Sb	0.2–0.5	0.5
Silver	Ag	0.015	0.015
Aluminum	Al	0.005	0.005
Arsenic	As	0.03	0.03
Bismuth	Bi	0.25	0.25
Cadmium	Cd	0.001	0.001
Copper	Cu	0.08	0.08
Iron	Fe	0.02	0.02
Indium	In	N/A	N/A
Nickel	Ni	N/A	N/A
Zinc	Zn	0.005	0.005
Gold	Au	N/A	N/A
Sulphur	S	N/A	N/A
Phosphorous	P	N/A	N/A
Tantalum Oxide	TaO	0.08	N/A

Aluminum (Al) Aluminum is classified as an oxide-promoting element, and levels of 0.0005 to 0.001 percent can create a significant amount of oxidation in the solder, leading to a dulling of the surface of the solder joints as well as preventing wetting. A maximum limit of 0.001 percent is recommended.[12]

Antimony (Sb) Antimony decreases solderability very slightly up to 0.5 percent. Up to 2.5 percent, it produces a progressive reduction in the area of spread of solder and up to 5 percent deterioration in wetting properties. It is, however, reported as beneficial if aluminum is present as a contaminant, because the antimony will tie up this element in the form of an intermetallic compound, which can be removed during the main melt of the alloy as dross.[13]

Arsenic (As) Some dewetting on copper surfaces can be caused by as little as 0.08 percent, and more severe dewetting at 0.1 percent, but it is primarily of concern because of its reaction with zinc, now more rarely found as a component of brass in electronic assemblies.[14]

Bismuth (Bi) The wetting properties of tin-lead solders have not been deleteriously affected by as much as 3 percent, and as has been described, larger amounts are added for particular applications.

Cadmium (Cd) Like aluminum, cadmium is an oxide-producer and can begin to cause dulling of the solder surface at a level of 0.01 percent. A maximum of 0.002 percent has been proposed to avoid the possibility of bridging and icicling during mass-soldering. Its drawbacks as a contaminant do not apply to alloys such as those described under the heading "Cadmium" previously in this chapter. Those alloys are not employed for soldering but as fusible alloys.

Copper (Cu) Above 0.3 to 0.4 percent copper content, the appearance of the solder can be affected, evidencing grittiness (FIG. 1-12). At these and higher levels, particles of tin-copper intermetallic will be formed (FIGS. 1-13 and 1-14). These can cause sluggish flow and an increase in the melting point of the alloy due to depletion of the tin, requiring higher reflow temperatures.

Gold (Au) Levels in tin-lead of 0.02 percent can cause sluggishness and dullness of appearance without any apparent effect on solderability.

Iron (Fe) Grittiness of the solder and a deterioration in flow characteristics can occur at 0.02 percent (FIG. 1-15).

Nickel (Ni) Solubility in tin-lead at normal reflow temperatures is negligible, and levels of above 0.02 percent can cause grittiness.

Phosphorus (P) As little as 0.01 percent can cause dewetting and grittiness.

Silver (Ag) Not considered a deleterious contaminant.

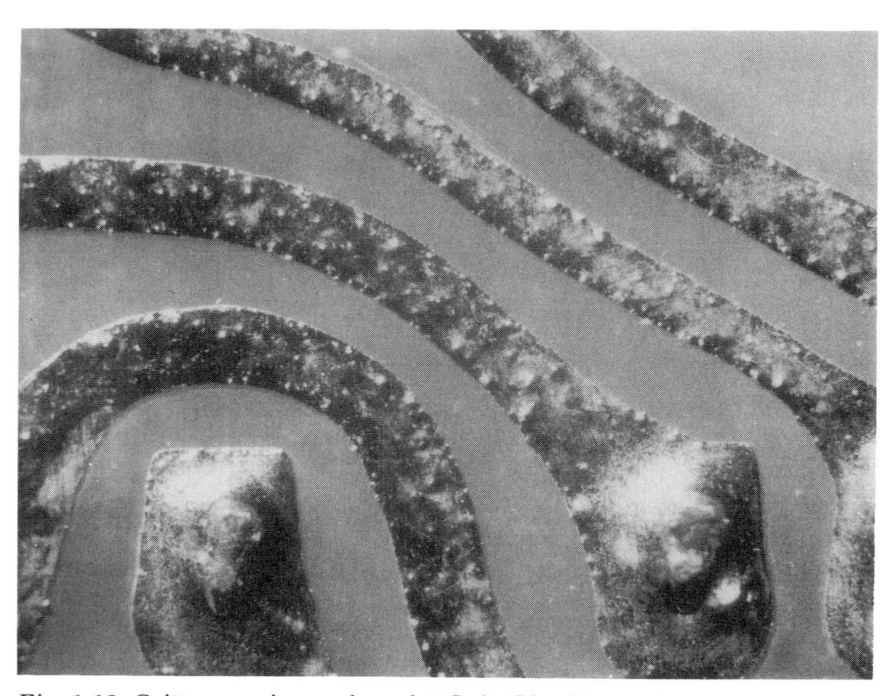

Fig. 1-12. Gritty or grainy surface of 63Sn/37Pb solder surface due to copper-tin intermetallic in solder.

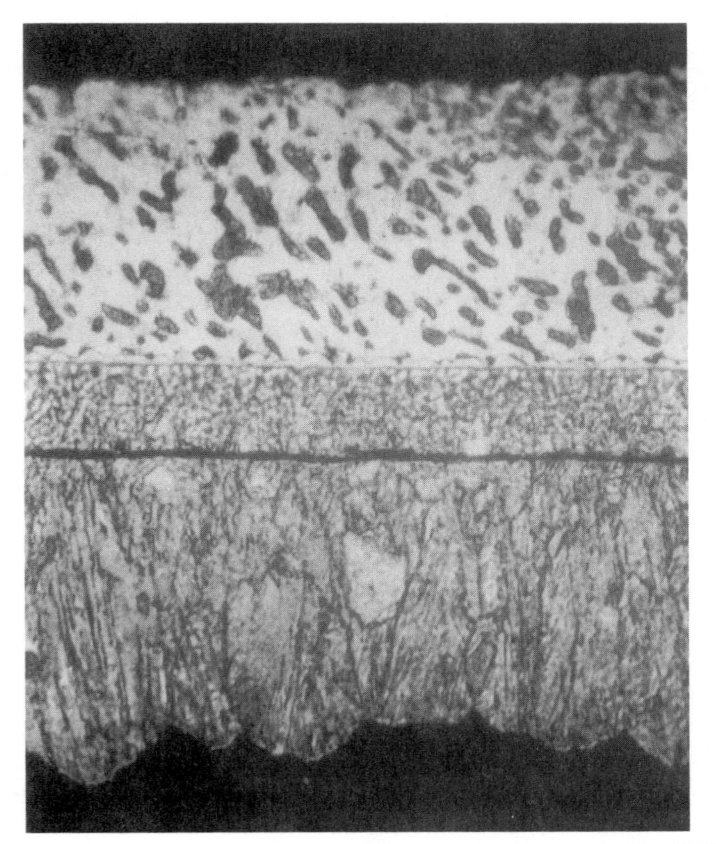

Fig. 1-13. Copper-tin intermetallic shown at interface of 63Sn/37Pb solder and copper metallization on printed wiring board.

Sulfur (S) Causes grittiness at 0.0015 percent due to the formation of tin and lead sulfides. Silver will react with sulfur to form "whiskers," which can produce electrical shorts.[15] However, such whiskers are generally regarded as the result of growth from pure tin coatings due to internal stresses.[16,17] Coatings with less than 70 percent tin content are not normally prone to whiskering.[17]

Zinc (Zn) Like aluminum, it's an oxidizer, and between 0.2 and 0.5 percent, zinc considerably decreases spreading capability of the alloy, also imparting a gritty appearance to the solder.

PARTICLE SHAPE

Particle shape plays an important part in determining the ultimate performance of the solder paste. It can have an effect on the flow or rolling

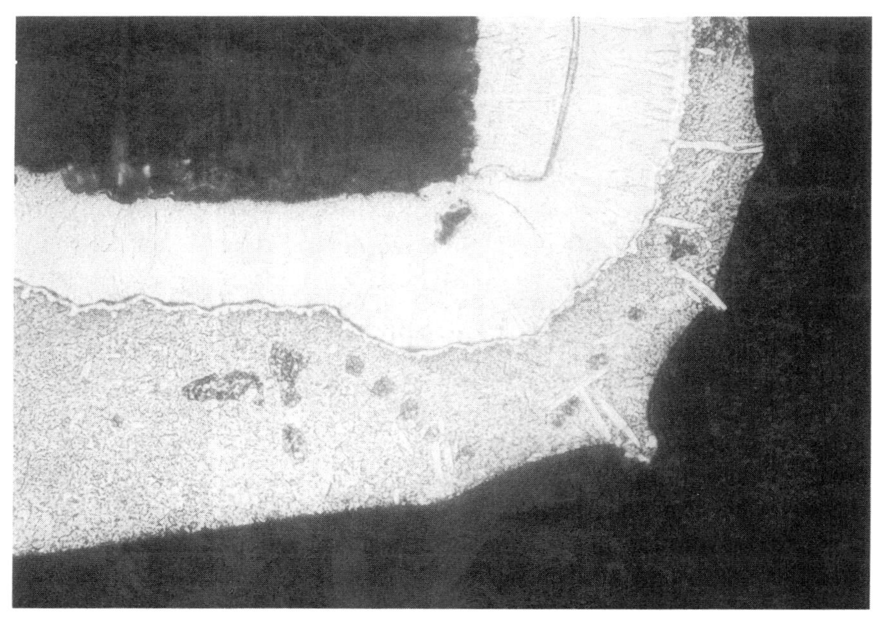

Fig. 1-14. Copper-tin intermetallic within solder coating at the corner of a plated-through hole in a printed wiring board.

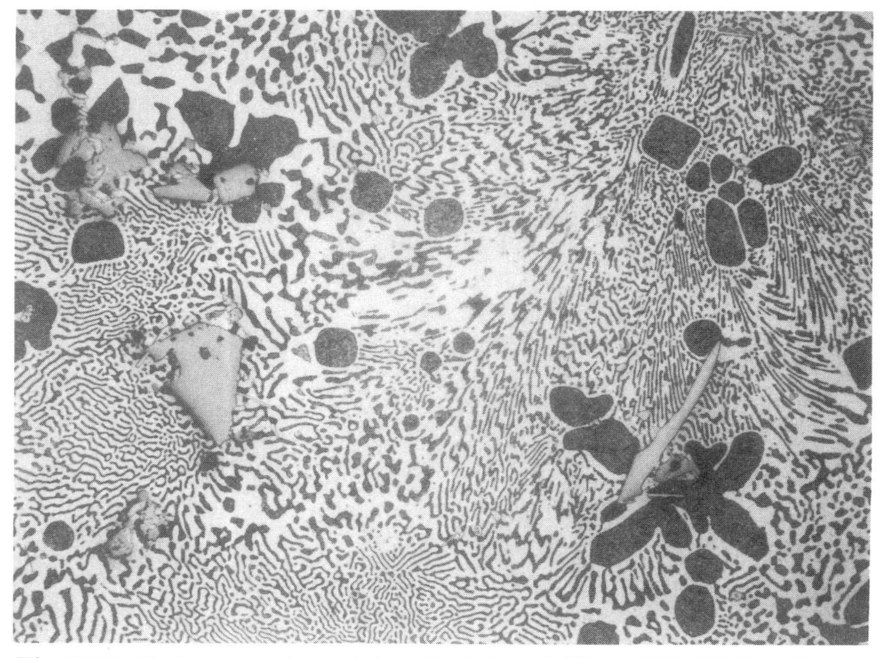

Fig. 1-15. Dark areas represent iron-tin intermetallic in 60Sn/40Pb solder.

characteristics of the product on a screen or stencil and on its ability to be pushed through an opening in a mesh, stencil, or needle orifice in a dispensing application. It also influences the apparent density of a solder paste and therefore its rheology, as well as viscosity.

The solder paste industry, with the concurrence of all the national and international bodies involved in drawing up standards for the product, is virtually unanimously agreed that spherical powder is the ideal to be pursued with regard to particle shape. In the correct processing conditions, a sphere can be fairly repeatably produced (FIG. 1-16), because surface tension naturally causes molten metal to assume a spherical shape. The other advantage is that a sphere always retains the same form and can pass through an opening, whatever its orientation. The sphericity is not always perfect because the shape can be altered by a number of factors, especially temperature. A spinning disk, for instance, typically imparts a slight tail to the sphere as it leaves the disk edge. This can become particularly pronounced with high-lead-containing alloys if the skin of oxide that forms on the sphere's surface is thick enough to prevent coalescence of the molten metal before it cools (FIG. 1-17). Frequently, the particles appear

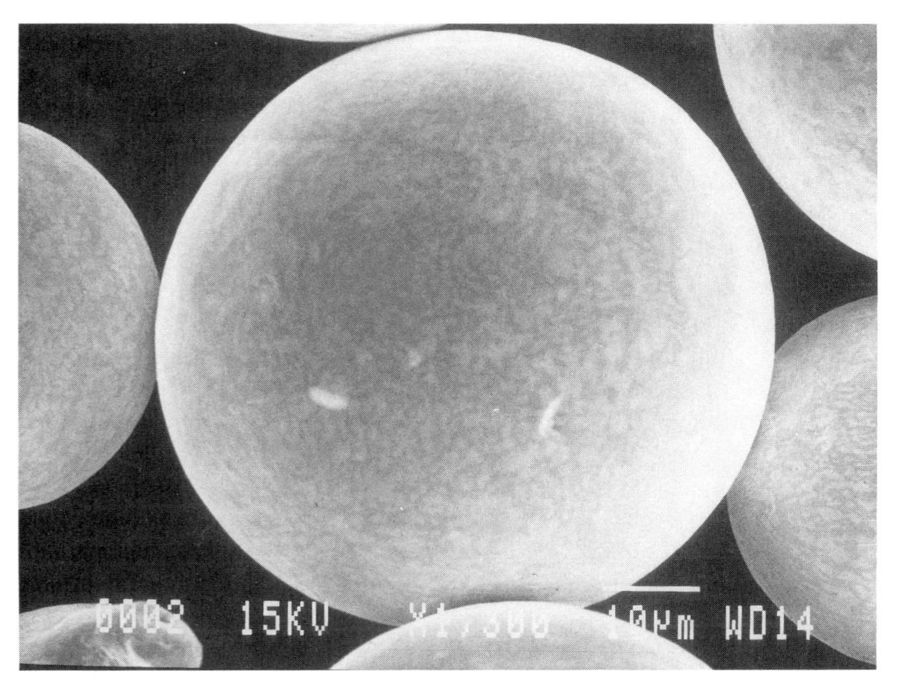

Fig. 1-16. 62Sn/36Pb/2Ag −200 mesh solder powder atomized in nitrogen chamber.

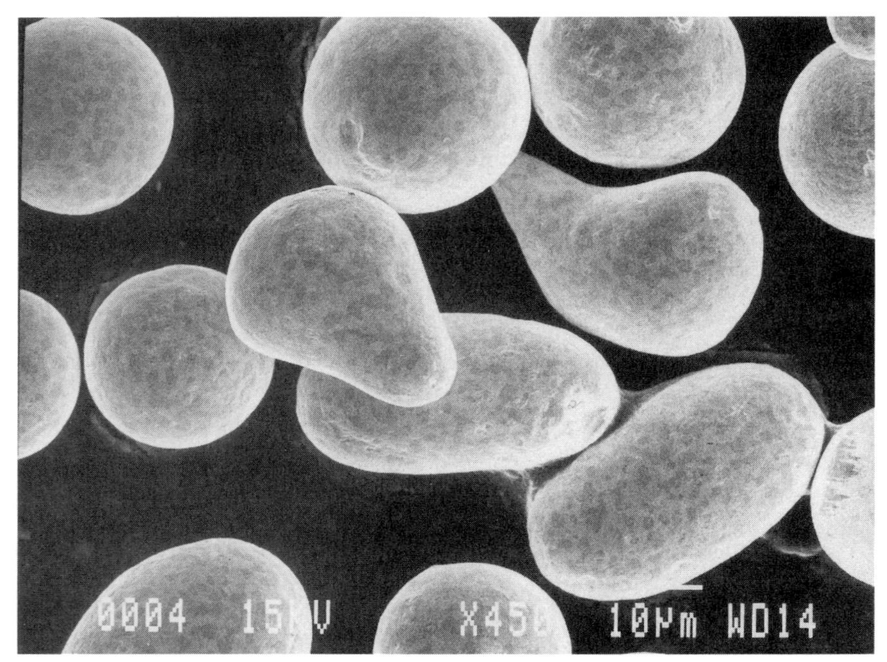

Fig. 1-17. 43Sn/43Pb/14Bi – 200 mesh solder powder atomized by spinning disk in nitrogen.

slightly flattened, or in extreme cases, much more so. Severe elongation (FIG. 1-18) can result in a particle length of over 0.015 inch (0.381 mm), and if the particle is lying horizontally, that size is more than enough to cause blockage of an 0.0088 inch (0.224 mm) opening in the ubiquitous 80 count mesh screen.

Powder particle shape, or its morphological characterization, can be inspected by means of scanning electron microscopy, but a quantitative determination is difficult. It should be sufficient to randomly inspect the paste under an ordinary microscope with, say, 40X magnification and an ocular scale with 0.001 inch (25 micron) divisions to verify that an acceptable proportion of the powder is spherical. Specification IPC-SP-819 requires that powder should be classified as spherical if, at sufficient magnification, it can be determined that at least 90 percent of the powder has particles in a length to width ratio of 1.5 to 1 or less. Figure 1-19 illustrates a near-perfect distribution of spheres, while FIG. 1-20 shows a sample that would fail the IPC test.

One author of a paper on solder paste claimed that each batch of powder bore a "fingerprint" by which its manufacturer could be identified.[18] This was true to some extent a few years ago, especially in regard to particle shape and surface quality, but the contrasts between the powders of different producers have now become much less sharply defined.

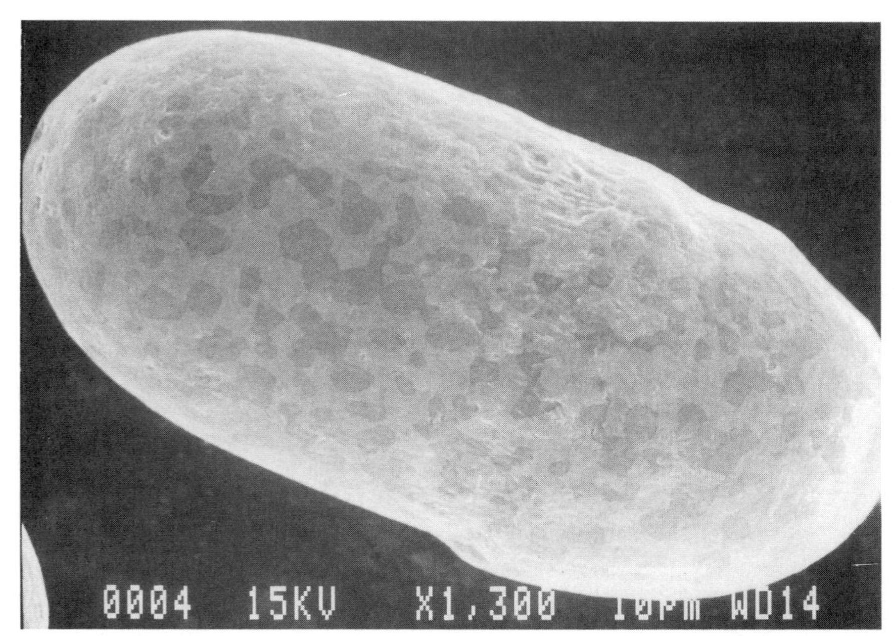

Fig. 1-18. 43Sn/43Pb/14Bi –200 mesh solder powder atomized by spinning disk in nitrogen.

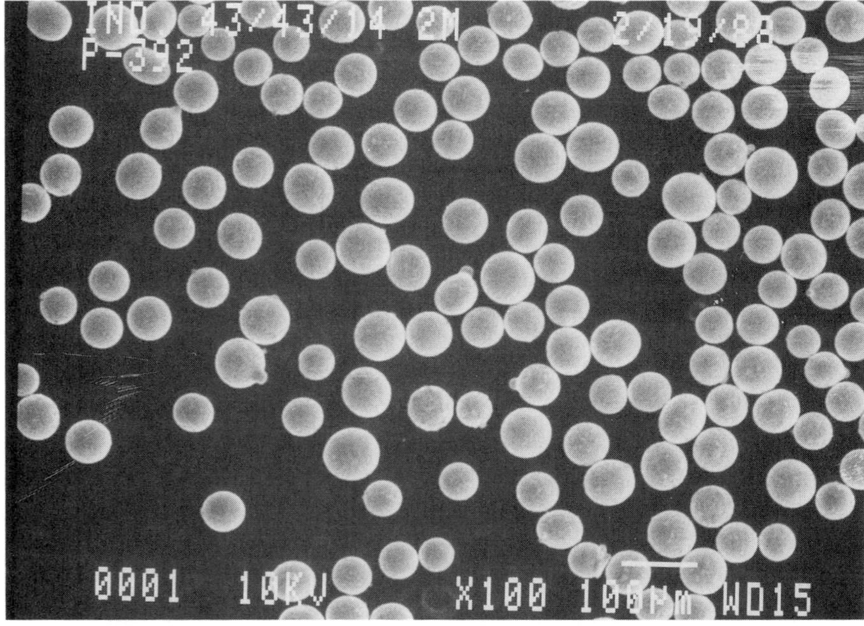

Fig. 1-19. 63Sn/37Pb –200 mesh solder powder atomized in nitrogen chamber.

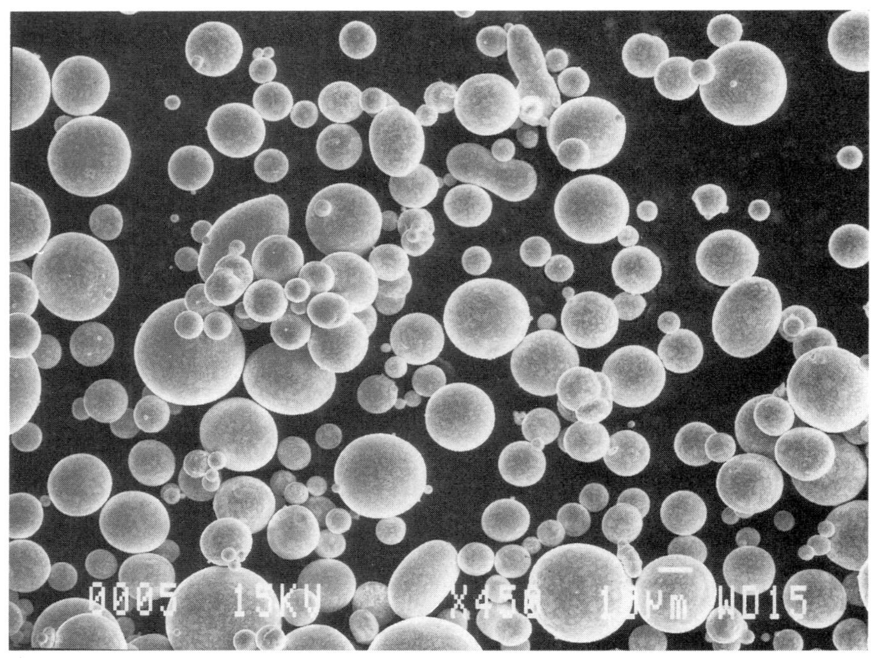

Fig. 1-20. 62Sn/36Pb/2Ag −325 mesh solder powder atomized in nitrogen chamber.

Apart from the tailing phenomenon already alluded to, the only other aspect of powder-making that could really demonstrably affect shape would be a water-atomization process, which generally produces irregular-shaped particles; in gas or air they are more rounded.[19]

PARTICLE SIZE

The powders currently used in solder pastes are classified under three main categories in terms of particle size: 100, 200, and 325 mesh. At least one leading solder powder and paste manufacturer in the United States has introduced an intermediate size as a standard commercial product for specific applications: 270 mesh. All of these powder size classifications are listed in TABLE 1-6 . For the purpose of electronics work, particularly surface-mount technology, the 100 mesh powder can be ignored: this is normally too coarse for any but the crudest joining tasks.

The accuracy of the method of measurement is essential if particle size distribution is to be controlled within reasonably close limits. In production, the powder should be properly mixed to ensure the uniform distribution of all particles

Table 1-6. Powder Particle Size Classification

U.S. Sieve No.	Tyler Sieve No.	Diameter Microns	Diameter Inches
100	100	149.0	0.00587
120	115	125.0	0.00492
140	150	105.0	0.00413
170	170	88.9	0.00350
200	200	73.7	0.00290
230	250	63.5	0.00250
270	270	53.3	0.00210
325	325	44.5	0.00175
400	400	38.1	0.00150

Note: The 200 mesh particle size is popularly referred to as being 75 microns while the 325 powder is generally termed 45 microns.

within the batch container(s) before it is added to the solder paste flux and other ingredients. Segregation can otherwise be caused by dumping, shaking during transit from the powder-making facility or supplier, and other vibrations, with a preponderance of small sizes accumulating at the bottom of the receptacle.

Recognized test methods exist for measuring powder particle size distribution, and these are used by both powder manufacturers and, in rather exceptional cases, by the actual solder paste user. The procedures used are normally ASTM B 214 "Sieve Analysis of Granular Metal Powders" in conjunction with B 215 "Sampling Finished Lots of Metal Powders." Improper sampling techniques are a major cause of inconsistent test results, so B 215 is useful in offering vendor and customer correlation in this respect.

Under the terms of B 214, which relates to powder sizes of 80 to 325 mesh, sieves of different sizes are placed one below the other in descending order of mesh size (with the largest largest mesh opening at the top) and a collecting pan at the bottom. The sample is placed on the top sieve and covered. The stack is then agitated by shaking, rotating, or tapping for a specific period of time. The powder fractions afterwards remaining on each sieve are then weighed separately and expressed as percentages of the weight of the sample retained on, or passed through, each sieve. Only standard ASTM or Tyler sieves (TABLE 1-7) according to ASTM E 11 "Wire Cloth Sieves for Testing Purposes" should be used, and it is usually necessary to arrange periodical calibration of both the vendor and customer sieves.

ASTM B 330 is used to provide a rapid approximate measurement of average particle size of sub-sieve (below 325 mesh) powder, employing the Fisher Subsieve Sizer. This device measures the permeability or flow of air through compressed powder.

Table 1-7. ASTM and Equivalent Tyler Standard Sieves

ASTM Sieve Opening			Tyler Standard Opening		
No.	Microns	Inches	No.	Microns	Inches
80	177	0.00697	80	175	0.00689
100	149	0.00587	100	149	0.00587
140	105	0.00413	150	104	0.00413
200	74	0.00291	200	74	0.00291
325	44	0.00173	325	44	0.00173

Another technique is to utilize light scattered by a laser beam through a stream of particles in circulating water, as in the Microtrac analyzer supplied by Leeds and Northrup. Still another method is sedimentation, which calculates particle size by the rate of settling in a liquid.

The required size of the particles of solder powder in a solder paste is determined by a number of important factors. Firstly, in the case of deposition by means of printing or pressure-dispensing, the major concern is that all of the particles will pass through either (a) the openings in the screen or stencil; or (b) the orifice at the needle or syringe opening. In the former case, the opening might be as small as 0.004 inch (0.102 mm); in the latter, 0.016 inch (0.406 mm).

In other applications, such as roller-coating, oversize particles can seriously interfere with even deposition of the solder paste onto the surface to which it is being applied. In pin transfer, where very small dots of 0.010″ (0.254 mm) and less are often being put down, larger particles can cause very uneven placement of paste.

It is therefore apparent that the way in which the paste is to be employed is an important starting parameter in the selection of the solder paste. There are, however, a number of other constraints with regard to both the maximum and minimum particle sizes, and these should also be borne in mind by the potential paste user.

Size is important not only with regard to the passage of a particle through an opening onto a part beneath, but there is often a need to apply a coating of paste to create a thin (0.001 to .002 inch or 0.025 to 0.050 mm) reflowed layer of solder on a pad or land area of a board or substrate (especially applicable in printed circuit fabrication). Such a process is described in Chapter 5. In this example, a maximum powder particle size would be 325 mesh or 0.0018 inch (45 microns) maximum, as anything larger would simply not pass consistently through the screen.

In another example, if a very thin deposit of wet solder paste were to be printed down onto a substrate and the solder paste contained larger particles (either by virtue of being oversized, or elongated, or otherwise misshapen), there would also be the danger that if left exposed to the air for an extended period of time, the fluxed surface of that part of the particle protruding above the surface of the printed deposit could become very dry and difficult to solder, depending upon the alloy being used (FIG. 1-21).

To a much lesser degree, the same observations apply to paste being printed onto pads for component mounting and reflow, because in practice, much thicker deposits are involved.

The danger of solder powder with excessively wide particle size distribution is obvious, but there have been instances where a paste with a very narrow particle size distribution at the upper range (−200 + 270 mesh) has caused similar problems. In these cases, because the particles are all similar in size, they tend to stack upon each other in columns (FIG. 1-22) instead of interlocking together in layers. With a reasonably wide range of sizes, the smaller particles press through the screen or fall into the gaps between the large particles (FIG. 1-23). This promotes printability and enhances tack retention.

In summary, select solder paste containing main sizes of powder particles that fall within a fairly narrow range but with a sufficient number of small particles to fill the spaces that might otherwise attract flux and promote separation. TABLE 1-8 shows the existing IPC-SP-819 Standard and notes the revisions with respect to an increase in the percentages of minimum particle sizes.

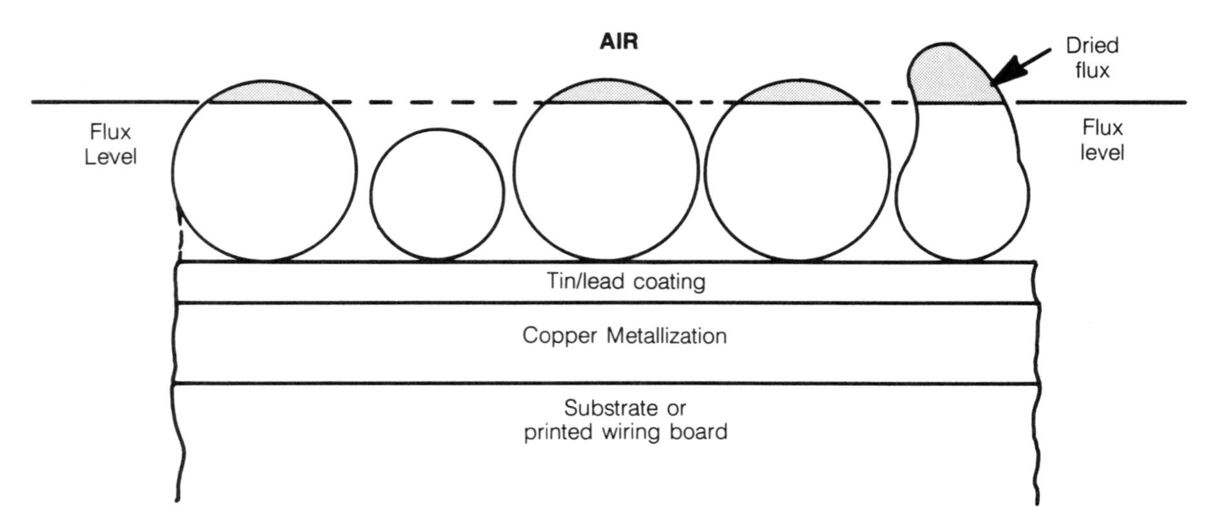

Fig. 1-21. The solderability of oversize solder powder particles can be affected by drying-out of their flux coating.

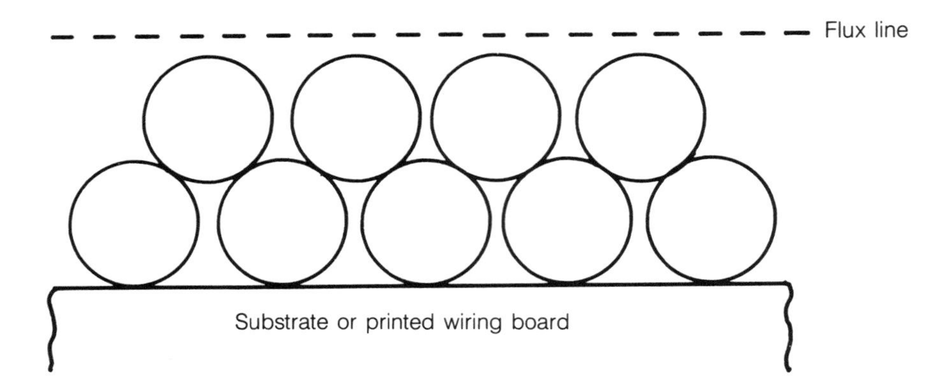

Fig. 1-22. Too uniform a particle size, especially for −200 mesh powder, will reduce the thickness of the flux layer at the surface, thereby reducing tack retention.

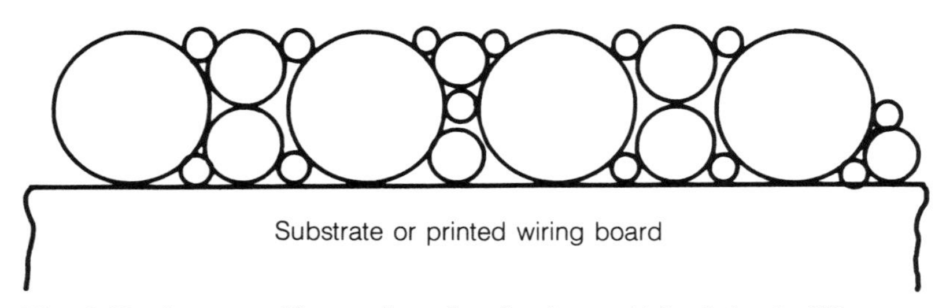

Fig. 1-23. A reasonable number of undersize particles helps to fill gaps, moving flux to the surface to provide essential tack retention. This wider particle size distribution also improves print definition and reduces flux spreading.

Table 1-8. Specification IPC-SP-819: Powder Particle Size Distribution

	Percent of Sample by Weight—Nominal Sizes		
	Less than 1% Larger than	**90% Minimum Between***	**10% Maximum Less than**
Type 1	150 microns	150–75 microns	75 microns**
Type 2	75 microns	75–45 microns	45 microns**
Type 3	45 microns	45–20 microns	20 microns
Type 4	38 microns	38–20 microns	20 microns

*Revised to 80%
**Revised to 20 microns

The larger surface area can also lead to more rapid deterioration of the powder in the presence of excessive humidity levels when, especially in alloys containing high (greater than 90 percent) lead, the moisture reacts with the lead oxides present, causing extreme discoloration and frequently almost total loss of solderability. For this reason, powders made from alloys such as 1 percent tin/97.5 percent lead/1.5 percent silver are often stored at sub-freezing temperatures to eliminate moisture contamination. This deterioration occurs even more rapidly when the powder has been incorporated into a paste flux, because any activators present accelerate the reactive process.

Expect a paste incorporating 325 mesh powder to exhibit more viscosity instability than the same formulation based on the larger 200 size. This is because the finer powder increases the apparent density of the paste.

HANDLING

Solder powder requires special handling because of the deterioration that can so easily occur in its surface solderability. This is especially true in the case of the oxide-prone alloys containing indium and high lead percentages.

The eutectic or near-eutectic tin-lead materials, as the 62 percent tin/36 percent lead/2 percent silver alloy, and tin-silver and tin-antimony compositions can normally be stored for many months in plastic-lined, sealed containers in a reasonably controlled environment (where temperature and humidity levels do not rise much above 35 degrees C (95 degrees F) and 50 to 60 percent, respectively). Caution must be exercised with any silver-containing product so that it is not exposed to sulfur, which destroys its solderability.

The indium and high-lead alloys have under these conditions much shorter shelf lives, and inventory control has to be much stricter. Storing such alloys in an inert atmosphere is very expensive, and refrigeration or freezing to minimize humidity effects is not advisable because of possible reactions as the result of condensation forming when they are returned to room temperature.

Any re-screening or other processing must be carefully handled. The oxide film first formed during atomization shields the particle surface against oxides that are normally created during its exposure to air at room temperature. When this barrier is removed (by abrasion or other means), the clean surface beneath reacts immediately to exposure to air, where a new, much more tenacious oxide layer is deposited.

2

Flux Classification

The choice of which solder paste to use for a given application depends on many factors that dictate which alloy and which flux to use. There are some analogies between liquid fluxes (as used in wave soldering, for example) and paste fluxes with regard to classification and terminology. Solder paste flux, however, must have additional attributes to ensure successful bond formation.

One of the factors that must be considered is "flux activity," a somewhat nebulous term that is indicative of the wetting ability and the corrosiveness of the flux. In general, the nomenclature used to specify flux activity takes into account a wide variety of activator systems. More precise definitions of activity are spelled out in various specifications. You might find differences as to what constitutes a particular type of flux activity depending on the particular specification used. This chapter discusses some of the properties and characteristics of fluxes that are generally used to assign each to a particular class. Selection criteria and details on the application of various flux types is discussed in more detail in Chapter 8.

SOLDER PASTE FLUX

The function of a flux as it is used in a soldering process has been reviewed in literature many times, especially with regard to "liquid" fluxes as used in wave or dip soldering. (See the first three references in the Notes section under

Chapter 2.) Regardless of the exact soldering process—wave, hot iron, paste reflow, etc.—it is generally accepted that a flux is a substance that

- prepares the surface to be soldered by removing surface oxides and tarnish to provide a nascent metallic surface that is conducive to wetting.
- protects the surfaces that are being soldered from further oxidation during the soldering process.
- leaves a residue composed of the flux ingredients and reaction products that is non-corrosive or removable by standard cleaning methods.

A flux can be a single substance or, as is usually the case, a mixture of substances. Depending on the soldering process, a flux can be a liquid, solid, molten salt, or viscous dispersion. What distinguishes a solder paste flux from all other fluxes? The first obvious difference is that solder paste flux must itself be a "paste," i.e., a structured substance that can suspend the alloyed solder powder. This is usually accomplished by adding substances that are specially designed to impart a thick, gel-like structure. The paste flux, of course, must be able to "clean" the substrate and perform the functions discussed above. However, for solder paste the source of the solder alloy is a finely divided powder dispersed in the flux. This means that the mechanism of forming a solder joint is somewhat different as compared to using wave, wire, or other soldering processes. Ultimately, the individual solder particles must melt and combine to form a uniform metallic bond.

The melting and coalescing of the individual solder particles, to some extent, is dependent on the surface condition of the powder. Expect oxides of tin and lead to be present on the powder surface. Oxide coatings inhibit the melting and flow of the solder, and the flux plays an important role in removing these oxides from the powder in addition to cleaning the substrate. The surface condition of the solder powder is of paramount importance in determining the quality of the solder paste.

ANATOMY OF A PASTE FLUX

To get a better understanding of what a paste flux is, use the following generic formulation. (Chapter 3 discusses some aspects of flux ingredients and chemistry in more detail.) Generic solder paste flux contains

Rosin/resin
Solvent
Activator
Thickeners/rheological agents
Other additives

This formula represents the barest essentials for an activated paste flux. The number of different ingredients in a formulation can easily increase: the use of multiple resins, solvents, mixed activators and other additives to achieve just the right properties can lead to complex formulations.

A solder paste usually contains rosin or some resin, natural or synthetic. *Resins* are usually solids or viscous liquids and impart certain physical properties to a flux. The resins play a role in the flow properties of the paste, especially at high temperatures, and they can contribute to imparting certain physical properties such as "tack." They provide the flux with "body" that contributes to the ultimate rheological properties of the paste. Resins can also contribute to the "activity" of the flux, as is the case with rosin-based fluxes. *Solvents* are needed to dissolve the resins and activators.

Activators are substances that give the flux its "cleaning" ability, that is, the ability to remove oxides from the surfaces being soldered. Activators cover a wide variety of chemical substances including inorganic acids and salts, amine hydrohalides, organic acids and bases, and any other substance that has the ability to reduce or remove oxides and tarnish. The activator is usually a small percentage of the formula, although the amount varies depending on the particular chemical species used and the desired "strength" of the flux. Choosing an activator system requires many aspects of the formulation to be taken into account; many times it is the synergistic effect of two or more substances that spells success or failure for proper activity. Of major importance is the corrosiveness of the activator residues after soldering.

Thickeners, viscosity modifiers, or *rheological control agents* (they are usually referred to as the latter) are usually a small percentage of the flux system. These substances are dispersed into the solvent/resin mixture and ultimately thicken the flux so that it obtains its characteristic paste-like consistency. The purpose is to suspend the powder so that it does not settle or separate to the bottom of the container. The thickening agent must also impart certain flow properties to facilitate consistent application of the paste (screening, stenciling, or dispensing).

Other ingredients might be added either to enhance the function of some of the above or provide other properties, for example, surfactants or tackifiers may be added.

The apparent simplicity of the formula given above is deceiving. Choosing and combining the proper functional materials to produce the desired result is a considerable challenge for the solder paste formulator.

CLASSIFICATION OF CHEMICAL FUNCTIONAL GROUPS

The nomenclature used to classify the flux types that are available is fairly well standardized, and most manufacturers of solder paste use similar terminology. Fluxes are usually classified as either R (rosin), RMA (rosin mildly activated), RA (rosin activated), OA (organic acid), WS (water soluble), IA (inorganic acid), and SA (synthetic activated).

This terminology is somewhat broad in scope, because a flux designation might imply something about its activator system (e.g. SA for synthetic activated), or it could imply something about its functional properties (e.g. WS for water soluble). In order to avoid confusion when discussing flux terminology, it is helpful to be familiar with some of the types of chemical functional groups commonly found in paste flux. The following discussion is presented, not as a comprehensive study of flux chemistry, but as an outline for those less familiar with chemical terminology.

Inorganic Salts

The use of inorganic salts are usually limited to the more active fluxes, such as the IA type. However, salts, particularly of organic compounds, are frequently found in other flux types. A salt is a compound consisting of a positively charged atom or radical such as K^+ (potassium ion) or NH_4^+ (ammonium ion) with a negatively charged atom or radical such as Cl^- or NO_3^-. These oppositely charged ions can combine to form ionic bonds and are held together by electrostatic attraction, thus forming compounds such as potassium nitrate (KNO_3) and ammonium chloride (NH_4Cl).

A salt might form by the displacement of the hydrogen in an acid by a metal. The reaction of metallic zinc (Zn) with an aqueous solution of hydrochloric acid (HCl) produces gaseous hydrogen and the salt zinc chloride according to the following equation:

$$Zn + 2HCl \rightarrow ZnCl_2 + H_2$$

Salts are ionic compounds, many of which dissolve readily in water and yield positive metallic ions (cations) and negative nonmetallic ions (anions). For example, ammonium chloride dissolves in water to produce ammonium and chloride ions.

$$NH_4Cl + H_2O \rightarrow NH_4^+ + Cl^-$$

These aqueous solutions may conduct an electric current due to the presence of the charged ions and are of concern in the soldering process. If not removed, they are a potential source for degradation of the insulating properties of the substrate assembly and/or a possible source of corrosion of the soldered metallic bonds.

The dissolution of a salt in water produces a solution that can be either acidic (have an excess of hydrogen ions—H^+), basic (have an excess of hydroxyl ions—OH^-), or neutral (the concentration of both H^+ and OH^- are equal). The reaction of an anion or cation of a salt with water is called *hydrolysis*. For example, the ammonium ion of ammonium chloride acts as an acid relative to water and donates its hydrogen ion (H^+) to the water molecule:

$$NH_4^+ + H_2O \rightleftarrows NH_3^+ + H_3O^+$$

The result is a slight excess of hydronium ion H_3O^+, and the resulting solution is slightly acidic.

On the other hand, some ions act as bases towards water and accept a hydrogen ion from it, leaving an excess of hydroxyl ions (OH^-), thus producing a basic solution. For example, the phosphate radical $PO_4^=$ of sodium phosphate acts as a base relative to water:

$$PO_4^= + H_2O \rightleftarrows HPO_4^- + OH^-$$

With some compounds such as sodium chloride (NaCl), neither the cation Na^+ nor the chloride anion Cl^- reacts with water, and the resulting solution is neutral.

Organic Acids

By definition, organic compounds are those that contain the element carbon. The term *organic acid* usually means those compounds that contain the carboxyl group, denoted by

$$\begin{array}{l} C=O \\ | \\ OH \end{array}$$

For example, acetic acid has the formula CH_3COOH. These compounds are acidic because under certain conditions, they donate the hydrogen ion of the OH group. In particular, acetic acid dissolves in water to yield an acidic solution of hydrogen ions and acetate ions:

$$CH_3COOH \rightarrow CH_3COO^- + H^+$$

You may write a carboxylic acid as R—COOH, the letter "R" being an abbreviation for the remainder of the molecule. There are numerous structures possible for the group R, so there are a wide variety of acids with different chemical and physical properties. Acetic acid, the main constituent of ordinary vinegar, is completely soluble in water and relatively acidic as compared to stearic acid—$CH_3(CH_2)_{16}COOH$—a waxlike substance that is insoluble in water. The number of carboxylic groups in a molecule can vary, and you often encounter organic acids in which two COOH groups are present; these are called *dicarboxylic acids*. Examples of dicarboxylic acids are HOOC—COOH (oxalic acid) and $HOOC(CH_2)_4COOH$ (adipic acid).

Amines

Organic amines are nitrogen-containing compounds of the general formula RNH_2, R_2NH, or R_3N where "R," as before, represents some chemical group. Organic amines can be considered as derivatives of the inorganic compound ammonia (NH_3), where the hydrogen atoms have been successively replaced by other groups denoted by R. Depending on the number of subs̆tuient groups, amines are classified as primary, secondary, or tertiary. For example, aniline, dimethylamine, and triethylamine are primary, secondary, and tertiary amines, respectively. See FIG. 2-1.

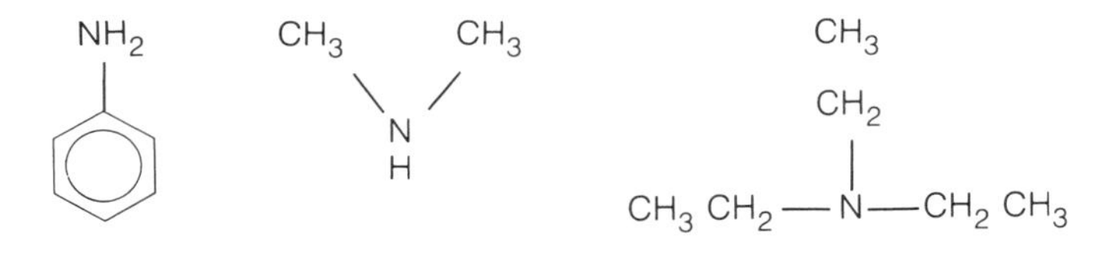

Aniline **Dimethylamine** **Triethylamine**

Fig. 2-1. Aniline, dimethylamine, and triethylamine are primary, secondary, and tertiary amines, respectively.

Organic amines are weak bases. The basic nature of an amine is attributed to its unshared electron pair that tends to attract a positive ion as in FIG. 2-2.

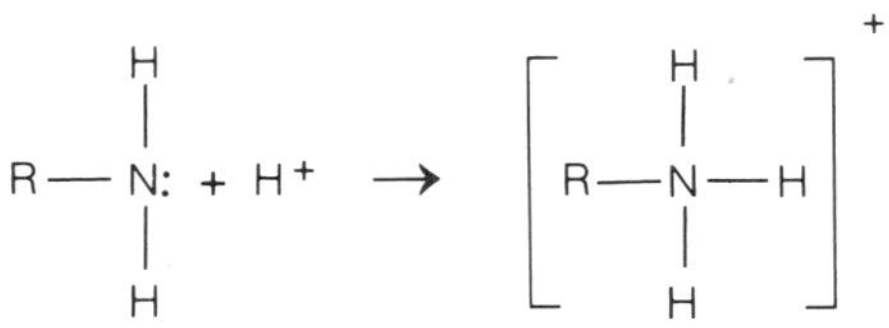

$$R{-}\underset{H}{\overset{H}{N}}{:} + H^+ \rightarrow \left[R{-}\underset{H}{\overset{H}{N}}{-}H\right]^+$$

Fig. 2-2. Unshared electron pairs in amines tend to attract positive ions.

Amine Salts

From the above reaction, note that amines may react with mineral acids such as HCl and HBr and organic acids to form salts. For example, dimethylamine and hydrochloric acid can form dimethyl amine hydrochloride:

$$(CH_3)_2NH + HCl \rightarrow (CH_3)_2NH_3^+ + Cl^-$$

The amine hydrochloride formed is often written as $(CH_3)_2NH_3 \bullet HCl$. Amine hydrohalides are ionic in nature and are analogous to the salts discussed previously, because an amine hydrohalide can also dissociate or ionize under the appropriate conditions to produce positive amine radicals and anions such as chloride, bromide, etc. Some other amine hydrohalides are shown in FIG. 2-3.

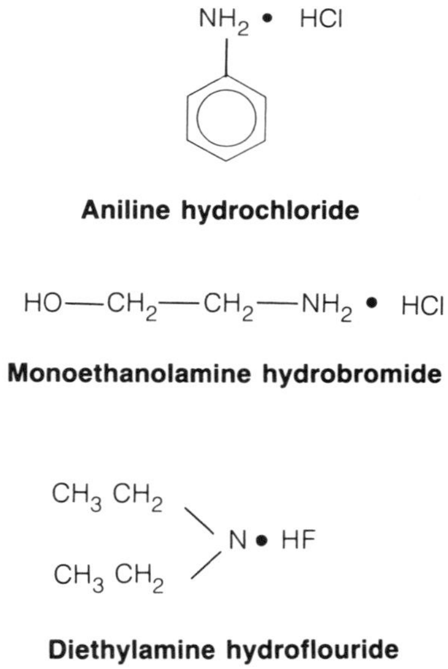

$NH_2 \bullet HCl$

Aniline hydrochloride

$HO—CH_2—CH_2—NH_2 \bullet HCl$

Monoethanolamine hydrobromide

$(CH_3CH_2)_2N \bullet HF$

Diethylamine hydroflouride

Fig. 2-3. Amine hydrohalides.

THE PHYSICS OF WETTING

The term *wetting* is often used when discussing the soldering process. The ability of molten solder to flow or spread during the soldering process is of prime importance, because the formation of a proper metallic bond is necessary. The phenomenon of spreading is usually referred to as wetting. By definition, wetting is a measure of the ability of a substance to spread over another substance and requires that at least one of the phases be a solid or liquid, that is, there must be at least one condensed phase. A measure of wetting is indicated as the change in area of, say, the liquid phase, or it is measured by the angle that the liquid makes with the solid surface. This is usually referred to as the *contact angle* and is shown in FIG. 2-4. Contact angles of 180 degrees and 0 degrees correspond to the limiting cases of no wetting and complete wetting, respectively.

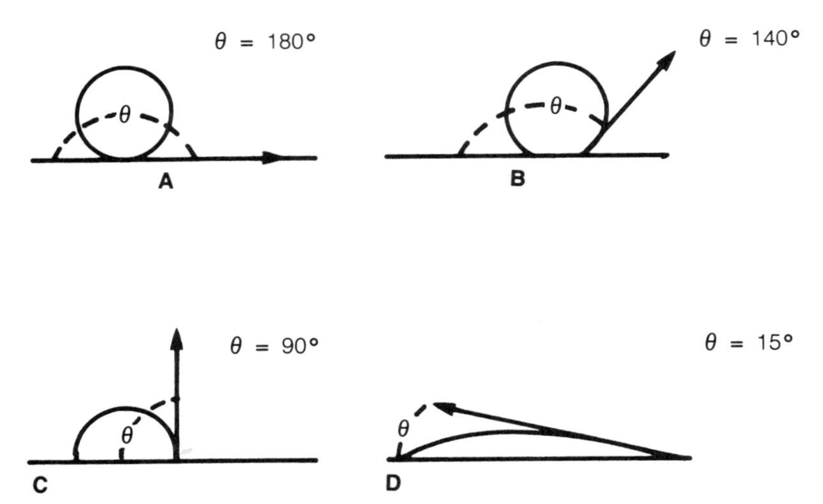

Fig. 2-4. Various degrees of wetting as indicated by the contact angle θ. The extreme case of complete nonwetting is shown in A with a contact angle of 180 degrees. Different degrees of wetting are shown in B, C, and D.

Ultimately, it is the nature of the surface forces that determine the ability of a liquid to wet a surface. As far as the soldering process is concerned, it is of interest to know what the conditions are that promote wetting. Wetting refers to how well a liquid spreads on a solid surface and the more the spread, the better the wetting. In soldering, this depends on the alloy and surface being soldered. Wetting might also refer to the ability of the solder alloy to form an intermetallic compound at the surface. For example, tin-lead alloys form an intermetallic of formula Sn_5Cu_6 on copper. However, sometimes it is desirable to have the apparently contradicting situation of a molten alloy with good wetting characteristics to produce a good bond but also have restricted flow.

The most fundamental characterization of wetting is given by considering the thermodynamics of the wetting process. Consider two branches of thermodynamics: *equilibrium* and *non-equilibrium* thermodynamics. The laws of equilibrium thermodynamics apply to systems in thermal equilibrium; that is, the observable properties of the system are not undergoing any change over time. Actually, the system must be in thermal, chemical, and mechanical equilibrium.

The state of a system can be characterized by the thermodynamic variables P, V, and T (pressure, volume, and temperature) and any composition variables, denoted by n_i. Because P, V, and T are usually related by some function, you only need to specify any two of the variables. Likewise, if the composition variables are related by some formula, you only need to specify the independent composition variables. The minimum number of variables that must be specified is referred to as the *number of degrees of freedom* of the system. Note that in equilibrium thermodynamics, you may consider the change in energy, work, etc., between two different states as long as each state is at equilibrium. The actual process of going from one state to another may or may not be an equilibrium process.

To characterize the thermodynamic properties of a changing system, requires the use of non-equilibrium thermodynamics. In such cases, time becomes a variable that must be included in describing the system so that the system is characterized by the variables (V,T,n,t). For example, if a chemical reaction is occurring, then the composition and possibly P, V, and T might all change simultaneously. You must then consider the rate of change of these quantities, that is, the kinetics of the process. If one has a very slow reaction, then the changes that occur for the thermodynamic variables can be carried out very slowly in a series of equilibrium steps, and thus equilibrium thermodynamics can be applied to the continuous change in the process.

The melting and wetting of molten solder on a metallic substrate is not an equilibrium process. It is an irreversible, non-equilibrium process. The solder cannot "dewet" or contract from small changes in the variables such as pressure, volume, or composition. Furthermore, for a tin-lead solder spreading on copper, the formation of the tin-copper intermetallic is a kinetic process and depends on time. However, an initial and final state of the solder-flux-substrate at some instant of time can be considered at equilibrium, and thus we can apply the laws of equilibrium thermodynamics to these two states. The initial equilibrium state could be the molten alloy, flux, and substrate at some temperature, pressure, and composition etc. What does thermodynamics tell us about the properties of the system with respect to wetting?

According to thermodynamics, a quantity called the *free energy*, denoted by G, can tell us whether a process will occur spontaneously under some given conditions. In particular, the sign of the difference in free energy between the two states (final minus initial) is an indication of the process occurring spontaneously. If the change in free energy is negative, then the process can occur spontaneously. In this situation, thermodynamics tells us that the process is possible but gives no information about the rate at which the process may occur. If the free energy change is positive, the process cannot occur spontaneously under the given conditions. If the change in free energy is zero, then the system is at equilibrium and remains so unless the conditions are changed.

The above is all fine and well if we know the free energy change of the system. It would be useful to relate the free energy change to some more familiar quantity. Without going into details, for a process at constant temperature and pressure, the free energy change can be related to the heat change during the process. An equivalent way of saying that the free energy must decrease for a spontaneous process is to say that the process must be exothermic, that is, heat must be liberated. As related to the soldering process, if there is a net result of a natural attraction between the surface atoms of the solder and the surface atoms of the substrate, then this is a stabilizing process providing liberation of heat. In terms of free energy, good wetting will occur if there is a great difference in the surface free energy between the two substances, because the result is a net lowering of the total free energy. That is, the surface energy of the solder lowers by forming an interface that is at a lower surface energy.

An elementary result of thermodynamics is that at constant temperature and pressure, the free energy represents the net work of the system and this net work can be related to other more basic variables of the system. For example, if the surface area of a liquid changes by an amount ΔA, then the work is W = force x distance or force per unit length x area, and W = dyne cm = [dyne/cm]cm^2. This force per unit length is denoted by γ and it is called the *surface tension*. Note that if $\Delta A = 1$ cm, then $W = \gamma$ so that the surface tension is the work required to expand a surface 1 cm^2. By using the fact that the free energy change represents work, you can relate the free energy change to surface tension as $\Delta G = \gamma \Delta A$ for a single pure substance.

For a system that consists of molten alloy (A), a substrate (B) and an interface (AB), one defines the work of adhesion as $W_{AB} = \gamma_A - \gamma_B$ and the work of cohesion $W_{BB} = \gamma_{BB}$. For spontaneous spreading, the work of adhesion should be greater than the work of cohesion. This is obtainable if the liquid B has a lower surface tension than the surface. Thus, it is not sufficient just to say that there should be a significant difference in the surface energy of the liquid and solid. A liquid with a high surface tension does not promote wetting.

Hence, thermodynamics provides some insight into the phenomenon of wetting. However, it is still difficult to explain or predict the exact properties desirable in a flux, metal, or substrate. Oxides and metal surfaces have surface free energies of the order of 500 to 5000 erg/cm^2. Based on surface energy alone it would be difficult to account for the wetting behavior of solder on metal surfaces. For the wetting of solder on copper, it could be that the dominant force is the formation of the intermetallic. The section on flux chemistry shows that the flux plays an important part in enhancing the ability of the solder alloy to form an intermetallic. The flux may also play a role in reducing the surface tension of the solder and thus reduce the work of cohesion, i.e., a flux can act as a surfactant. Note also that the purpose of a flux is to clean the surfaces to be soldered of any oxides that may have formed, but its purpose is *not* to overcome poorly prepared substrates or components. Grease or any other contamination should not be present to begin with.

Wassink claims that the mutual solubility of the solder alloy and the substrate is related to the wetting process.[4] The greater the solubility of the metal in the solid state, the lower the interfacial energy will be when these metals are brought into contact with one another. It is probable that a solid metal will be wetted by another liquid metal if the two metals are fairly soluble in one another.

Finally two other terms, *nonwetting* and *dewetting*, should be defined. Nonwetting is simply the inability of the molten solder to spread over the substrate. Dewetting, however, is the phenomenon where the molten alloy spreads over the substrate then retracts, leaving a "halo" of solder. This is illustrated in FIG. 10-4C. Dewetting is often observed and is not limited to solder. It has been found that liquid-to-liquid interfaces can show dewetting due to the occurrence of mutual saturation. The resulting saturated layer has a much higher surface energy than that of the original pure surface liquid. Dewetting of solder on copper is accounted for by Bondi by the formation of an outer zone of Cu_3Sn and an inner zone of Cu_6Sn_6.[5]

In summary, it is desired to have a molten solder alloy with a low surface energy (surface tension) and a substrate with a high surface tension so that the interface is stabilized by having a net lower surface energy. Some claims are made that oxides have low surface energies, so wetting would be expected to be poor and hence would explain the need for flux in order to remove the oxides. If oxides are high-energy surfaces, then other mechanisms such as intermetallic formation might also be important, and this again is influenced by the action of the flux.

FEDERAL SPECIFICATION QQ-S-571E

Flux selection depends on what is being soldered. If you are developing low-reliability consumer items or structural assemblies, then relatively active fluxes might be called for. On the other hand, high-reliability assemblies might call for less corrosive fluxes. The question to be answered is how do you distinguish flux activity?

In a broad sense, activity is related to the ability of the solder to wet the substrate. The amount of spread and time for this to occur are controlled to a great degree by the flux. As a general rule, fluxes that contain activators that are highly reactive to oxides of tin-lead and copper usually enhance the ability of the molten solder to spread. You might then expect activity to be based on some criteria of spread and time; indeed, such measurements can be carried out, and various devices exist that are available commercially to determine wetting rates.[6, 7] However, the ability of the molten solder to wet a substrate is enhanced by more active fluxes, and the possibility of corrosion is also proportionately much greater. Thus, when speaking of activity, concern is with many aspects of the chemical and physical properties of the flux.

The first attempts to classify flux activity arose from government specification. In particular, the specification QQ-S-571E, which is under the aegis of the U.S. Army Electronics Research and Development Command, contains specifications for solder alloy in the form of bar, wire, and preforms, and it also includes specifications for solder paste. With respect to flux activity, this document gives criteria for distinguishing between R, RMA, and RA fluxes. It also allows for a flux classification denoted by "AC," which refers to a non-rosin or non-resin flux that could contain acid and inorganic salts. This is what is usually referred to as IA (inorganic acid flux); the abbreviation AC is rarely used. The QQ-S-571E document provides specifications and test methods for alloy impurities, mesh size for powders, and other physical properties. However, flux activity is based on the following tests:

1. Resistivity of water extract
2. Chlorides and Bromides
3. Solder Pool
4. Spread factor
5. Effect on copper mirror

Details of some of these test methods are discussed in Chapter 10. Here is a brief summary of each test.

Tests 1, 2, and 5 might be considered tests that relate directly to the activity of the flux, because they evaluate the reactivity and potential corrosiveness of the flux. The solder pool and spread factor are indirect indications of flux activity, because they evaluate the wetting properties of the paste.

To carry out tests 1, 2, and 5, it is necessary to separate the flux from the solder powder. A detailed procedure is given in QQ-S-571E for the extraction of the flux so that you obtain a solution of about 35 percent by weight of flux in 99 percent IPA (isopropyl alcohol). This solution is used for the three tests mentioned above. Water extract resistivity (WER) provides an indication of the ionic content of the flux. The concern for the presence of ionic material is that it could be a potential source of corrosion and/or a source of electrical failure (electrical shorts) in the final assembly. The conductivity of a water extract of the 35 percent flux solution is measured and is expressed as resistivity, the reciprocal of conductivity, in units of ohm-cm. The mean of the specific resistances of the water extracts must be equal to or greater than 100,000 ohm-cm in order to qualify as either R or RMA. An RA flux must have an average specific resistance of at least 45,000 ohm-cm.

The chlorides and bromides test is a more specific test of the chemical nature of the flux. It tests for the presence of ionic chlorides and/or bromides. The presence of these ions is of concern, because they are potential sources of corrosion. They can contribute to a self-sustaining corrosion cycle in the presence of air (a source of carbon dioxide) and moisture to form lead carbonate. This test is also referred to as the "chromate test" because it is carried out by placing a drop of the extracted test solution on a piece of silver chromate test paper. A white to off-white discoloration of the brick red paper indicates the presence of these ions. Fluxes classified as R or RMA show no color change. The QQ-S specification provides visual standards. Examples of the chromate test can be found in FIG. 10-19.

The copper mirror test is another test of flux activity. A copper mirror consists of a vacuum-deposited film of copper on a glass substrate. The flux extract solution is placed on the mirror and stored for 24 hours at 23 ± 2 degrees centigrade and 50 ± 5 percent relative humidity. At the end of the 24-hour period, the flux is washed off the mirror and the surface is inspected for removal of the copper. For a flux to be classified as R or RMA, there should be no indication of any complete removal of the copper. Visual standards are supplied in the specification, and FIG. 10-20 shows examples of copper mirror tests.

The solder pool test is listed under fluxing action in the specification. This is more of a functional test and evaluates the ability of the solder paste to uniformly wet a copper substrate. No visual standards are supplied for this test. This test is commonly referred to as solderability on copper. This is also discussed further in Chapter 10, and FIG. 10-4 gives examples of different types of wetting that might be obtained in this test.

ROSIN FLUX CLASSIFICATION

As previously mentioned in the section concerning chemical classification flux designations such as R, RMA, and OA etc. might not necessarily reflect the true nature of the chemistry and/or activity of the flux. For rosin-based fluxes, more precision is used here in defining the difference between different activity levels. Again fluxes are classified as R, RMA, or RA if they conform to the requirements of the QQ-S-571E specification.

Fluxes classified as being of type R do not contain any added activators, and they depend on rosin as the main source of activation. For most applications with solder paste, these fluxes are not active enough to provide good solderability. However, high-melting alloys, such as high-lead alloys are sometimes used with an R flux because the rosin becomes active at the higher reflow temperatures and is able to provide wetting.

Type RMA flux is the most widely used flux system. RMA fluxes contain additives to enhance solderability and wetting. The specific chemical composition of the activators added to RMA type fluxes is not restricted by the QQ-S-571E specification. However, the flux must meet the test requirements of the QQ-S document. RMA solder pastes find a wide range of use in the electronics industry, especially for high reliability electronic assemblies. RMA-type activity is typically used in surface-mount assemblies. In general, an RMA flux is not corrosive or conductive before or after the reflow process. RMA pastes can also be used for most thick-film hybrid substrates. The telecommunications industry also makes great use of RMA paste, although they have developed their own specifications regarding the requirements for this type of flux.

Under the heading of RMA is the so-called "no-residue" solder paste. These fluxes contain substances that sublime and/or volatilize, leaving no residue. This represents the ideal situation with respect to the problem of cleaning and of residues. At this time there are no special classifications or test methods for this type of flux.

Fluxes designated as RA (meaning rosin-activated) contain activators that are more aggressive towards oxides and provide better wetting. This type of flux is used when the surfaces to be soldered are highly oxidized or difficult to wet, such as nickel and some thick-film substrates. RA flux might typically contain ionizable chloride compounds. The residues of such fluxes are corrosive, and in the presence of air and moisture can cause corrosion of the soldered joint. The use of RA fluxes necessitates cleaning the flux residue in those cases where reliability is of concern. RA fluxes also find use in some structural soldering applications.

IPC SPECIFICATIONS

The advent of the Institute for Interconnecting and Packaging Electronic Circuits (IPC) specifications IPC-SF-818 (General Requirements for Electronic Soldering Fluxes) and IPC-SF-819 (General Requirements and Test Methods for Electronic Grade Solder Paste) represents the most significant change in flux classification since the establishment of the military specifications. IPC-SP-819 addresses the properties of the paste, which include viscosity, tack, wetting, metal content, particle size, slump, and solder balls. For evaluating flux activity IPC-SP-819 refers to IPC-SF-818, and it should be noted that this specification also covers liquid flux, cored solder and solder preform flux. It is SF 818 which proposes a new and more general classification system for flux activity. The classification scheme is not restricted to rosin type fluxes. Any flux type may be classified, although inorganic fluxes are not covered by this document because their use in electronics is not common.

Flux activity is determined by the outcome of three classification tests: copper mirror, silver chromate, and corrosion. These tests can be performed on either the flux itself or the flux residue, that is, the residue left after carrying out a reflow of solder with the flux. One of the main differences between this specification and QQ-S-571E is that the IPC document is somewhat more flexible, since it allows for options as agreed by the supplier and user. For instance, as mentioned above, flux activity may be carried out on either the flux or the flux residue. The copper mirror and halide (Silver Chromate test) are similar in principle to the tests as described previously for the QQ-S-571E specification. The IPC document also gives explicit procedures for carrying out these tests. The corrosion test is not included in QQ-S-571E. This test gives direct evidence of the effect of the flux on a copper substrate after being stored in a controlled environment for a given period of time. At the end of the test period, the test panels are observed for any signs of corrosion. Figure 10-21 shows some examples of the flux corrosion test. Flux activity is denoted by the letter L, M, or H.

Flux activity is designated by the letters L, M or H, denoting low, medium, or high activity. The determination of the activity class depends on the results of the above three corrosion tests as described in TABLE 2-1. The flux is further characterized by evaluating its conductive properties using a surface insulation resistance test. This test is discussed in more detail in Chapter 10. Briefly, the test evaluates the electrical resistance of a test board. Typically a "comb" conductor pattern is treated with the flux and stored in controlled conditions of temperature and humidity. The resistance between the conductor pattern is measured and must meet certain requirements, as shown in TABLE 2-2, in order to be classified as either 1, 2, or 3. The numbers refer to the following assembly classes:

CLASS 1: Consumer Products
CLASS 2: General Industrial
CLASS 3: High Reliability

A further designation of "X" is used, which means that the insulation resistance failed in both the cleaned and uncleaned test modes. Hence, a series of flux designators are used: L, M, or H, and 1, 2, 3 (denoting flux type) and C

Table 2-1. Flux Activity Classification Tests

Flux Type	Corrosion Resistance Tests		
	Copper Mirror	**Silver Chromate**	**Corrosion**
	(Section 4.4.3.1)	(Section 4.4.3.2) or Halide Test (Section 4.4.3.3)	(Section 4.4.3.4)
L	No evidence of complete copper removal (No white background visible)	For Class 1 and 2 less than 0.5% halides (based on flux solids). For Class 3 must pass silver chromate paper test.	No evidence of corrosion. If blue/green peripheral border is visible flux must pass silver chromate paper test
M	Partial or complete copper removal	<2% Halides	Minor corrosion acceptable. Flux must pass halide requirement
H	Complete copper mirror removal allowed	>2% Halides allowed	Major corrosion evident

IPC, Lincolnwood, IL

Table 2-2. Surface Insulation Resistance Requirements

Flux Type	Assembly Classes Test Conditions		
	1	**2**	**3**
	50°C 90% RH 7 days	50°C 90% RH 7 days	85°C 85% RH 7 days
L	100 Megohm Residues not cleaned (N) *or* cleaned (C)	100 Megohm Residues not cleaned (N) and cleaned (C)	100 Megohm Residues not cleaned (N) and cleaned (C)
M	100 Megohm Residues not cleaned (N) or cleaned (C)	100 Megohm Residues not cleaned (N) and cleaned (C)	100 Megohm Residues not cleaned (N) and cleaned (C)
H	100 Megohm Residues cleaned (C)	100 Megohm Residues cleaned (C)	100 Megohm Residues cleaned (C)

IPC, Lincolnwood, IL

or N, meaning cleaned or not cleaned. If desired, the letter R may be used to denote a rosin-based flux. For example, a flux might be designated as MR2C. This describes a moderate flux or flux residue activity that passes the SIR requirements for a class 2 assembly when cleaned.

OA, IA, SA, AND WS FLUXES

OA Flux

The abbreviation OA refers to organic acid flux. With regard to solder paste flux, this term is somewhat misleading, because OA implies an organic acid activated flux, that is, a compound characterized by the carboxylic group. The term OA is used very loosely, and activators might contain carboxylic type compounds or amine hydrohalides. Typically, OA fluxes have had the reputation of being highly activated—more active than RA fluxes. In many cases, OA fluxes are water soluble. OA pastes have found use in "bumping" on printed wiring boards, repair work, and in soldering structural assemblies.

IA Flux

IA designates an inorganic acid activated flus. These are among the most active fluxes available. They are used in soldering assemblies that have very poor solderability. They are extremely corrosive, containing mixtures of ammonium chloride and zinc chloride. These fluxes might contain water as a solvent and have a tendency to spatter. These are typical plumbing fluxes. They are also used for automobile radiators and other structural assemblies.

SA Flux

SA denotes synthetic activated flux. Again, this is a somewhat nebulous term. Specifically, this flux type was designed to have residues that were soluble in chloro-fluorocarbon solvents. Specially designed activators were synthesized to have this property; hence the name synthetic activated. In general, these fluxes are also quite active and residues must be cleaned. With the concern of CFCs and atmospheric pollution, the use of this type of paste is expected to decline in the future.

WS Flux

In light of the current problems just mentioned, i.e., the desire to eliminate CFCs, it would be advantageous to have a solder paste that is acceptable for high-reliability electronic assemblies, yet be cleanable in water. This has the advantage of using non-polluting solvents and the possibility of cost savings also. There are fluxes in the market place that are water cleanable and can be classified as being analogous to the RMA type in flux activity or acceptable for high-reliability electronics, meeting the requirements of IPC-818. WS solder pastes are available that are halide free and cleanable in ordinary tap water without the need for any detergents or saponifiers. The availability of these pastes is only fairly recent, and their use in electronic assembly is quite limited at this time. A direct comparison of a water soluble paste and a solvent cleanable RMA paste has been carried out by Grunwald and Lowell.[8] Surface-mount assemblies were constructed with both types of paste. The RMA paste residues were cleaned with Freon TMS and the WS paste was cleaned with hot, deionized water. Surface insulation resistance was used as a measure of cleanliness. The result of their comparison was that aqueous cleaning was comparable to solvent cleaning.

3

Flux Chemistry

A discussion of flux chemistry should include some of the details of the mechanism of the fluxing process. This is important since it provides insight into what the best activator might be for a given type of flux. An activator should be inactive at ambient temperatures and active near the reflow temperatures of the solder. The activator must be dormant at room temperature or temperatures of up to about 120 degrees Fahrenheit (those that might be reached during transit). Any reaction at these temperatures could cause instability in the paste that could result in viscosity changes, crusting of the surface, or other adverse effects.

As discussed in Chapter 2, there are many classes of activator types such as halides, organic acids, and amines. It is expected that the ability of these compounds to react with metallic oxides of tin-lead and copper, for example, depends on the type of activator. Mixtures of activators cause even further complications. At this time, the approach to choosing an activator system is somewhat empirical, and much of the work is based on trial and error. Studies of activation energies or wetting rates for given activator types do not always necessarily show an obvious trend or correlation with molecular structure or other molecular parameters, and detailed studies of ''activator reactions'' are relatively scarce.

While the substrates and parts that are soldered in electronics applications are usually metals or alloys such as tin, tin-lead, palladium-silver, and nickel, the principle metal used to elucidate the mechanism of ''fluxing'' has been copper. Many of the standard test methods used to classify flux activity and corrosion use some form of copper, for example copper mirror, IPC flux corrosion test, and solderability on copper.

The following sections present a survey of some of the work that has been published on this subject. Also covered are other raw materials used in paste flux formulations and how they affect the chemical and physical properties of the flux.

ROSIN

Aside from the metal alloy itself, of all the materials that play a part in the soldering process, rosin must be placed in the forefront. Rosin has a long history of use in all forms of soldering, whether it is related to cored wire, wave soldering, or solder paste. With respect to solder paste, the single most widely used flux type is "rosin mildly activated" (RMA).

The use of rosin as a fluxing medium is no accident. Its properties and interactions with molten alloys of solder are quite unique. Even though rosin itself is not usually considered a "strong" activator, the duplication of the wetting and flow characteristics of rosin-containing fluxes by non-rosin fluxes is difficult. The chemical and physical properties of rosin give it a unique status as a fluxing material. The chemistry of rosin allows the cleaning of residues in both solvent and aqueous systems. Solvent cleaning systems that might prove to be a problem due to increased concern for health and atmospheric hazards do not prohibit the use of rosin as a continued viable raw material for fluxes.

Source and Composition

Rosin is a natural product obtained from the wood of pine trees. When discussing flux chemistry, it should be noted that rosin is classified as a natural resin. By definition, a natural resin is an organic solid that breaks with a conchoidal fracture (having a shell-like surface when fractured). Natural resins cover a broad range of chemical classes including balsam, copal, and amber. The term resin is also used in a general sense to cover various natural, amorphous organic substances or synthetic substances such as polymerized PVC.

Chemically, rosin is a mixture of organic acids (about 90 percent), the remainder consisting of non-acidic material. The rosin acids are monocarboxylic acids derived from alkylated hydrophenanthrene nuclei mainly of the abietic and pimaric type, as shown in FIG. 3-1.

Rosin can further be classified into three main types: gum, wood, and tall oil. A typical analysis of rosin is given in the following table.

Abietic acid	34%
Dehydroabietic acid	24%
Palustric acid	9%

Isopimaric acid	6%
Dihydroabietic acid	5%
Pimaric acid	5%
Neoabietic acid	3%

Gum rosin is obtained from trees that are individually tapped to yield the gum and then distilled to separate gum rosin and turpentine. Wood rosin is obtained from solvent extraction of wood stumps that undergo distillation. Tall oil rosin is a by-product of the pulping of resinous woods by the sulfate process. An alkaline "cook" converts the resin acids to soaps, separates them, then acidifies them to yield tall oil. The relative amounts of the rosin acids in wood and gum rosin are essentially the same so that the main differences are in the neutral fractions. Tall oil rosin is similar to both gum and wood rosin in composition and physical properties, and all three undergo virtually the same chemical reactions.

Being a natural product, expect variations in composition from lot to lot. In particular, the geographic location and species of the tree play a role in determining the resulting composition of the rosin. Commercial rosin is available from many sources throughout the world. For example, Joy and Lawrence[1] have analyzed rosins from widely varying geographical locations and give a quantitative breakdown of the constituents of the rosin acid fraction. The fact that there can be some variation in the composition raises some concern as to its use as a raw material, because any variation in the rosin might be reflected in the final paste flux. This fact has given some impetus to consider completely synthetic flux vehicles that, in theory, should be more reproducible from lot to lot. While minor variations in composition are probably insignificant as far as flux activity is concerned, monitoring some simple characteristics such as softening point, color, solubility, acid number, and tendency to crystallize can signal any potential problems.

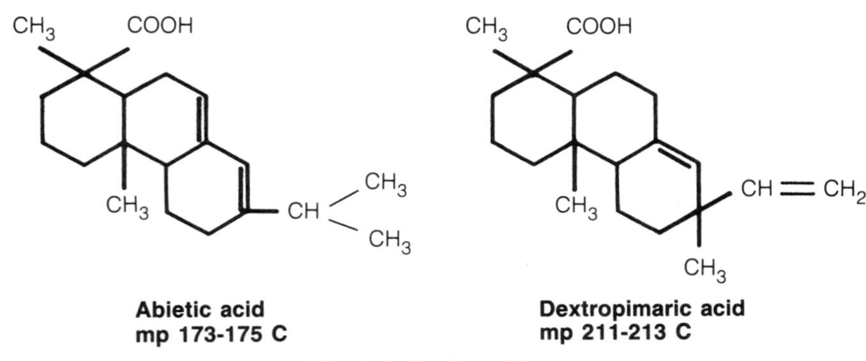

Fig. 3-1. Rosin acids.

Chemistry

The chemistry of rosin is determined essentially by its functional groups, the carboxylic group, and the double bonds. There are an enormous number of reactions and transformations that rosin can undergo, but only those reactions that relate to the purpose of this book are considered.

Oxidation. The double bonds in rosin are subject to oxidation. This oxidation can take place at room temperature and is enhanced if the rosin is in powdered or granular form. Processes using rosin at 150 degrees centigrade or above are normally protected against oxidation by using an inert atmosphere such as nitrogen. A possible product of oxidation would be for example a peroxy dihydroabietic acid, as in FIG. 3-2.

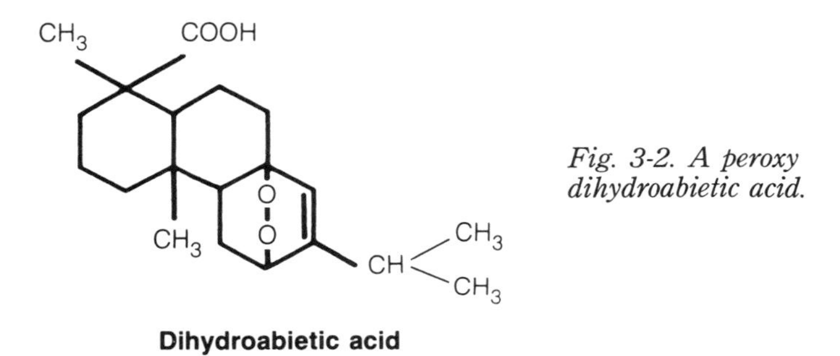

Fig. 3-2. A peroxy dihydroabietic acid.

Knecht and Hibbert[2] studied the absorption of oxygen by rosin quantitatively and obtained a result equivalent to the addition of two atoms of oxygen. Veitch and Sterling [3] studied the changes of powdered rosin in storage in closed containers. They found that in six weeks, the acid value fell 12 percent, while the saponification value increased by 8 percent and the iodine value decreased by 55 percent. They concluded that the change was wholly due to oxidation.

G. Dupont and J. Levy studied the rate of absorption of oxygen by a 33 percent solution of abietic acid in xylol.[4] They concluded that the oxidation is autocatalytic, the first stage being the formation of an oxide that acts as a catalyst.

Mitra studied the oxidation of molten rosin in the presence of various inorganic oxides.[5] Rosin was heated to 200 degrees centigrade with an air flow of 739-44 cc/min. The oxidizing effect on rosin of various oxides (0.01 percent oxide) was evaluated. Of the many oxides used, it was found that PbO, SnO and SnO_2 yielded a rosin that was oxidized less than the control.

Minn studied the oxidative stability of rosin products using high-pressure differential scanning calorimetry.[6] Results of this study indicate that gum rosin was more stable than wood and tall oil rosin with regard to oxidation.

Isomerization. The abietic acid isomers (abietic, neoabietic, and levopimaric acids) are unstable in the presence of acids and/or high temperatures because shifting of the conjugated double bonds can occur. Since, under reflow conditions, temperatures will be reached where isomerization can occur, consider these transformations. For example, when heated to about 200 degrees centigrade or treated with a solution of acidic ethanol, abietic acid is an equilibrium mixture of both levopimaric and neoabietic acid, as shown in FIG. 3-3.

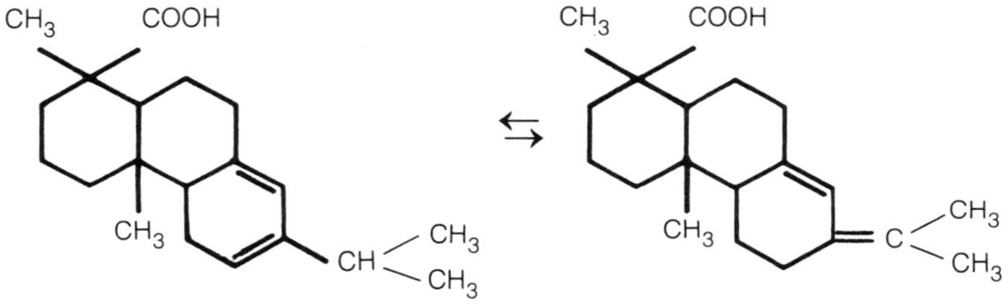

Levopimaric acid **Neoabietic acid**

Fig. 3-3. Isomerization of abietic acid yields an equilibrium mixture of levopimaric and neoabietic acid.

In the treatment with acid and heat, some disproportionation can also take place. *Disproportionation* is a simultaneous dehydrogenation-hydrogenation reaction. The net reaction results in the removal of two atoms of hydrogen from the two double bonds of abietic acid and subsequent rearrangement to form dihydro and dehydro abietic acid. Disproportionated rosins are less sensitive to atmospheric oxidation. In the study of Minn, it was also found that disproportionation of rosin was more effective than hydrogenation in providing stability toward oxidation.[6] Disproportionated rosins are commercially available.

Another method to produce stability against oxidation is to carry out a *dimerization* that occurs during a condensation reaction involving the double bonds, catalyzed by inorganic acids. The dimer molecule no longer has a conjugated double bond and is thus more stable against oxidation. These dimerized rosins are also commercially available. They have higher softening points and are less subject to charring at high soldering temperatures.

Carboxyl group. The carboxyl group of abietic acid is also a site for chemical transformations. One typical reaction of the acid group is esterification. *Esterification* is difficult to carry out due to the steric effects of the carboxylic group. However, once formed, rosin esters are extremely resistant to hydrolysis.

Esterification is the reaction of a carboxylic acid group with an alcohol, usually carried out in acidic or basic conditions. Using abietic acid as the source of the carboxyl group and combining it with ROH (some alcohol) gives the result shown in FIG. 3-4.

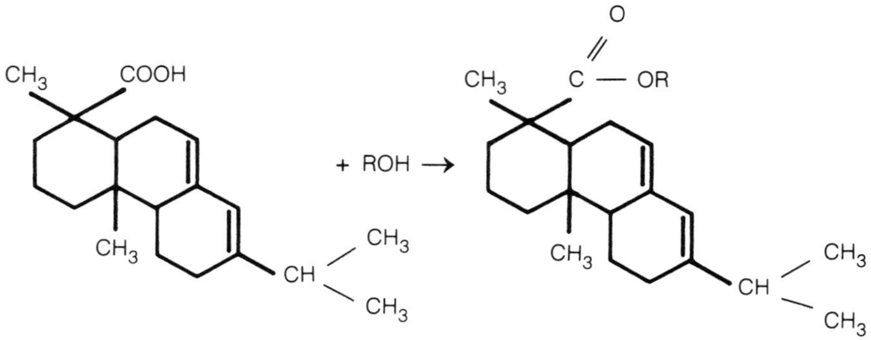

CH_3 COOH + ROH → CH_3 C(=O)—OR

Fig. 3-4. The resultant reaction of abietic acid and an alcohol.

Due to the vast choice of alcohols, numerous rosin esters are possible. A wide variety of these esters are available commercially. Rosin esters find use in paste flux as viscosity modifiers and tackifiers.

Salt formation can lead to many different metal resinates. These find use in various industrial applications. However, our main concern with resinates is that tin, lead, and copper resinates can form during the soldering process and can be the source of undesired residues. Rosin acids also form amine salts, i.e., amine soaps. These soaps are water soluble and thus form the basis of aqueous cleaning with saponifiers. See FIG. 3-5.

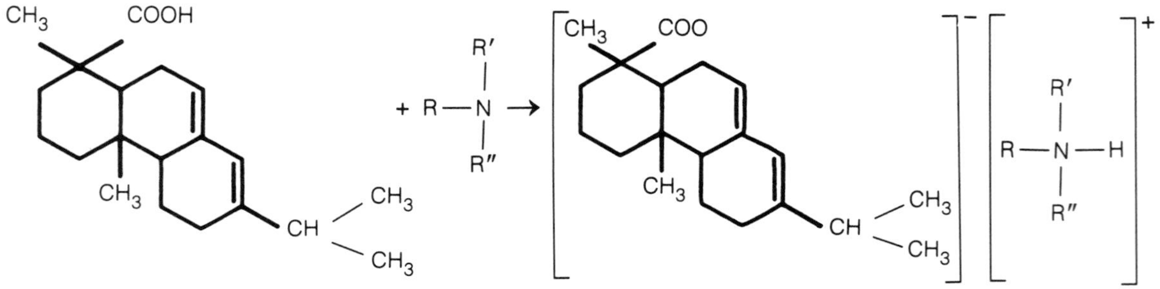

CH_3 COOH + R—N(R′)(R″) → [CH_3 COO]$^-$ [R—N(R′)(R″)—H]$^+$

Fig. 3-5. Reaction of rosin and an amine to form an amine soap.

Rosin Residues

Of a more practical nature, especially for the user of the solder paste, is the concern over white residues that can appear on the reflowed solder after defluxing. These white residues vary in color from white to various shades of tan and might appear as distinct particles or lumps or show up as a haze or film on the reflowed solder joint. The term "white residue" is commonly used in a generic sense and can refer to residues from any type of flux. Depending on the flux formulation and circumstances of reflow and cleaning, the compositions of the residues are usually difficult to analyze. In the case of rosin-based fluxes, some studies have been undertaken to identify the nature of these residues.

In particular, Lovering carried out an analysis of rosin residues.[7] The residues consisted of an ethanol portion containing ordinary rosin isomers and another insoluble component that is most likely a mixture of tin and lead resinates. This is supported by infrared and microprobe analysis.

Archer and Cabelka studied this problem further.[8] Their work indicates that while there is a substantial isomerization of the rosin isomers under soldering conditions, the majority of the rosin is unchanged. They also confirmed that the residues left after defluxing are a mixture of lead and tin salts of the rosin acids. It is much more common to obtain tan residues using fluorocarbon solvents as opposed to chlorinated solvents. This is accounted for by the fact that the chlorinated solvents can remove the rosin isomers more efficiently and quickly and carry along with it any tin and lead resinates.

While rosin might be a possible source of white residues, keep in mind that paste flux can contain other additives to enhance or impart other desired properties, and these can also be a potential source of residues. Also, the substrate and the parts being soldered must not be ruled out as a possible source of white residue.

ACTIVATORS

Inorganic Acid Activators

Inorganic acid fluxes are quite different from electronic grade fluxes in that they contain as a typical activator system a mixture of zinc chloride ($ZnCl_2$) and/or ammonium chloride (NH_4Cl). Vehicles often contain glycols, and the flux might also contain water as a substantial part of the formulation. This is the single outstanding difference. Water cannot be tolerated in electronic grade fluxes because when the flux is heated, the water tends to volatilize rapidly and cause a spattering of the flux and molten solder over a large area of the assembly being soldered. Furthermore, the residues after soldering are quite corrosive.

Studies of the fluxing action of this type of flux have been carried out over the years, and Latin concluded that a "tinning" mechanism was involved in the phenomena of wetting.[9] He observed narrow grayish areas around the solder alloy. These metallic areas are due to the formation of a thin deposit of tin on the copper where the tin is alloyed with the copper. It was shown that the deposition takes place from the flux. Oyama studied this further and confirmed that the mechanism involved the formation of stannous chloride formed by the reaction of the chloride flux and the molten solder, resulting in the deposition of the tin on the copper.[10]

A tinning sequence has been proposed by Lewis, which is shown in the following series of reactions.[11]

$$ZnCl_2 + H_2O \rightarrow Zn(OH)Cl + HCl$$

The HCl thus formed dissolves copper oxide.

$$2HCl + CuO \rightarrow CuCl_2 + H_2O$$

and the HCl might also react with the molten alloy to form tin chloride.

$$Sn + 2HCl \rightarrow SnCl_2 + H_2$$

A replacement reaction can also occur.

$$SnCl_2 + Cu \rightarrow CuCl_2 + Sn$$

The tin then forms an alloy with the copper surface.

Amine Hydrohalides

Amine hydrohalides have been used extensively as activators for rosin based fluxes. An amine hydrohalide of the general formula shown in FIG. 3-6 has associated with it a molecule of the acid HX, (X represents Cl^-, Br^-, I^- or F^-). The liberation of this moiety under the appropriate conditions accounts for the activity of the flux. The halides most frequently used are those of either chlorine or bromine. The relative activity of amine hydrochloride depends on the particular molecular structure of the amine, however, the more active rosin activated fluxes (RA) usually contain amine hydrochlorides and rosin mildly activated fluxes (RMA) typically contain the hydrobromide. The amine hydroflourides, while less corrosive towards copper mirror and undetected by

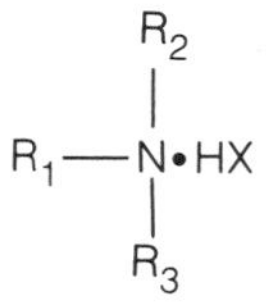

$$R_1 - \underset{\displaystyle R_3}{\overset{\displaystyle R_2}{N}} \cdot HX$$

Fig. 3-6. Formula of an amine hydrohalide.

the chromate test, are proportionately less active towards wetting and are not found to any great extent as activators in solder paste.

It has been demonstrated that halides play an important role in the wetting mechanism. Okamoto et al., in an extensive investigation of flux reactions, studied the wetting of tin-lead solder on copper using amine hydrochlorides.[12] They concluded that the hydrochloride reacts with the metallic copper to give a cupric halide, CuX_2, and a copper complex. These in turn react with the molten tin-lead alloy to form metallic copper and tin and lead chlorides. The copper formed immediately dissolves into the molten solder, forming a copper-rich layer at the outer surface that enhances the wetting of the molten alloy. For example, with aniline hydrochloride, both cupric chloride and a copper amine chloride complex could be isolated as indicated by the following reactions:

$$2C_6H_5NH_2 \bullet HCl + HCl + Cu \rightarrow CuCl_2 + 2C_6H_5NH_2 + H_2$$

$$2C_6H_5NH_2 \bullet HCl + CuCl_2 \rightarrow Cu[C_6H_5NH_2]\ Cl_4$$

It was also verified that the following reactions readily occurred at soldering conditions.

$$CuCl_2 + Sn \rightarrow SnCl_2 + Cu$$

$$2CuCl_2 + Sn \rightarrow SnCl_4 + 2Cu$$

$$CuCl_2 + Pb \rightarrow PbCl_2 + Cu$$

Other studies of the effect of various amine hydrochloride activators showed that there was no relation between base strength and activity as determined by Meniscograph wetting tests.[13, 14] Energies of activation did not correlate with base strength.

Organic acids

Organic acids can be used as activators, although in solder paste flux, they are quite often found in combination with other activators such as amine hydrohalides or amine salts. Onishi et al also claim a fluxing mechanism for organic acid similar in principle to what is proposed for amine hydrohalides.[15] Studies with stearic acid and oxidized copper indicate that stearic acid reacts with copper oxide to form copper stearate.

$$2C_{17}H_{35}COOH + CuO \rightarrow Cu(O\overset{\overset{\displaystyle O}{\|}}{C}C_{17}H_{35})_2 + H_2O$$

The copper stearate, in turn, reacts with the tin-lead alloy to form stearic acid and elemental copper.

The most common organic acids to be found in solder paste fluxes are the organic acids of rosin. These acids can react with oxides of tin, lead, and copper. Studies by Manko showed that rosin fluxes (rosin dissolved in isopropyl alcohol) reacted with copper oxide at room temperature to form green copper abiete salts.[16] Like rosin, these salts are inert and have high values of surface insulation resistance. Under similar conditions, no reaction was observed for tin-lead alloys. Metal salt rosinates can also be formed by fusion of molten rosin and metal oxides.

Formation of these salts was observed by Disque when a rosin-copper oxide mixture was heated to 400 degrees F, whereupon the resulting mixture, when cooled, showed a green color of copper abiete.[17] At higher temperatures, (about 500 to 510 degrees F), the copper oxide was reduced and a copper mirror was formed. When a similar experiment was carried out with an activated rosin flux containing amine hydrochlorides or organic acids no mirror was observed. However, green colorations were observed on cooling. This behavior is attributed to the more rapid reaction of the added activators with cupric oxide. It can then be expected that at temperatures below and near the reflow temperature of solder, oxides are removed by both the rosin and the added activators.

SOLVENTS

The function of solvents, as used in solder paste, are to dissolve the resins and, in particular for rosin based fluxes, a suitable solvent is desired so that a stable non-crystallizing solution is obtained. For non-rosin fluxes, you usually need a solvent in conjunction with some resin(s) and thickening agents to produce a gel-like flux. As mentioned in Chapter 2, water, if usable as a solvent, is used only in IA-type fluxes. Water is not desirable as a constituent of typical electronic-grade fluxes.

The choice of solvent is critical in formulating paste flux because there are many constraints that must be met. As with all raw materials used in the flux, the solvent must be acceptable from the toxicological standpoint. Non-hydroscopic solvents are desirable because absorption of water might accelerate reactions of activators with the tin-lead alloy powder. This could result in viscosity changes, crusting of the surface of the solder paste, or could cause a general deterioration of the ability of the molten alloy to wet the substrate during reflow. The solvents must be of low volatility at ambient temperatures; otherwise evaporation during storage could transform the paste into a solid-like crystallized mass of rosin and metal powder.

Typically, the boiling points are high; however, the solvent should be volatile enough so that the solder paste can be "cured" if necessary, before reflow.

Curing means evaporation of the solvent from the paste after deposition and before reflow. For example, this is done with vapor phase reflow so that there is no violent eruption of the solvent during the sudden heating of the paste by the condensing reflow liquid. Boiling and eruption of the solvent could cause movements of the parts on the substrate resulting in misaligned components. Finally, the solvent should be compatible with the activator system used and with any rheological additives that might also be present.

Some of the chemical classes of solvents that are used quite frequently are alcohols, glycols, and glycol ethers. A glycol ether can be considered a derivative of a glycol, and a glycol, in turn, can be considered a derivative of an alcohol. Alcohols are compounds that contain the hydroxyl group (OH), which in general can be written as R—OH. The hydroxyl group is polar (there tends to be a separation of charge between the oxygen and the hydrogen atom producing a dipole.) The degree of polarity depends on the structure of the remainder of the molecule. In particular, a class of alcohols known as *terpineols* ($C_{10}H_{18}O$) are effective solvents for rosin, and they have found extensive use in paste solder flux formulations for many years. This is not surprising because terpineols comprise some of the principal components of pine oil, which is obtained from an extract of the pine tree. Synthetic terpineols are available also. There are three terpineol isomers (molecules with the same molecular weight and same chemical composition but differing in their structure) that are denoted as alpha, beta, and gamma terpineol (see FIG. 3-7). The alpha and beta isomers are the most used of the three. These solvents have a characteristic pine-like odor and are of low volatility.

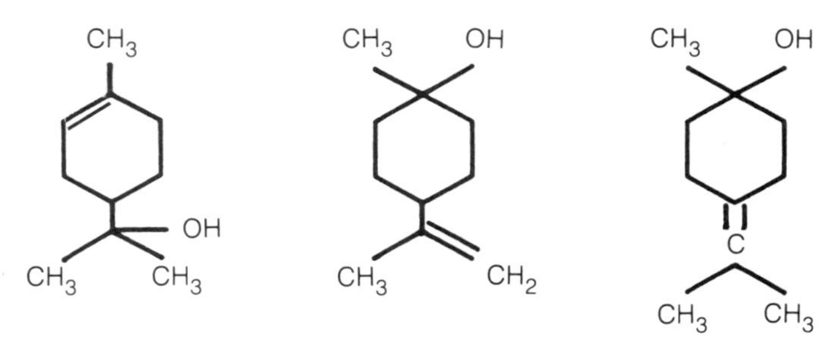

Fig. 3-7. Alpha, beta, and gamma terpineol isomers.

Glycols are dihydroxy alcohols, as exemplified by ethylene glycol and hexylene glycol, as shown in FIG. 3-8.

If more than two hydroxyl groups are present, one obtains compounds referred to as polyols, such as glycerine as shown in FIG. 3-9.

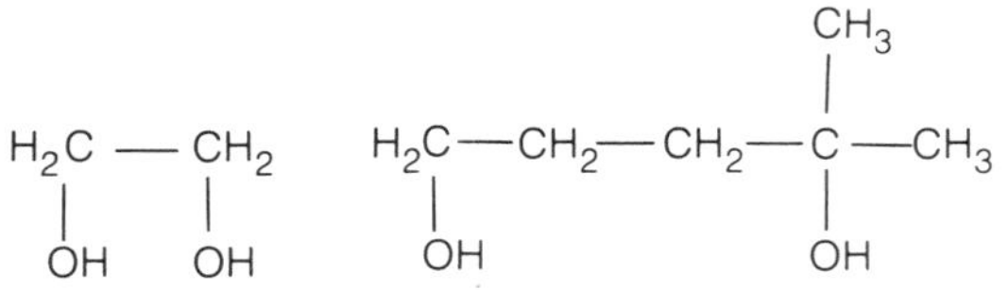

$$H_2C-CH_2 \qquad H_2C-CH_2-CH_2-C(CH_3)-CH_3$$
$$\text{(OH on each carbon)} \qquad \text{(OH on C1 and C4)}$$

Ethylene glycol **Hexylene glycol**

Fig. 3-8. Glycols are dihydroxy alcohols.

Low-molecular-weight glycols are odorless, white liquids and, because of the presence of the polar hydroxyl group, tend to be hygroscopic. Solder paste can contain a range of glycols. They are frequently used in conjunction with other solvents. The low volatility of glycols helps to prevent the solder paste from drying out. Glycols have been used to control or restrict the flow of the molten tin-lead alloy. Because the lower molecular weight glycols are water soluble, it's conceivable that they would be useful as solvents for water-cleanable paste flux. In actuality, their use in such formulations is limited because of the hygroscopicity of these compounds.

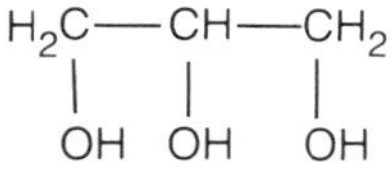

$$H_2C(OH)-CH(OH)-CH_2(OH)$$

Fig. 3-9. Glycerine is a polyol.

A slight modification of a glycol in which one of the hydroxy groups is converted to an ether linkage (R—O—R) yields a class of compounds known as glycol ethers, also referred to as carbitols and cellosolves. Simple mono- and di-ether derivatives of ethylene glycol can be represented as shown in FIG. 3-10. More complex molecules that include multiple ether linkage are produced to yield a wide array of commercially available solvents with varying properties.

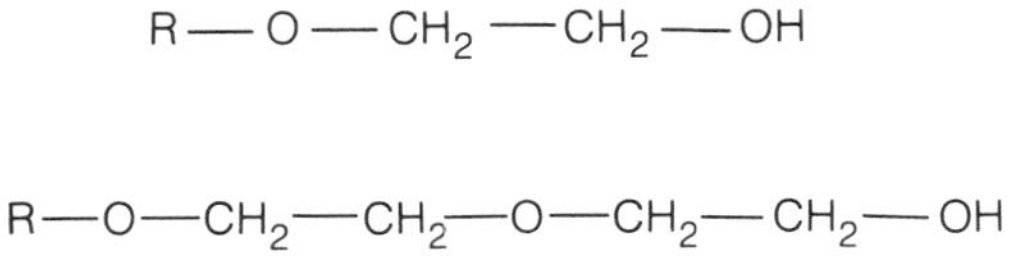

$$R-O-CH_2-CH_2-OH$$

$$R-O-CH_2-CH_2-O-CH_2-CH_2-OH$$

Fig. 3-10. Mono- and di-ether derivatives of ethylene glycol.

RHEOLOGICAL ADDITIVES

The term *rheological additive* is an all-encompassing term when referring to paste flux. Substances used to impart rheological properties might truly be an additive, because the amount of substance present is a relatively small part of the flux; 10 percent by weight or less, for example. On the other hand, substances might be present in substantial amounts and act not only to control the rheological properties but to act as the major resin or vehicle for the flux.

In any case, they all have the purpose of imparting certain physical properties to the flux so that it will be a paste or gel. The term ''gel'' is used loosely.

Specifically, a gel refers to a disperse system in which some substance provides a three-dimensional "structure" to give rigidity or body to a system; the remaining components of the system fill the space between the structural units (intermolecular network).

One of the principal reasons for incorporating these additives is to provide enough of a gel-type structure so that the alloyed solder powder, when mixed with the flux, remains suspended. If solder powder is placed in a flux that lacks sufficient structure, the powder slowly settles, and after a long enough time, becomes compactified at the bottom of the container; redistribution of the powder can be accomplished only with difficulty. Addition of powder to the flux causes changes so that the rheological properties of the paste are quite different than that of the starting flux. This means that the flux must be formulated so that after addition of the solder powder, the desired properties are obtained. Details of the rheological effects of adding powder to flux are discussed in Chapter 4. Rheological agents are also needed so that the paste will perform properly in actual use. Cartridge dispensing, screening, and stenciling all require specific rheological properties.

Deciding which rheological substances to use is dictated to a great extent by the class or type of flux: rosin, non-rosin, solvent, or aqueous based. For aqueous systems, there are an enormous variety of rheological additives to choose from. These include natural gums (guar and xanthan gums) and cellulose derivatives (sodium carboxymethyl cellulose, ethoxylated cellulose, etc.), of which there are numerous derivatives. Silicates, clays, and acrylic acid polymers provide even further choices. Unfortunately, aqueous systems are limited to IA type fluxes, so these materials are essentially unusable for the majority of fluxes used for solder paste. Some of the above materials can be dissolved in certain non-aqueous solvents, but even in these cases they are difficult to incorporate.

The thickening of solvent systems to produce a suitable gel of the desired viscosity is much more difficult than for aqueous systems. For non-aqueous solvents, the number of commercially available thickeners is quite limited. One of the more commonly used thickeners is based on a modified castor oil derivative. These are available from various manufacturers under different trade names. The exact modifications and structures of these compounds are not disclosed by the manufacturers. Castor oil itself is classified as a non-drying oil and has a long history of use in alkyl resin paints. Castor oil is a triglyceride ester of fatty acids. Approximately 90 percent is ricinoleic acid, an 18 carbon acid containing a double bond in the 9-10 positions and a hydroxy group on the 12th carbon.

The castor oil can be modified by hydrogenation or possibly amine modified. The resulting substances exhibit oxidation stability, and are inert non-hygroscopic. These materials must be incorporated into the flux under very specific conditions of time, temperature, and shear during mixing. The fully dispersed state of these substances is that of a colloid gel. Picture this by thinking of the controlled swelling of deagglomerated particles until a fully activated state of a continuous network forms.

Another class of substances sometimes used as rheological control agents are waxes. This covers an extremely broad range of materials. In simple terms, a wax is a substance that is a plastic solid at room temperature and tends to soften and become a liquid at higher temperatures. When waxes are melted in a compatible solvent at an appropriate concentration, the mixture, when cooled to ambient temperatures, can have a gel-like appearance. This is not as easy to achieve as it sounds. The wax must have a high enough softening temperature for the flux to withstand extreme storage conditions, especially thermal cycling, where softening and resolidification of the wax could result in a permanent change in the paste rheology. Wax-based fluxes are more difficult to formulate for satisfactory screening and stenciling properties.

Waxes can be classified according to their source: insect and animal, petroleum or synthetic. Insect-animal waxes are mixtures of long-chain acid esters of fatty acids and hydrocarbons. Petroleum waxes are composed of hydrocarbons. Various grades of petrolatum have been used in fluxing materials. Probably their oldest use is in the inorganic acid plumber-type fluxes.

Synthetic waxes offer another source of these materials. They might be based on different starting materials. Molecular weights and functional groups can be varied to control melting points, solubility etc. For example, a large number of synthetic waxes are based on amide derivatives of fatty acids, polyol ether-esters, and polyoxyethylene glycols. The general formula for a polyoxyethylene glycol is

$$HO-(CH_2CH_2O)_N-H$$

Some waxes, such as those based on a polyoxyethylene glycol, might be water soluble. However, be careful because these molecules are also polar and tend to be hygroscopic. The properties of synthetic waxes are related to their molecular weights. In general, as molecular weight increases, solubility in water, hygroscopicity, and vapor pressure decrease. Melting point and viscosity increase with increasing molecular weight.

4

Rheological Considerations

The rheological behavior of solder paste relates directly to the methods used in applying it to an assembly. *Rheology* is the study of the flow and deformation of substances when subjected to a stress. Application methods such as screening, stenciling, dispensing, and roller coating subject the paste to some particular "stress history" that is unique to that process. Properties such as (1) quality of print definition and "paste life" in printing, (2) the tendency of paste deposits to clog and shape in dispensing and (3) uniformity of coverage in roller coating all depend on the rheological properties of the paste. Paste formulations must be tailor made so that performance is optimized for any given process.

With respect to rheological considerations, there are two main questions to consider:

1. What are the best methods and instruments to determine and study the rheological properties of solder paste?
2. What are the rheological properties that are desired in a solder paste for a given application?

The subject of rheological instrumentation and methodology is probably one of the most controversial areas of paste technology. This is due, partly, to the fact that the process methods using solder paste have become more demanding and, as a result, more meaningful information is needed from rheological measurements.

In particular, the importance lies in measuring rheological properties under conditions that are similar to the actual operating conditions of the production process. The standard method of measuring paste viscosity uses the Brookfield viscometer. While this instrument is simple to use and affordable, it provides limited information in predicting the overall paste behavior. One then looks to other, more sophisticated instrumentation, such as an absolute viscometer, which can provide shear rates that are comparable to those found in the production process. While these instruments are valuable research tools, they are more costly and require more attention in obtaining and interpreting data. Even with these instruments there are pitfalls, and their applicability may be limited in certain cases. This is an active area of study and there are many avenues of approach open for investigation.

With respect to the second question, there is surprisingly little in the literature that deals with solder paste and its rheological properties, especially under the conditions of printing and/or dispensing. Most of the efforts in this regard appear in studies on the screening properties of thick-film inks. Although the results of these studies are valuable, solder paste has its own unique characteristics that must be elucidated further. The typical solder pastes of today, with 89 to 91 percent by weight (about 50 percent by volume) of alloyed powder are concentrated dispersions in "non-Newtonian" flux vehicles. In general, the properties of dispersion systems are complex and far from being completely understood, especially on a theoretical basis; therefore, one must rely on empirical and semi-empirical methods to obtain results of practical value.

The remainder of this chapter provides an elementary overview of rheology to make you familiar with the common terms and definitions encountered in the study of this subject. The types of viscometers/rheometers being used in the solder paste industry today and how paste rheology can be related to process conditions are also discussed. Finally, there is an introductory discussion of *viscoelasticity*, a subject of greater complexity that provides information that cannot be obtained from the analysis of flow curves alone. Any complete account of the behavior of solder paste must ultimately consider the effects of both viscosity and elasticity.

RHEOLOGY DEFINED

The introduction to this chapter stated that rheology is the study of the deformation and flow of matter. To be more specific, the fundamental goal of the rheologist is to be able to predict the behavior of a substance (i.e. determine its velocity distribution) when it is subjected to external forces. The fundamental equation of motion for a deformable substance is given by

$$\varrho(d\mathbf{v}/dt) = \varrho\mathbf{g} + \nabla \bullet \mathrm{T} \qquad [4\text{-}1]$$

where ϱ is the density, **g** is the acceleration due to gravity, **v** is the velocity and $\nabla \bullet$ T, the divergence of T, represents the internal stresses. This equation is applicable to both solids and liquids. The feature that distinguishes liquids and solids is the exact nature of the internal stresses. Solids are rigid bodies and internal stresses oppose any deformation due to shearing. A solid can only undergo a limited amount of deformation beyond which fracture will occur. Fluids undergo irreversible deformation and continue to deform as long as they are subjected to a shearing stress. Any attempts to solve equation [4-1] require a definite statement about the internal stresses, that is, an expression that relates the stress to the deformation, or rate of deformation (shear rate). Such a relation is called a *rheological equation of state.*

With the concern being mainly in fluid-type behavior, the interest is in equations of state that relate stress components (τ) to rate of shear ($\dot{\gamma}$). In general, the stress is considered to be a function not only of the rate of deformation, but also of time t, temperature T, pressure P, and composition variable C_i. To simplify things, assume a given set of conditions and express the stress as some function of the shear rate:

$$\tau = f(\dot{\gamma})$$

or in the inverse form $\dot{\gamma} = g(\tau)$, depending on which is more convenient.

Even if the equation of state is known, it is only in the simple case when stress is proportional to shear rate that solutions to equation [4-1] can be obtained. These are known as the *Navier-Stokes equations*. These equations play a fundamental role in the theory and application of fluid mechanics.[1,2] The solution of these equations yields the stress and velocity distribution for the fluid for some given boundary conditions. The Navier-Stokes equations become extremely difficult to solve if non-linear equations of state are used. Unfortunately, most of the fluids in the real world are "non-linear." Predictably, solder paste falls into this category. Thus, any hope of predicting its behavior based on a theoretical framework is far from practical. However, what we can study and formulate is the relation between stress and shear rate. Rheological equations of state are usually empirical or semiempirical equations that result from an experimental plot of stress versus shear rate; such a plot is called a *rheogram*.

KINEMATICS

Stress Defined

The kinematical state of a body is specified if the position and velocity of all points in the body are known at some time t. In the case of deformation and flow, suitable kinematic variables are the strain and rate of strain. The dynamical state of a body can be determined when the forces acting on the body are known. The forces that act on a body can be divided into two categories—body forces and surface forces. *Body forces* are those that act throughout the substance such that the body is not in physical contact with the source of the force. Examples of body forces are gravitational and electromagnetic forces. *Surface forces* arise by direct contact with a boundary of the body.

Figure 4-1 shows an arbitrary element of surface area ΔS of a body where a force F is being applied. The area ΔS is small enough to assume the force is constant over the surface ΔS. The force F can be resolved into normal and tangential components. The body itself is not accelerating so that the forces represent tensions that tend to deform the body. The stress is defined as the ratio of the force F to the area ΔS:

$$\text{Stress} = \tau = \frac{F}{\Delta S}$$

The units of stress are $N(\text{Newton})/m^2$ = Pa(Pascal). Dividing the normal and tangential component of force by ΔS gives the normal and tangential stress respectively:

$$\tau_{\text{normal}} = \frac{F_{\text{normal}}}{\Delta S} = \tau_{ii}$$

$$\tau_{\text{tangential}} = \frac{F_{\text{tangential}}}{\Delta S} = \tau_{ij}$$

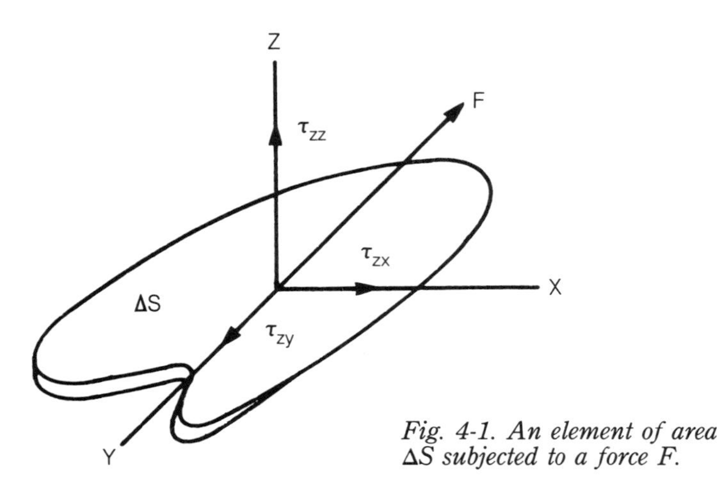

Fig. 4-1. An element of area ΔS subjected to a force F.

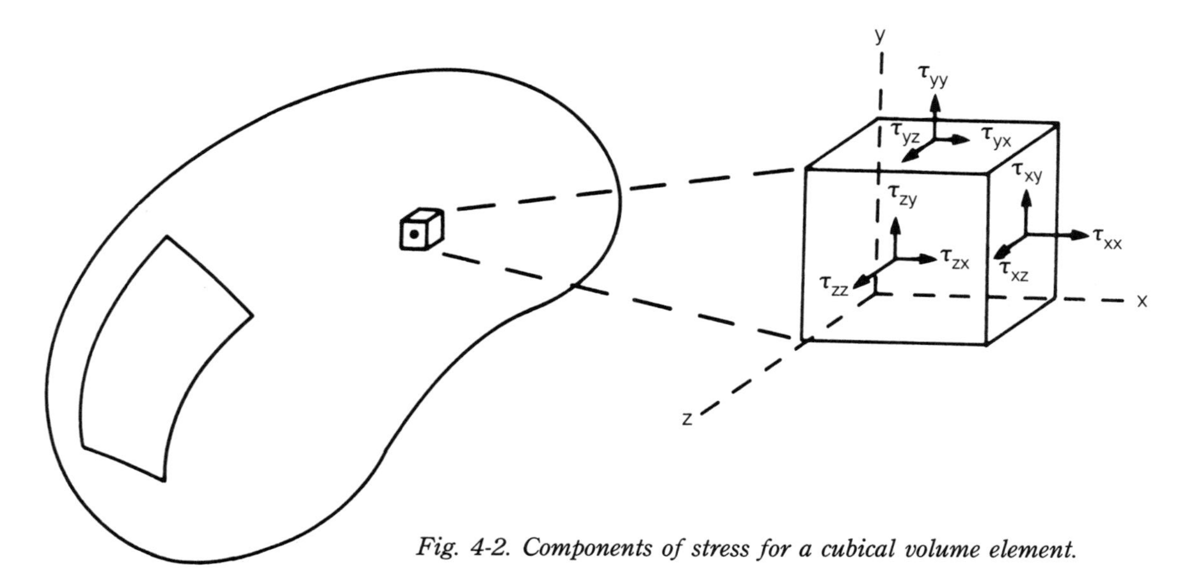

Fig. 4-2. Components of stress for a cubical volume element.

The general state of stress in a body can be characterized as follows. Consider a point P of a body surrounded by a cubical volume element. A surface force is acting on the body so as to cause some state of tension. For every surface element of the cube, there is one normal and two tangential components of stress as shown in FIG. 4-2. The stress components are characterized by a double-variable subscript. The first subscript indicates the direction of the normal to the plane associated with the stress, while the second subscript denotes the direction of the stress itself.

There will be a stress on each face of the cube so that there will be 18 components. However, if the cube is taken to be small enough, you can consider stresses on opposite faces as equal to each other. This follows from Newton's third law—for every action there is an equal and opposite reaction. The result is the need to only consider nine distinct components, and these can be arranged in an orderly array in the form of a matrix.

$$\tau = \begin{bmatrix} \tau_{xx} & \tau_{xy} & \tau_{xz} \\ \tau_{yx} & \tau_{yy} & \tau_{yz} \\ \tau_{zz} & \tau_{zy} & \tau_{zz} \end{bmatrix}$$

The above matrix is called the *stress tensor*. The diagonal terms are the normal components and the off-diagonal terms are the shear components. The normal components represent the hydrostatic pressure P of the fluid, and $\tau_{xx} = \tau_{yy} = \tau_{zz} = P$. For substances of a more complex nature such as those that exhibit elastic properties, normal forces can arise from sources other than hydrostatic pressure. This phenomenon is examined further when discussing viscoelasticity later in this chapter.

The purpose in considering the stress tensor is simply to use it as a convenient ''bookkeeping'' device. Depending on the values of the components, in particular which components are non-zero, you can classify different states of stress. A further simplification results for a body at rest (stated without proof). At equilibrium, the sum of the moments about each axis must be zero. This leads to the result that $\tau_{xy} = \tau_{yx}$ etc. so that the preceding matrix is symmetric about the diagonal. The general state of stress in a body is represented by,

$$\tau = \begin{bmatrix} \tau_{xx} & \tau_{xy} & \tau_{xz} \\ \bullet & \tau_{yy} & \tau_{yz} \\ \bullet & \bullet & \tau_{zz} \end{bmatrix}$$

where the dots represent the corresponding terms below the diagonal.

Stress under Simple Shear

There is a specific type of deformation called *simple shear* that produces a great simplification in the state of stress of the body. Simple shear is used as a model for many physical processes and is the basis of many rheological measurements. Suppose that a rectangular solid body is subjected to a force that is acting parallel to the top surface, which has an area A. The result is that the rectangular body is transformed into a parallelogram of the same volume, as shown in FIG. 4-3. Because F is the only force acting on the body, the stress tensor becomes

$$\tau = \begin{bmatrix} \tau_{xx} & \tau_{xy} & 0 \\ \bullet & \tau_{yy} & \bullet \\ 0 & 0 & \tau_{zz} \end{bmatrix}$$

Thus, for the shearing component, you need only consider one component and simply write $\tau_{xy} = \tau = F/A$.

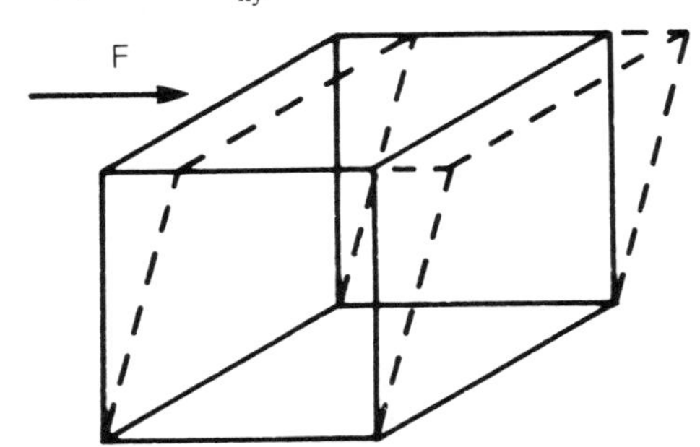

Fig. 4-3. Simple shear of a rectangular body.

Shear Rate

The distinguishing feature of liquids, as mentioned before, is that a liquid will continue to deform as long as it is subjected to some external force. The inherent resistance to flow at any time depends on the rate of change of the deformation, that is, the gradient of the deformation. In general, the rate of deformation of a fluid can be treated in a similar manner as it was done for the stress. The rate of deformation tensor is written as follows

$$\dot{\gamma} = \begin{bmatrix} \gamma_{xx} & \dot{\gamma}_{xy} & \dot{\gamma}_{xz} \\ \bullet & \dot{\gamma}_{yy} & \gamma_{yz} \\ \bullet & \bullet & \dot{\gamma}_{zz} \end{bmatrix}$$

The exact derivation and interpretation of all the components requires some lengthy mathematical derivations and is beyond the scope of this discussion. (For details, refer to notes 3 and 4 for this chapter.) Fortunately, this matrix does not have to remain as it stands. Considering again the case of simple shear, the shear rate tensor reduces to

$$\dot{\gamma} = \begin{bmatrix} 0 & \dot{\gamma}_{xy} & 0 \\ \bullet & 0 & 0 \\ 0 & 0 & 0 \end{bmatrix}$$

because deformation occurs only with respect to the y axis. The term $\dot{\gamma}_{xy}$ has a simple interpretation and it is this quantity that is usually referred to as the shear rate.

Consider a liquid subjected to simple shear as shown in FIG. 4-4. The liquid is placed between two parallel plates of area A. Under the specification of simple shear, the plate is subjected to a tangential force F resulting in a shear stress $\tau = F/A$. This stress is the same throughout the liquid once it reaches a steady-state of flow. Picture the liquid as being composed of "layers" of thickness (Δy) piled upon each other, so that as the top plate moves, these layers or laminae slide over each other, producing the deformation shown in FIG. 4-4. This model, although apparently naive, is found to be justified for describing flow as long as there is no violent motion or turbulence. The fluid velocity will range from zero to that of the moving boundary at the top. The shear rate is defined as the rate of change of the velocity with respect to the distance y, so for a given layer,

$$\text{shear rate} = \dot{\gamma} = \frac{\Delta v}{\Delta y} = \frac{(L/T)}{L} = T^{-1} \qquad [4\text{-}2]$$

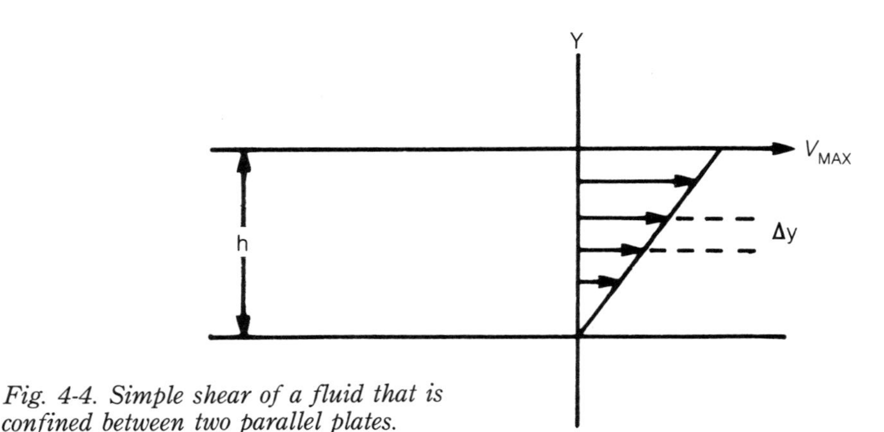

Fig. 4-4. Simple shear of a fluid that is confined between two parallel plates.

Shear rate has units of reciprocal time, usually expressed as $\sec^{-1}$. The shear rate will be constant and the velocity profile will be linear. This is a completely general result and holds for any fluid undergoing simple shear. Consider the general rheological equation of state $\dot{\gamma} = g(\tau)$. This specifies how the shear rate varies with the stress, which does depend on the properties of the liquid. However, imposing some given constant stress, denoted by k, gives

$$\dot{\gamma} = g(k) = K$$

and the shear rate is constant, denoted by K because a function of a constant is also constant. Using the definition of shear rate gives $\Delta v/\Delta y = K$, which is the slope of the straight line $v = Ky$ (which passes through the origin). This shows that the velocity is a linear function of the distance between the plates. A convenient formula for the shear rate can be written by substituting the velocity and distance of the upper and lower plates (v_{max}, $y = h$, $v_0 = 0$, $y_0 = 0$) into equation [4-2] so that

$$\dot{\gamma} = \frac{v_{max} - v_0}{h - y_0} = \frac{v_{max}}{h}$$

and the shear rate can easily be calculated by knowing the velocity of the top plate and the thickness of the liquid.

The situation just described for a fluid between parallel plates is idealized. End effects have been neglected but can be justified if the areas of the plates are large compared to the thickness of the fluid. It is also assumed that the velocity of the fluid, at the solid boundary of the upper plate, is equal to velocity of the plate, i.e., there is no slipping. Finally, to actually achieve a steady state so that the stress is constant throughout the liquid requires a very thin separation between the plates.

The definition of shear rate as given above is fairly simple and can be calculated if information is available for the velocity and thickness of material being sheared. In the following examples, assume that the situation is as shown in FIG. 4-4—that is, a constant shear rate is maintained. The velocity is the maximum velocity that occurs, and all calculations are based on $\dot{\gamma} = v_{max}/h$, where h is the thickness of the substance being sheared.

Slump on a Vertical Surface. Suppose that a fresh layer of paint on a wall has a thickness of 4 mil (0.004 inch). The action of gravity causes flow down the wall. The rate of flow depends on the properties of the paint. Suppose that in this case the velocity is 0.05 in/min. Figure 4-5 illustrates this idealized slump. Under these conditions,

$$\dot{\gamma} = \frac{0.008 \text{ in/sec}}{0.004 \text{ in}}$$

$$\dot{\gamma} = 0.2 \text{ sec}^{-1}$$

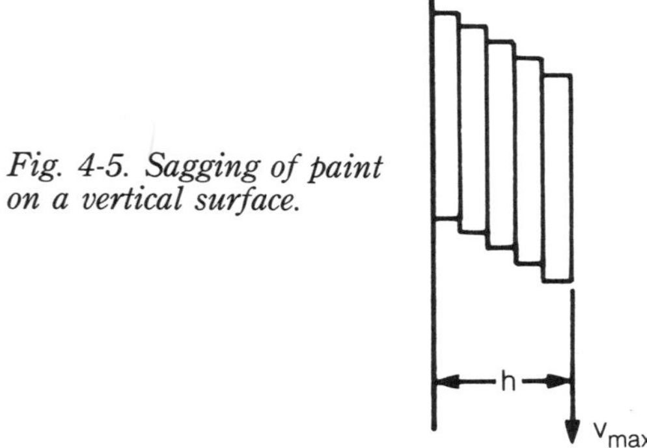

Fig. 4-5. Sagging of paint on a vertical surface.

Spreading Sealant with a Spatula. A sealant is being applied to fill a crack in a wall. For a 1/16-inch-thick layer of sealant moving at 10 in/sec,

$$\dot{\gamma} = \frac{10 \text{ in/sec}}{1/16 \text{ in}} = 160 \text{ sec}^{-1}$$

Brushing Paint onto a Surface. Paint is applied to a surface with a velocity of 20 in/sec and a thickness of 4 mil.

$$\dot{\gamma} = \frac{20 \text{ in/sec}}{.004 \text{ in}} = 5000 \text{ sec}^{-1}$$

VISCOSITY AND FLUID CLASSIFICATION

Defining the explicit mathematical relationship between shear stress and shear rate allows for accounting for various types of fluid behavior. One of the simplest behaviors is described by an equation of state for which the shear stress is directly proportional to the shear rate:

$$\tau = \eta\dot{\gamma} \qquad [4\text{-}3]$$

where η (eta) is a constant of proportionality. Fluids that obey this equation are called *Newtonian*. For Newtonian fluids, a plot of τ vs $\dot{\gamma}$ is simply a straight line as shown in FIG. 4-6. Rearranging equation [4-3] gives

$$\eta = \frac{\tau}{\dot{\gamma}}$$

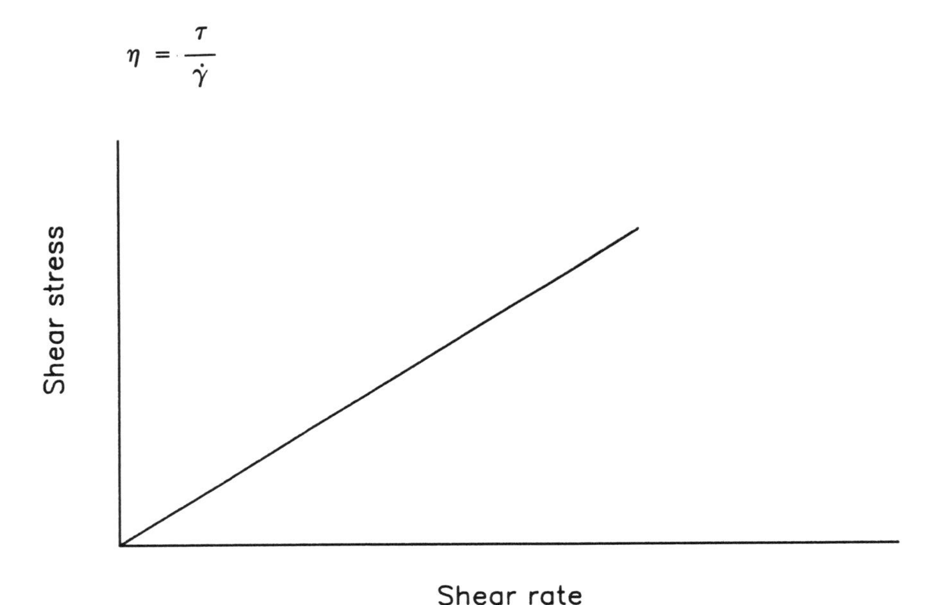

Fig. 4-6. Flow curve (rheogram) for a Newtonian fluid.

The ratio of shear stress to shear rate is constant, and this constant is called *viscosity*. This definition allows quantification of the notion that a liquid is "thick" or "thin," because it relates viscosity to the magnitude of the internal stresses. "High" viscosity means that high internal stresses must be overcome to produce flow. Thus, a Newtonian fluid has a constant viscosity: double the shear rate doubles the internal stress, etc. The viscosity of a Newtonian fluid is independent of shear rate so that the viscosity is simply a straight line parallel to the shear axis as illustrated in FIG. 4-7. Newtonian behavior is exhibited by water, low-molecular-weight organic liquids, dilute solutions, and some oils.

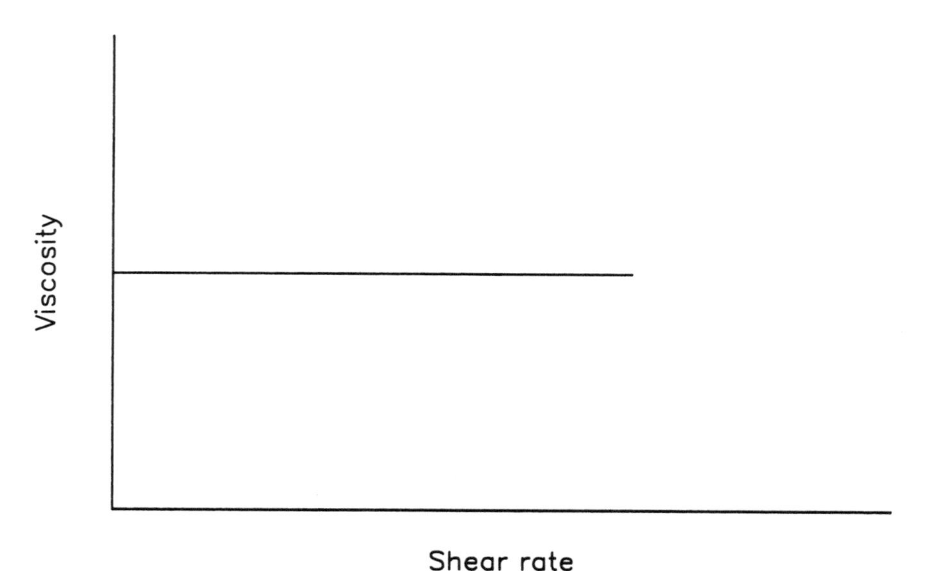

Fig. 4-7. A plot of viscosity versus shear rate for a Newtonian fluid.

The SI units of viscosity are pascal seconds since from our previous definition of stress and shear rate we have $\eta = \text{Pa/sec}^{-1} = \text{Pa} \cdot \text{sec}$. Although these units are used in scientific literature, it is standard practice to specify solder paste viscosity using the *centipoise* unit. These units are derived by using dynes as the unit of force:

$$\eta = (\text{dynes/cm}^2)\text{sec}^{-1} = (\text{dynes} \cdot \text{sec})/\text{cm}^2 = \text{poise}$$

and 1 poise = 100 cps. Note that 1 mPa • s = 1 cps. In all discussions in this book cps is used as the unit of viscosity. Note that in the flow curve in FIG. 4-6, the abscissa is labeled as the shear rate axis (independent variable) and the ordinate is the shear stress axis (dependent variable), or, as written previously, $\tau = f(\dot{\gamma})$. For our purposes, the choice of the shear rate as the independent variable is more natural because the interest is usually in the response of a substance to a given imposed shear rate. Such a process is known as being *shear-rate controlled*. Some rheometers are based on this principle. However, in certain cases it is advantageous to carry out measurements in a stress-controlled mode so that the shear rate is measured for some predetermined values of shear stress.

Non-Newtonian Fluids

Any fluid that has a non-linear relationship between shear stress and shear rate is called *non-Newtonian*. Non-Newtonian fluids can be classified into three main categories.

- Bingham plastic
- pseudoplastic
- dilatent

Bingham plastic. A fluid which exhibits a flow curve as shown in FIG. 4-8 is called a Bingham plastic. The Bingham plastic is similar to a Newtonian fluid except that it intercepts the stress axis at some value τ_0. The intercept τ_0 is called the *yield value* and is the minimum shear stress that must be exceeded before flow will occur. This usually means that the fluid has some internal structure. If the yield value is exceeded, the structure is destroyed and the liquid behaves as a Newtonian fluid under an effective stress $\tau - \tau_0$, with viscosity

$$\eta = \frac{\tau - \tau_0}{\dot{\gamma}}$$

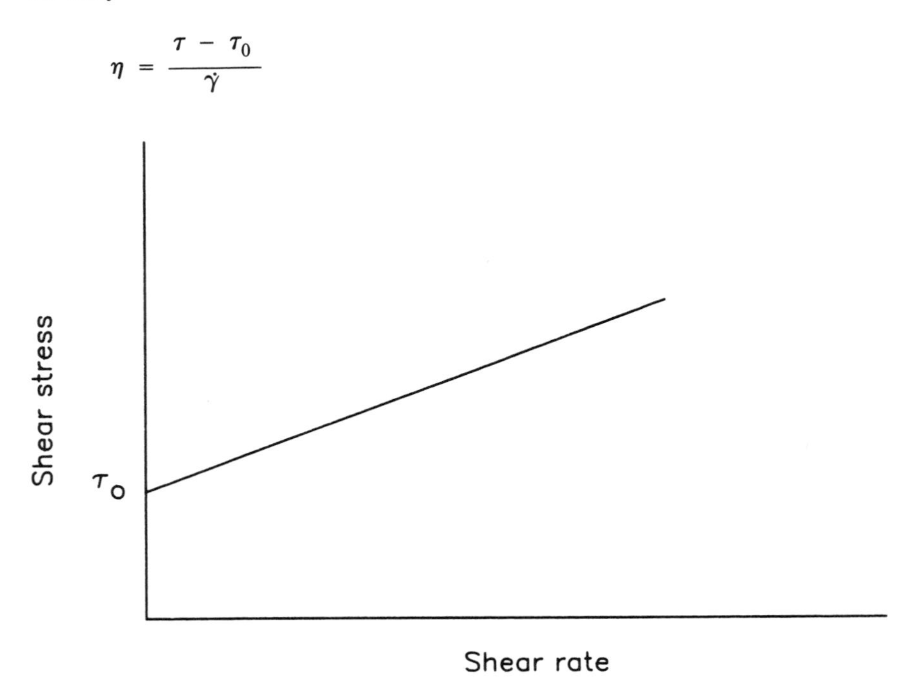

Fig. 4-8. Flow curve for a Bingham plastic showing the yield value τ_0.

The yield stress is a useful parameter for characterizing substances although it is the subject of some controversy in recent years. Barnes and Walters claim that if measurements are accurate enough, no yield stress exists.[5] Fluids that flow at high shear stress can also flow at lower stresses. This is based on measurements using extremely low shear rates down to 10^{-5} sec^{-1}. These studies indicate that all fluids will ultimately exhibit Newtonian behavior at low shear rates. Cheng reviews the measurement of yield stress in more detail and points out that the actual value obtained depends on whether you can determine if the sample actually has developed continuous flow or has ceased flowing.[6] The concept of yield point is helpful in discussing the properties of solder paste. Any reference to yield point will be taken as the minimum stress to cause flow.

Pseudoplastic fluids. Pseudoplastic fluids are characterized by a flow curve as shown in FIG. 4-9. As the shear rate increases, the internal stresses increase in a non-linear manner. A simple relation showing this is given by the power law (Ostwald equation)

$$\tau = k\dot{\gamma}^{n}$$

where k and n are constants ($n < 1$). What is the viscosity of such a liquid? Since the linear relation is no longer as in equation [4-3], the apparent viscosity is defined as the shear stress divided by a given shear rate.

$$\eta_{app} = \frac{\tau}{\dot{\gamma}} \qquad [4\text{-}4]$$

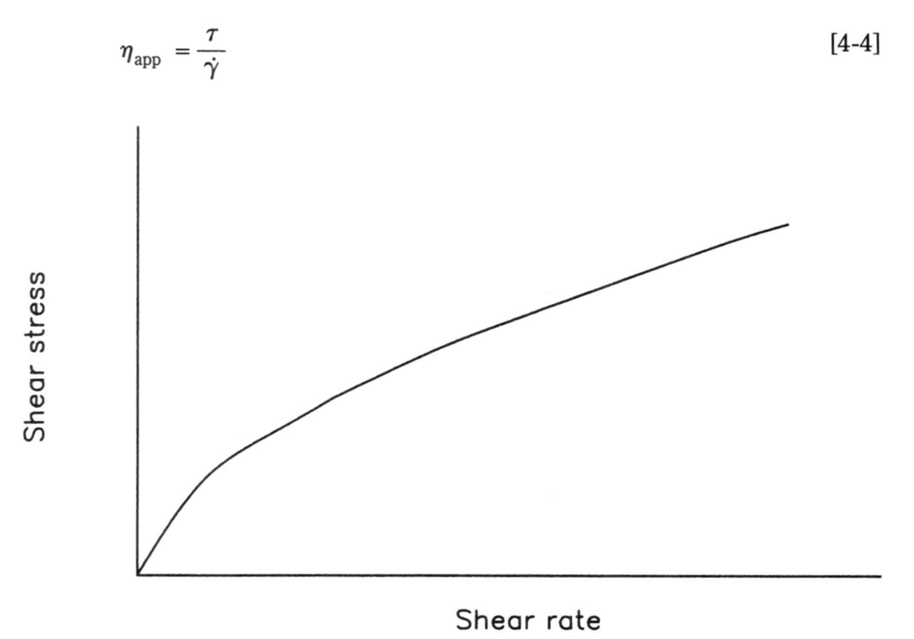

Fig. 4-9. Flow curve for a pseudoplastic fluid.

The apparent viscosity is the slope of the line connecting the origin and some point of the flow curve, as shown in FIG. 4-10. The figure shows that the slopes and hence the viscosities decrease as the shear rate increases. A plot of viscosity versus shear rate would appear as shown in FIG. 4-11.

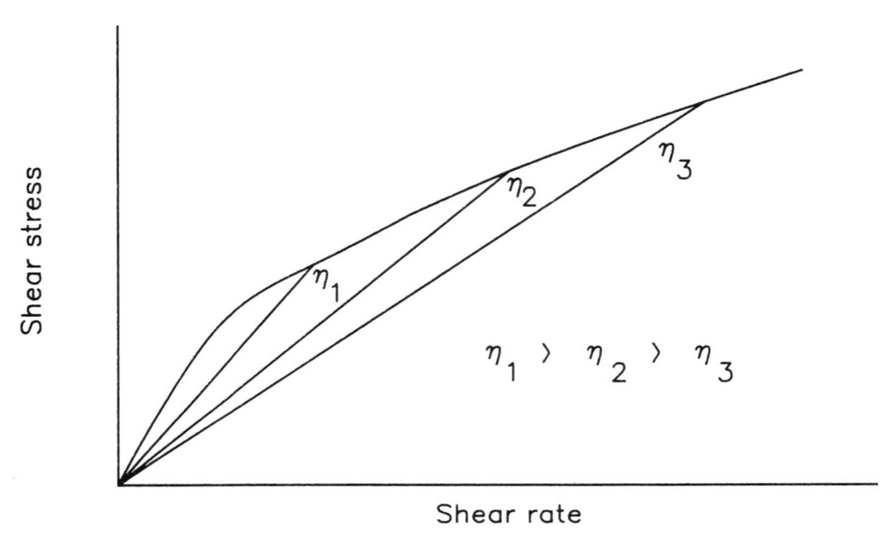

Fig. 4-10. Apparent viscosities for three different points of a flow curve. The apparent viscosity is determined by the slope of the line connecting the origin and the desired point.

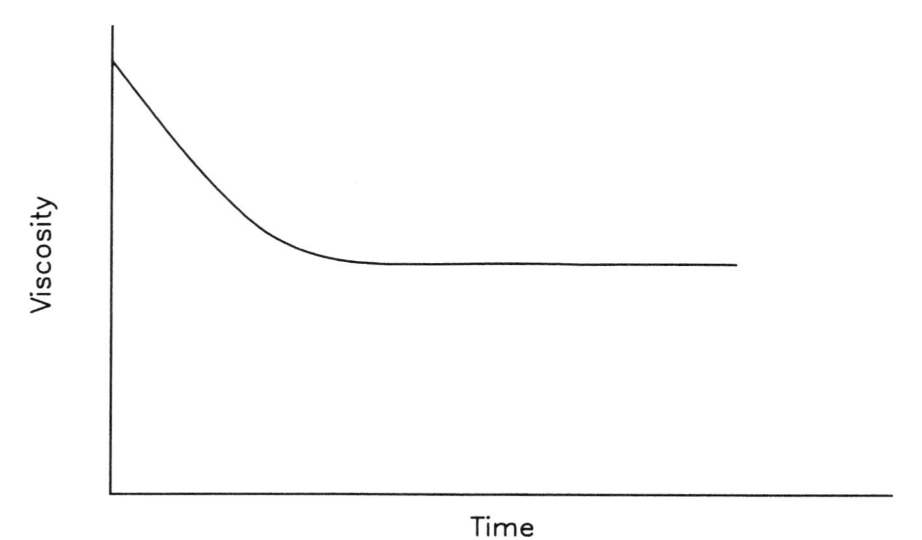

Fig. 4-11. Viscosity versus shear rate for a pseudoplastic liquid.

Pseudoplastic fluids are said to be *shear thinning*. Such behavior is typical of suspensions of particles and polymer solutions. Shear thinning occurs due to the alignment of the particles or molecules in the direction of flow. The apparent viscosity continues to decrease as shear rate increases until no further alignment is possible. Note that time independence implies an "instantaneous" recovery once shear is stopped or changed. The same flow curve would be obtained if the shear rate ran from 0 to some maximum shear rate $\dot{\gamma}_m$ or if it were run in the reverse manner, $\dot{\gamma}_m$ to 0.

Below are some standard equations that are used as models for pseudoplastic flow. They are usually fitted to some specific shear rate range. The value η_∞ denotes the viscosity at infinite shear, and A, B, and C are constants determined from a best fit of the data.

Ostwald $\tau = k\dot{\gamma}^n \ (n<1)$

Herschel-Bulkley $\tau - \tau_0 = k\dot{\gamma}^n$

Casson $\tau^{1/2} - \tau_0^{1/2} = \eta_\infty^{1/2}\dot{\gamma}^{1/2}$

Eyring $\tau = \dot{\gamma}/B + C\sin(\tau/A)$

Williamson $\tau = A\dot{\gamma}/(B + \dot{\gamma}) + \eta_\infty\dot{\gamma}$

Dilatent Fluids. *Dilatent behavior* is the opposite of pseudoplasticity in that the apparent viscosity increases as shear rate increases, with a resulting flow curve as shown in FIG. 4-12. This behavior is observed only rarely. The

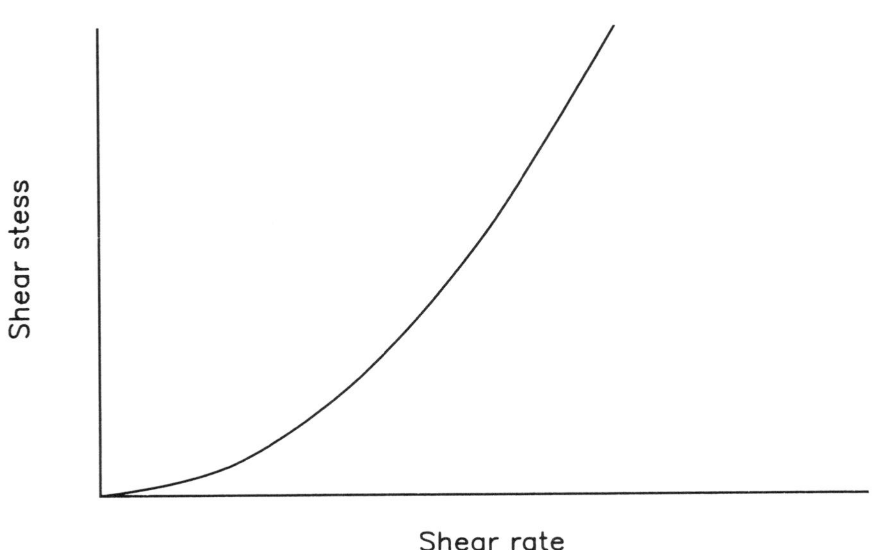

Fig. 4-12. Flow curve for a dilatent fluid.

most typical example is that of suspensions of solids at high concentration. At rest, the suspension has only enough liquid to fill the void volume. At high rates of shear, the dense packing is disturbed and suspended particles dilate and the void volume increases. A lack of liquid to fill the expanded voids requires an increase in stress in order to cause flow; this results in an increase in the apparent viscosity.

Thixotropy

In this case, the rheological equation of state also includes time as an independent variable; $\tau = f(\dot{\gamma},t)$. Time dependent fluids are classified as either *thixotropic* or *rheopectic*. *Time dependent* means that a thixotropic substance decreases in viscosity if subjected to a constant shear rate and rheopectic, if it increases in viscosity when subjected to a constant shear rate.

Thixotropy, which literally means "change to touch," is more common than rheopexy and was originally discovered in the study of dispersions of iron oxide in water. These dispersions were found to have the property that if left undisturbed, they would form a gel such that when inverted, no flow would take place. Upon mixing, the dispersion "liquefied" and was found to flow easily from its container. This behavior was found to be reversible, so that when left alone the dispersion would resolidify to form a gel.

From the time of this discovery in 1923, it has been found that thixotropic behavior is quite common and is more likely to be found than the simple behavior exhibited by Newtonian and pseudoplastic fluids. Thixotropic effects are most likely to be found in heterogeneous systems with a dispersed phase.

Figure 4-13 illustrates the dependence of time and shear rate. For a given constant shear rate, the viscosity decreases over time, ultimately reaching a constant value. If the experiment were repeated by subjecting the sample to higher rates of shear, then a series of curves would be obtained, each curve being successively lower. Thus, the decrease in viscosity depends on the time and the magnitude of the shear rate. To account for this behavior, the presence of some "internal structure" is assumed that can be broken down when subjected to shearing, so there is less internal resistance and the viscosity decreases. At a constant shear rate, only a certain amount of this structure is broken, finally reaching a constant viscosity.

The shear thinning behavior of a thixotropic substance is accounted for in the same manner as a pseudoplastic substance; however, for a thixotropic substance, the rebuilding of its structure is not instantaneous. As a result of this we can experimentally observe a hysteresis loop when the shear rate is ramped up to some maximum shear rate $\dot{\gamma}_m$ and then reversed back to 0 as shown in FIG. 4-14.

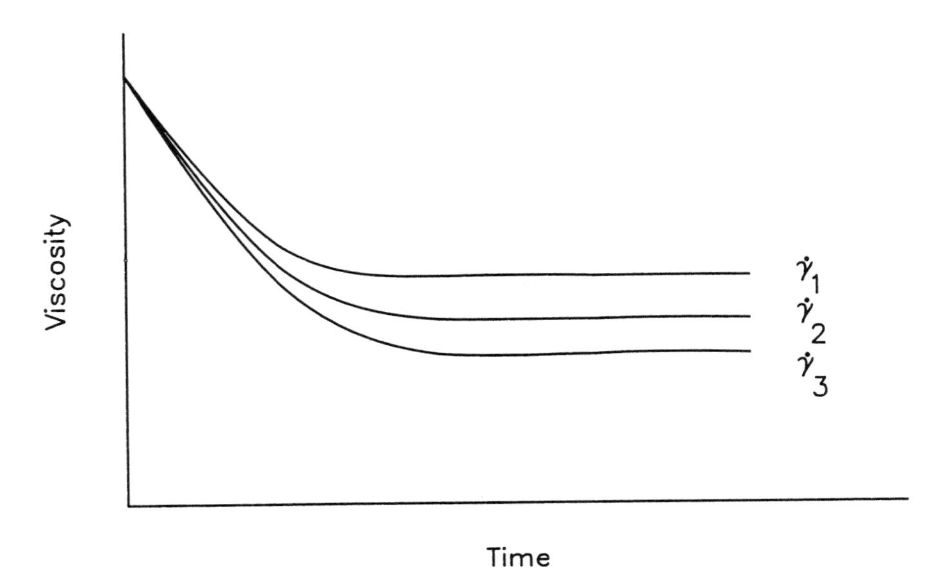

Fig. 4-13. Stress versus time for a thixotropic substance at three different shear rates where $\dot{\gamma}_3 > \dot{\gamma}_2 > \dot{\gamma}_1$.

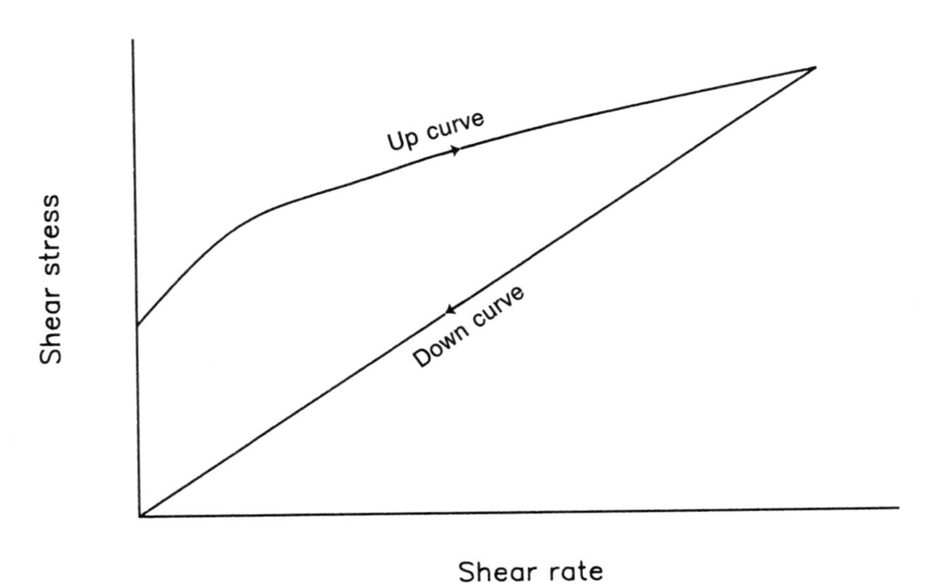

Fig. 4-14. Thixotropic loop obtained by ramping "up" and "down."

The actual loop obtained depends on the total time to reach $\dot{\gamma}_m$, i.e., the ramp time. The area of the thixotropic loop is sometimes used as an indication of the "amount" of thixotropy. The loop area can be interpreted as the energy to break the thixotropic structure for some particular volume of material. With the units of area for the rheogram given by $\tau \cdot \dot{\gamma}$,

$$\text{Area} = \frac{\text{N}}{\text{m}^2\text{sec}} = \frac{\text{N m}}{\text{sec m}^3} = \frac{\text{power}}{\text{volume}}$$

Shearing might not completely destroy the structure. Substances are said to have *false body* if they still exhibit a yield value after shearing. For example, the rheogram in FIG. 4-15 shows a substance with an initial yield value τ_0. After ramping up and down, the down flow curve intercepts the axis at $\tilde{\tau}_0$.

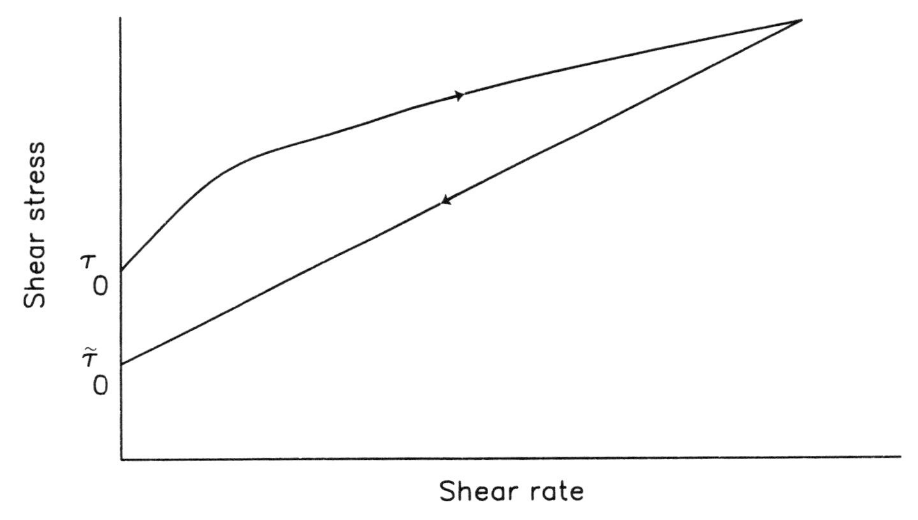

Fig. 4-15. Thixotropic loop of a false-bodied substance.

A thixotropic fluid is a complex substance, so you might find slight differences in the definition of this term depending on the source. The following summarizes the properties of a thixotropic substance; this is the working definition for the purposes of this book. A thixotropic fluid is one that has a

- yield point
- plastic or pseudoplastic flow curve
- decrease in viscosity on shearing over time at constant shear rate
- total or partial destruction of the yield value
- tendency to rebuild structure (viscosity) in the absence of shear to its original yield value. The recovery time depends on the particular fluid.

ABSOLUTE VISCOMETRY

What is the best instrument for evaluating solder paste? In choosing a viscometer, you must consider the purpose of the measurement, the cost, and the ease of operation. A research instrument might not be the best one for routine quality control testing. Both high-priced, sophisticated viscometers and simple, inexpensive instruments have their own particular attributes and shortcomings.

An extraordinary number of viscometers are available, and methods of measurement are based in a wide variety of principles including capillary flow, falling needles, vibrating spheres, and rotating cylinders, to name only a few. This discussion is limited to viscometers that are now being used in the industry by both users and manufacturers of solder paste. They are mainly of two types; absolute viscometers and relative viscometers. The term *absolute viscosometer* refers to the fact that the instrument is capable of subjecting a sample to known shear rates for which the corresponding stresses are determined; then, using equation [4-4], the viscosity can be calculated. To obtain definite shear rates, the viscometer measuring head must have a well-defined configuration. Viscometers that use various types of spindles placed in some container of an arbitrary size do not provide well-defined shear rates and are called *relative viscometers*.

The fact that absolute viscometers can be programmed for a wide range of shear rates accounts for one of their most important advantages, that is, they can produce flow curves (rheograms). Viscometers that give readings for only a limited number of points, especially at low shear rates, do not give a complete rheological picture. Figure 4-16 illustrates this. If a single or multi-point

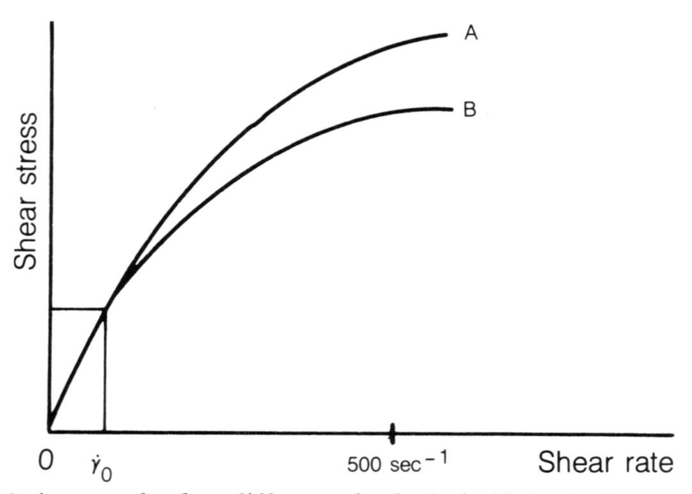

Fig. 4-16. An example where differences in rheological behavior become apparent at high shear rates.

measurement was made at shear rates of $\dot{\gamma}_0$ or less, it would be difficult to distinguish between the two samples. In practice, you might find that sample A performed satisfactorily while B did not. By ramping to higher shear rates, sample B shear thins to a greater extent than A and has lower viscosities at higher shear rates. Hence, rheologically, the two samples are quite different.

Rotational Viscometers

"Rotational" viscometer is a general term that simply means the viscometer contains some rotary mechanism capable of shearing the sample. Common configurations consist of either a system of rotating concentric cylinders (cup and rotor) a cone and plate, or variations of these designs, as shown in FIG. 4-17. The sample can be subjected to various shear rates by varying the rotational speed of the cone or cup (Searle viscometer) or rotor (Couette viscometer) and the resulting torque is measured. This data can then be transformed to an output of shear stress versus shear rate.

Cone and Plate. In the cone and plate system, the sample (about 0.1 cc) is placed on the plate. The cone is lowered onto the plate and makes contact at the vertex of the cone. If the angle of the cone is small enough, then the stress can be considered to be constant throughout the fluid so that the geometry of this set-up allows a constant shear rate to be produced. Figure 4-18 shows a schematic of a cone and plate where the cone is rotating with an angular speed of ω (rpm). The radius of the cone is R and the distance between the cone and plate z at some point r is $r \cdot \tan(\alpha)$. The velocity at r is ωr so that the shear rate is

$$\dot{\gamma} = \frac{\omega r}{r \cdot \tan(\alpha)} = \frac{\omega}{\tan(\alpha)}$$

and is constant for a given rotational speed and angle. A further simplification can be made: since the angle α is small, $\tan(\alpha) = \alpha$, and the equation reduces to

$$\dot{\gamma} = \frac{\omega}{\alpha}$$

Cup and Rotor. A schematic of a cup and rotor is shown in FIG. 4-17. It is also an inherent property of this geometry that the shear rate is not constant across the annular gap. In fact, for a Newtonian fluid, the shear rate across the gap is given by the Margules equation

$$\dot{\gamma} = \frac{2/r^2}{1/a^2 - 1/b^2}\ \omega$$

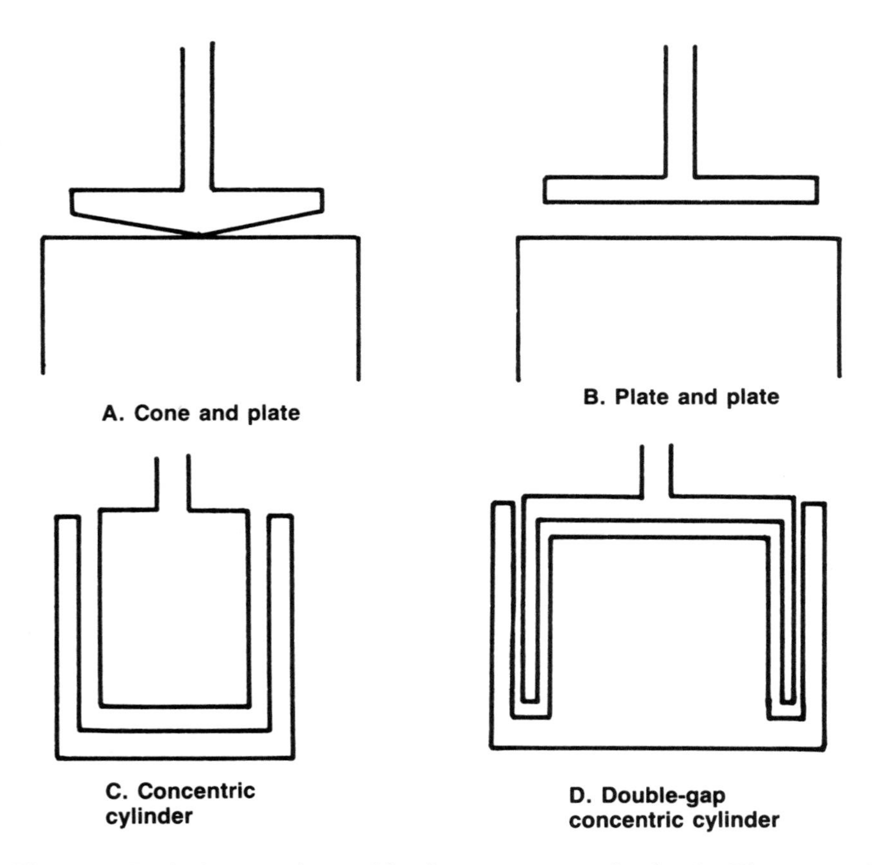

Fig. 4-17. Typical geometries used in viscometer measuring heads. The cone and plate provides a constant shear rate throughout the sample, making extremely high shear rates obtainable. The plate and plate might be necessary for samples containing particulate matter; the shear rate will not be constant throughout the sample. The concentric cylinder geometry is a general-purpose sensor system and can be used with a wide variety of fluids including many disperse systems. The double-gap concentric cylinder is designed for measuring low-viscosity fluids.

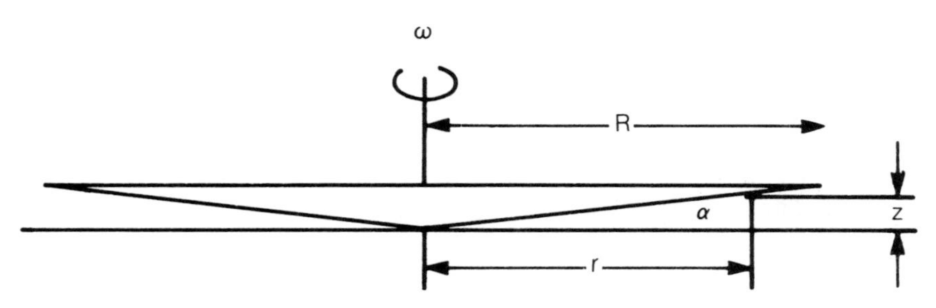

Fig. 4-18. Cone and plate geometry.

where $\dot{\gamma}$ varies as $1/r^2$, where r is the radial distance from the center of the cup, and a and b are the radius of rotor and cup respectively.[7] In practice, the gap size is reduced in order to linearize the velocity gradient to produce a constant or "almost constant" shear rate throughout the fluid as shown in FIG. 4-19. A DIN standard specifies limits for the ratio of the radius of the cup and rotor as

$$1.00 < \frac{r_{cup}}{r_{rotor}} < 1.10$$

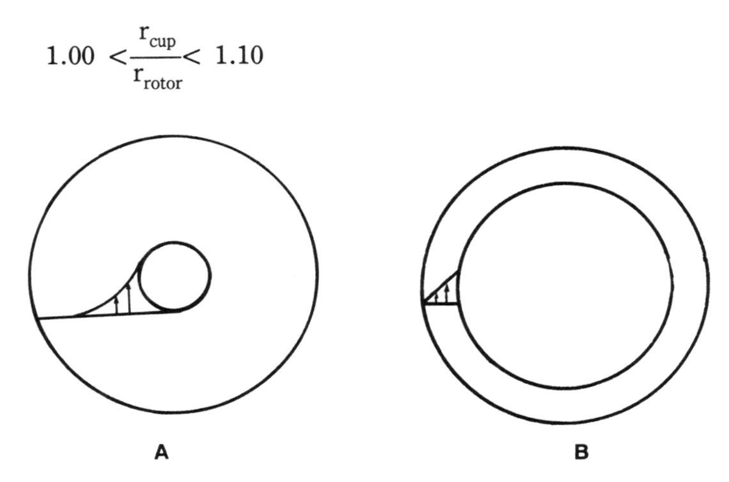

Fig. 4-19. Cup and rotor. In B, the gap has been reduced to linearize the velocity gradient to produce a constant shear rate.

Commercial Viscometers

Previously we have used the terms viscometer and rheometer without making any distinctions between them. In discussing commercial instrumentation, there is a distinction between these terms in that *viscometer* refers to those instruments that provide only a means to measure the viscosity of a substance (relative or absolute), while *rheometer* is an all-encompassing term meaning that not only viscosity, but also other rheological parameters such as viscous and elastic moduli or normal forces, for example, can also be determined.

There are a number of commercial instruments to choose from. Advances are continually being made and certain rheological measurements that were once obtainable only on costly and specialized pieces of apparatus can now be carried out routinely. Prices range from $2,000 to well over $100,000, depending on the unit and its capabilities. Many of these instruments are designed on a modular concept so the user can add other components if desired. Interfacing the instrument with a computer is common. This allows control of the measurement process, that is, to control which types of measurements are to be made and the sequences in which they are to be performed. The computer also allows storage and manipulation of the data, thus simplifying interpretation of the data. Oscillatory and/or controlled stress modes are other possible options; these modes are used to study the elastic properties of a substance. The following section provides a sampling of various instruments that are typical of what is available. Examples of the various types of measurements that can be made with these instruments is given throughout the remainder of this chapter.

The Haake RV 20 (FIG. 4-20) is a shear-rate-controlled instrument. It is based on a modular design that allows various types of measuring heads to be used and can be interfaced with a computer if desired. Viscosity can be measured from 0.02 to 10^5 mPa·sec. A wide variety of measuring sensors are available. Oscillatory measurements can also be carried out, allowing measurement of the storage and loss modulus, which provides information about the elastic and viscous properties of the substance, respectively. The CV 100 measuring system is based on the Couette system, while the M5-OSC is a Searle-type system and can measure higher viscosities (up to 500 times) than the CV 100.

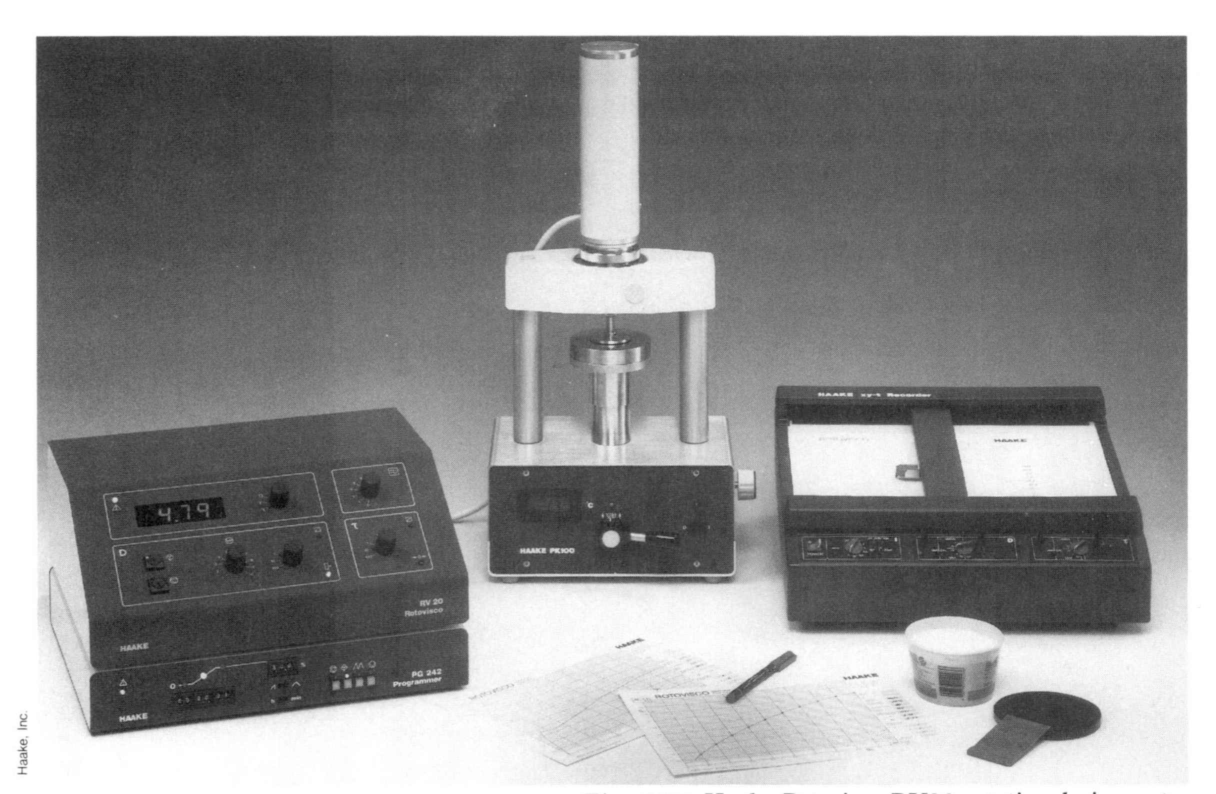

Haake, Inc.

Fig. 4-20. Haake Rotovisco RV20 rotational viscometer.

The Carri-med CS Rheometer (FIG. 4-21) is a controlled-stress instrument. Instead of forcing the sample being measured to undergo a predetermined shear rate, you control the stress that is applied. One of the most significant consequences of this is that you can observe the behavior of a substance at zero shear rate. This allows the study of many low-shear phenomena such as settling, leveling, and flow under gravity. With a stress-controlled rheometer, you can conduct a *creep analysis* (apply a force to the sample and measuring the strain). This is another means of obtaining information about the viscoelastic behavior of a substance. A stress-controlled mode of operation can provide extremely accurate determinations for the yield value. The Carri-med can also be used in an oscillatory mode and with the appropriate software, it can be used as a shear-rate-controlled instrument.

The Bohlin VOR provides a full range of measuring capabilities that include viscometry, oscillatory measurements, relaxation, and a "Job Stream" mode. The Job Stream mode is a software package that allows up to 20 different tests to be performed automatically. A stress-controlled rheometer is offered as a separate unit.

The Ferranti-Shirley is a cone-and-plate instrument and has a long standing reputation for an excellent instrument design for this type of measuring head. At this time, a computer/software package is not offered as standard equipment. The Rheometrics Fluids Spectrometer allows rheological characterization using both dynamic oscillatory and steady shear modes. The Rheometrics Stress Rheometer allows investigation of the properties of a substance under constant stress.

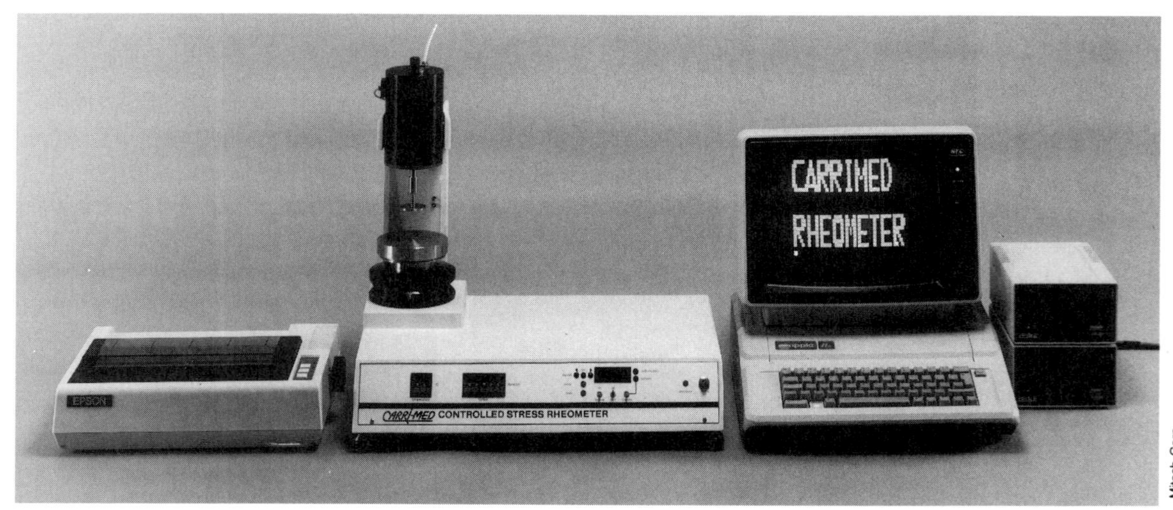

Mitech Corp.

Fig. 4-21. Carri-med CS rheometer.

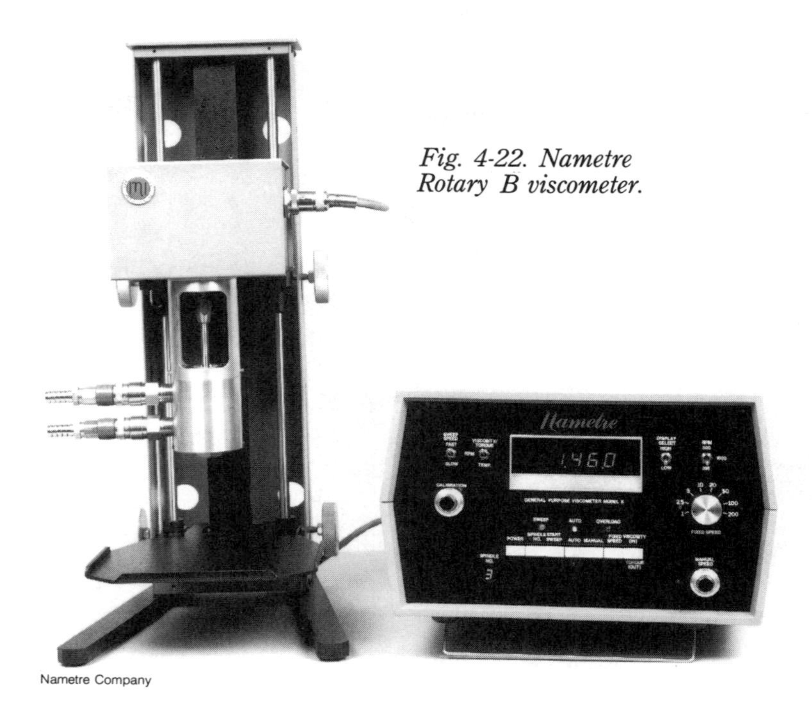

Fig. 4-22. Nametre Rotary B viscometer.

Nametre Company

The Nametre Rotary B (FIG. 4-22) is a unique instrument and is priced substantially lower than the previous rheometers. It can be used with a cone-and-plate or cup-and-rotor system. It does not have oscillatory capabilities, but this instrument does have an attractive feature of being readily adapted with spindles, which are used with relative viscometers. (The benefit of this is discussed in the next section.)

In a different category is the Malcolm PC-1 Paste Controller from Japan (FIG. 4-23). This is being marketed as a device designed especially for measuring the viscosity of inks and solder paste. The measuring head is in the form of a probe, and an inner, stationary, threaded cylinder is surrounded by an outer rotating cylinder as shown in FIG. 4-24. The measuring probe can actually be placed on a screen or stencil if desired. In operation, the solder paste flows up through the probe and emerges at the top and continues to flow out and down the side of the cylinder. Rotational speeds can vary in discrete steps up to 100 rpm (60 sec^{-1}).

Absolute Viscometry of Solder Paste

Be aware of some of the pitfalls of absolute viscometry when applied to solder paste. Complications can arise due to the complex nature of the paste, and in some cases it might be difficult or impossible to obtain any meaningful

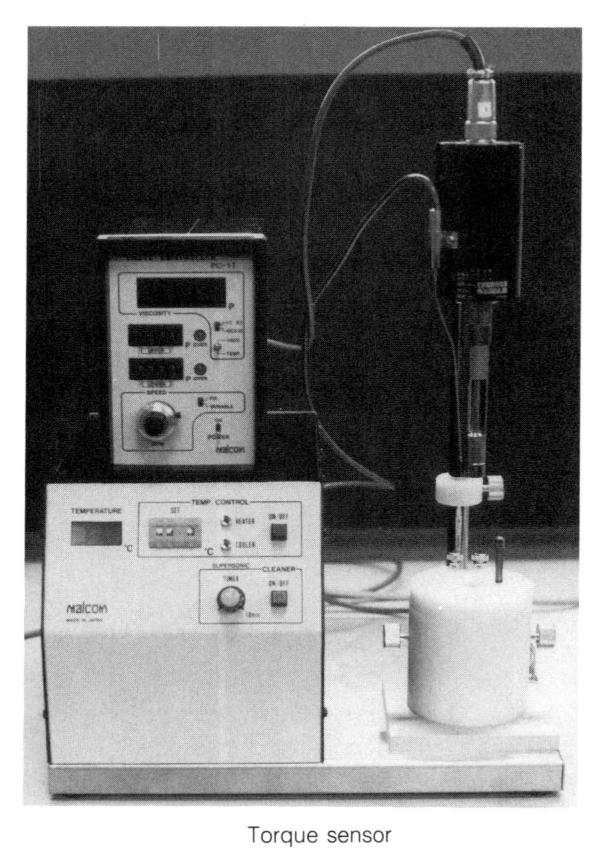

Fig. 4-23. Malcom PC-1 viscometer.

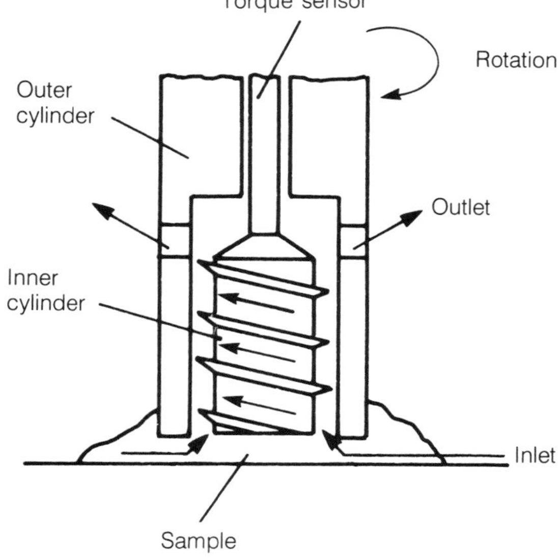

Fig. 4-24. Spiral pump measuring head of the Malcom PC-1 Paste Controller viscometer.

information from the viscometer. This depends on the particular solder paste formulation.

Cone and plate systems, while noted for their high accuracy and for being able to produce constant shear rates, might be of questionable use in the case of solder paste. The high metal-loading of the relatively "large" solder particles can cause problems in the narrow gap of the cone due to grinding and deformation. To alleviate these effects, wide-angle cones, truncated cones, or a cone-and-plate geometry are available. Most solder pastes are viscoelastic (they tend to move in a direction normal to the applied stress). The result of this is a creeping of the paste out of the cone and plate (FIG. 4-25) or, in the case of a cup and rotor, the paste might climb out of the cup leaving a space in the annular gap.

In many cases, due to the high loading of metal, solder paste tends to act more like a solid rather than a fluid. When placed in a rotor, the paste offers resistance at low shear rates but eventually the rotor will slip past the paste and a complete flow curve cannot be obtained.

While absolute viscometers are invaluable research tools, they are designed to study "fluid" behavior. The assumption that a solder paste is a fluid is not always satisfied, thus limiting the applicability of certain absolute viscosity measurements.

Fig. 4-25. Anomalous behavior of a sample of solder paste creeping out of a cone and plate viscometer.

Our discussion has been limited to viscometers for studying rheological behavior, there are other phenomenological properties such as penetration, texture, and flow characteristics that can provide some insight when viscometry alone does not suffice. Solder paste rheology is still in its infancy, and many avenues are open for investigation.

THE BROOKFIELD VISCOMETER

General Description of the Viscometer

The Brookfield viscometer is the single most widely used viscometer in the solder paste industry today. Most manufacturers of solder paste use some kind of Brookfield viscosity either as a parameter for classifying the intended use of the paste (screening, stenciling, or dispensing) or for establishing quality control specifications. The Brookfield has found acceptance in industry for more than 50 years as an inexpensive, reliable, and easily used piece of equipment. Different models are available, designated as LV, RV, HA, and HB, depending on the viscosity range to be measured. Further subclassifications are denoted by F or T, which refer the number of different rotational speeds that are available.

The Brookfield RVTD (FIG. 4-26) is the one recommended for solder paste measurements. RV signifies medium viscosity range, T means that there are eight rotational speeds (0.5, 1, 2.5, 5, 10, 50 and 100 rpm), and D means digital readout. The digital readout is recommended because with this model, the viscometer can be interfaced with a chart recorder that is necessary for the IPC (Institute for Interconnecting and Packaging Electronic Circuits) viscosity procedure. The viscometers are supplied with disk or T-bar spindles.

The T-bar spindles (FIG. 4-27) are designed especially for measuring viscosities of gels and pastes. They are designated by the letters A, B, C, D, E, and F, and refer to the length of the crosspiece of the "T," with A having the longest crosspiece. The particular T-bar used depends on the viscosity range of interest. Table 4-1 shows viscosity ranges and the recommended T-bar. If a rotating T-bar is placed in a thixotropic substance with a high yield value, the T-bar could create a channel (a disk shaped void) in the paste and the torque would quickly decay to some small value.

To overcome this difficulty, T-bars are used in conjunction with the Brookfield Helipath Stand. The helipath stand slowly raises or lowers the viscometer (7/8 of an inch per minute) so that the T-bar describes a helical path through the paste. By always cutting through undisturbed paste, channeling is eliminated. When using the helipath stand, specify when to record the viscosity reading, i.e., after one minute or two minutes of descent or after some up and down cycle. Since the choice is arbitrary, it is of utmost importance to obtain the exact procedure from the paste manufacturer. For a Brookfield viscosity to be meaningful, specify the viscometer head, rotational speed, time to record the measurement, and the temperature. For example, a typical solder paste viscosity could be specified as "500,000 cps, RVTD, 5 rpm, T-F spindle, 2 min., 77° ± 1°F." The RVTD viscometer does not give readings of viscosity directly.

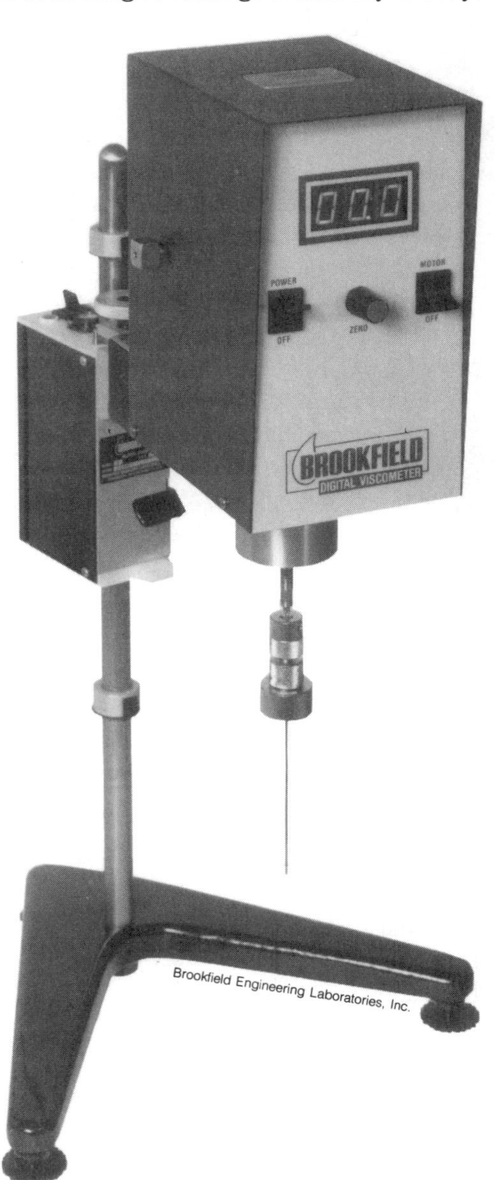

Fig. 4-26. Brookfield RVT DV-II viscometer.

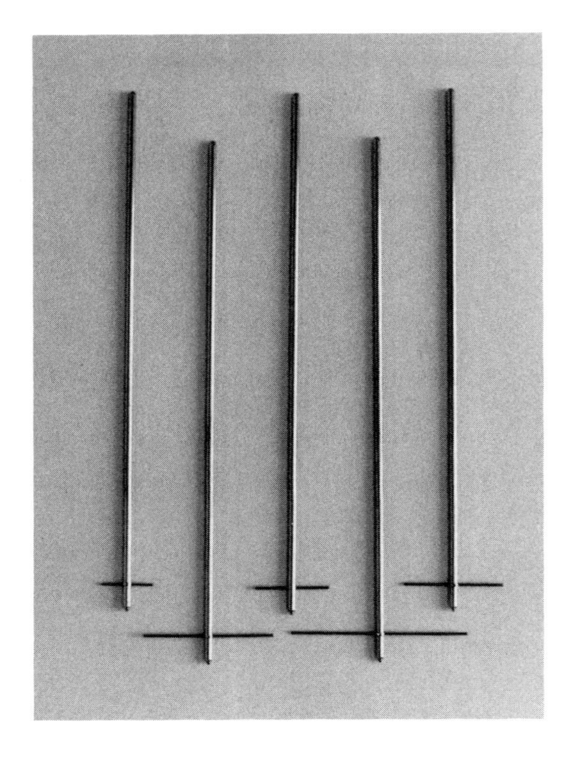

Fig. 4-27. T-bar spindles used in conjunction with the Brookfield viscometer and Brookfield Helipath stand allows viscosity measurement of highly viscous substances such as solder paste.

Table 4-1. Viscosity Ranges with Recommended T-bars

Spindle	Viscosity Range (cps)
T-A	2,000–400,000
T-B	4,000–800,000
T-C	10,000–2,000,000
T-D	20,000–4,000,000
T-E	50,000–10,000,000
T-F	100,000–20,000,000

It is calculated from the particular spindle speed combination in conjunction with a multiplication factor. (Further details of the procedure are discussed in Chapter 10 under "Viscosity.")

Brookfield now offers a "DV-II" series of viscometers that provide a direct readout of viscosity so that no calculations need be performed by the operator. These viscometers can also be interfaced with a chart recorder or computer if desired.

Limitations of the Brookfield Viscometer

A rotating T-bar following a helical path represents a complex flow problem that is not easily analyzed using a mathematical model as in the cases of the cone-and-plate or cup-and-rotor geometries. Thus, any information about the rheological behavior of a substance is based on undefined shear rates.

Accepting the fact that the shear rates are unknown, it is still necessary to deal with the severe restriction of making measurements at very low rotational speeds. At this point, there appears to be a dilemma. On one hand, there is a descending T-bar that overcomes the problems associated with high yield thixotropic materials, and on the other hand, there is the limitation to low shear rates. What is needed is a T-bar mechanism that can sustain high torque and therefore high rotational speeds. A viscometer that meets these requirements is the Nametre Rotary B. Because of an extended torque range, rotational speeds of up to 500 rpm are possible. The viscometer can be operated at pre-set speeds of 1, 2.5, 5, 10, 20, 50, 100, and 500 rpm or the rotational speed can be ramped from 0 to some given value. Thus a ''rheogram'' can be obtained with a T-bar spindle. Figure 4-28 illustrates the potential of such measurements. Two different solder pastes with similar nominal viscosities at 5 rpm (sample A 550,000; sample B 600,000 cps) have different behavior at higher rotational speeds.

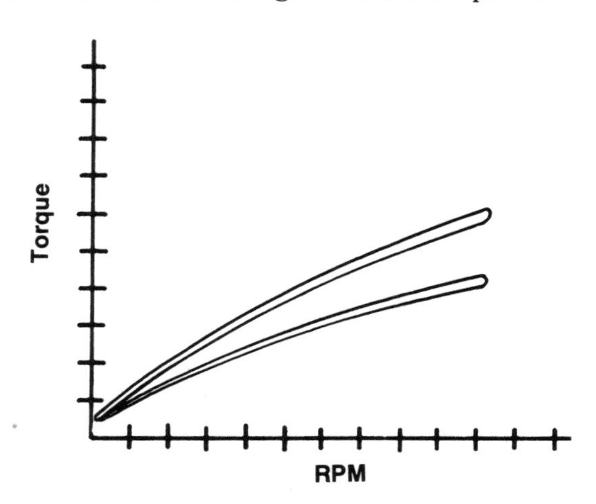

Fig. 4-28. T-bar rheograms of two samples of solder paste obtained with the Nametre Rotary B viscometer. A Brookfield T-E spindle was used. The total up/down ramp time was 2.5 minutes and the rotational speed was increased to a maximum of 200 rpm. In this experiment, the Brookfield Helipath stand was not used.

FURTHER ASPECTS OF PASTE RHEOLOGY

Screening and Stenciling

Screening and stenciling represent two of the major methods that are used for depositing paste onto a substrate. As its name implies, a screen consists of fine metallic (stainless steel) or polymeric (polyester) threads woven to a certain mesh size. An emulsion is applied to the screen to allow formation of a pattern to match the land areas where solder paste is to be placed; variation of the emulsion thickness along with the thickness of the wire determines the ultimate thickness of the paste deposit. A stencil or mask is simply a sheet of metal (usually brass) of a given thickness on which a pattern is etched.

The parameters that affect the outcome of the final print in both screening and stenciling are numerous. A discussion of the practical aspects of choosing a screen or stencil and the controlling factors in the actual mechanism of printing are discussed in Chapter 5. This chapter considers only the rheological aspects of the screening/stenciling process.

Screening makes more demands on a solder paste than stenciling, which could explain the increasing use of stencils. In order for paste to pass through the small mesh openings of the screen, the particle size of the solder powder becomes critical and must be controlled to avoid oversize particles from clogging the mesh openings. A general rule of thumb is that the average particle size should be ⅓ of the mesh opening. Such problems are minimized if a stencil is used. The openings are relatively large compared to the solder particles.

The success or failure of a solder paste in screening or stenciling depends on both rheological and physical properties. During the screening process, paste is placed on the screen and undergoes a continuous rolling movement. The paste is thus subjected to a shearing motion. Because solder pastes are generally thixotropic, they tend to decrease in viscosity during the printing process. Depending on the shear sensitivity of the paste, the print quality might deteriorate due to a spreading or slumping of the paste. On the other hand, the viscosity might increase, preventing the paste from passing through the mesh openings of the screen or stencil. An increase in viscosity can occur if there is loss of solvent due to evaporation. The choice of solvent is critical, especially for pastes with high metal loadings. This is usually less of a problem in stenciling, because the larger openings of the stencil allow a higher viscosity paste to be used.

The rheological properties that are desired in a paste for screening or stenciling are determined by the shear history that the paste experiences in each phase of the process. It is subjected to low shear rates when it is taken from its container and placed on the screen. During the printing process, the paste is moved back and forth by the squeegee and subjected to higher rates of shear. In a screening process, the paste is subjected to further shearing as it is forced through the small openings of the wire mesh.

You would think that a solder paste with a plastic or pseudoplastic (with yield point) behavior would be ideal for printing. For a plastic substance, the viscosity would remain constant during printing and would produce a well-defined print, because recovery to the original yield value would occur instantaneously, thus minimizing or eliminating any slump. If the paste were truly pseudoplastic, there would be some shear thinning during printing, but again, the yield value would be recovered instantaneously, producing no slump. However, it is unlikely that a solder paste can be formulated to be a true plastic or pseudoplastic substance. The paste flux itself is usually a complex mixture of substances including additives to impart a yield point. Vehicles and resins tend to be highly viscous, which inhibits or slows down any recovery or rebuilding of structure.

It is more typical that the solder paste flux itself is thixotropic and the dispersed solid phase (solder powder) also contributes to the thixotropic nature of the paste. For example, solids dispersed in a Newtonian fluid can produce a disperse system that might show shear thinning as well as thixotropic effects. The success of the printing process depends on the proper control of the shear thinning and yield characteristics of the paste. Solder paste must be able to withstand the rolling motion of being pushed back and forth by the squeegee. If too much shear thinning occurs during this process, the viscosity could become low enough so that after printing, rebuilding of the viscosity is too slow and slump occurs.

The rheology of the screening paste is more crucial because the viscosity must be low enough, but not too low, so that the paste can be pumped through the screen openings. Rheograms as shown in FIG. 4-29 indicate the general behavior desired for a screening and stenciling paste. A screening paste might have a lower nominal viscosity than the stenciling paste. Shear thinning is necessary to allow pumping through the mesh openings. Viscosity must rebuild enough to prevent slump. The stenciling paste can have much higher nominal viscosities because shear thinning is not as crucial and false body can be used to the advantage to prevent slump.

Further insight into the various stages of the printing process can be obtained from some simple models used to calculate shear rates. Such a model assumes that the time it takes for a squeegee to pass a mesh opening is equal to the time it takes the paste to go through the opening.[8] The time it takes for a squeegee with velocity v_{sq} to pass an opening of length z is,

$$t = \frac{z}{v_{squeegee}}$$

The velocity of the paste through the screen which has a wire thickness of d_{wire}, is

$$v_{paste} = \frac{d_{wire}}{t} = \frac{d_{wire}\ v_{sq}}{z}$$

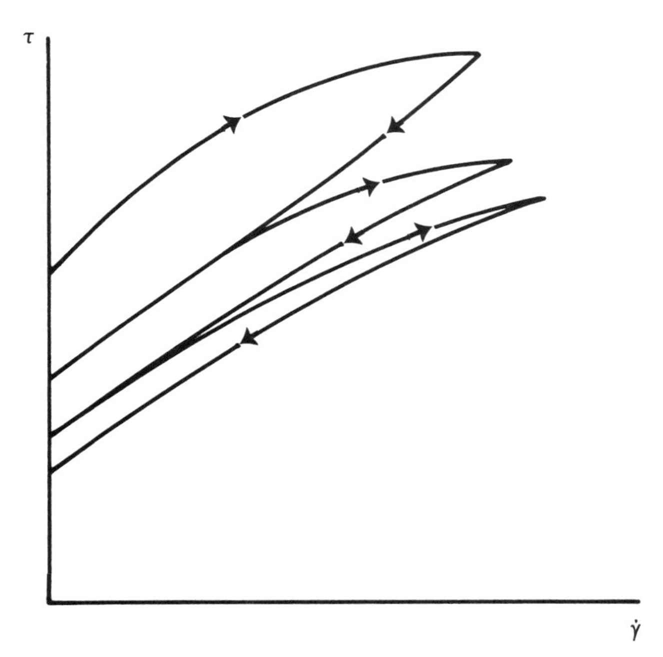

Fig. 4-29. "Idealized" rheograms for a screening/stenciling paste. After repeated cycling, viscosity decreases but some structure remains (false body), as indicated by the intercept of the down curve with the shear stress axis.

The paste is also assumed to flow through the wire opening as in capillary flow with a velocity distribution as shown in FIG. 4-30. A maximum velocity, v_{max} occurs at the midpoint of the velocity profile. The actual effective thickness being sheared is taken as z/2; thus the shear rate is

$$\dot{\gamma} = \frac{v_{paste}}{z/2} = \frac{2d_{wire}\, v_{sq}}{z^2}$$

Note the strong influence of the mesh size, because the shear rate is inversely proportional to the square of the screen opening.

A more detailed analysis of the screening process was undertaken by Riemer[9, 10, 11,12] and others[12, 13, 14]. The printing process is modeled by the motion of an infinitely wide, inclined, flat plate over a porous plane sheet. The inclination of the flat plate simulates the angle of the squeegee blade. Starting from the Navier-Stokes equations assuming a Newtonian fluid and that gravitational and inertial effects are negligible compared to any surface stress, the Stokes equation is stated as

$$\text{grad } p - \eta \Delta v = 0$$

where p and v represent pressure and velocity respectively.

With suitable boundary conditions, solutions of the Stokes equation can be used to represent streamlines of flow. The interpretation is that the squeegee acts as a hydraulic pump in that it generates hydraulic pressure to pump the substance through the screen openings. The necessary hydraulic pump effect that takes place in front of the squeegee depends on the fluid's ability to transmit shear stress. This ability is reduced if the viscosity of the fluid decreases suddenly. In the case of inks, this has been demonstrated by Stanton.[16] This model can provide an estimate of the shear rate during the rolling motion of the paste near the surface of the screen or stencil. According to the definition given in note 10, we have instead found the shear rate at the surface of the stencil to be given by the following expression.

$$\dot{\gamma} = \frac{2V}{r}\left[\frac{\alpha - \sin(\alpha)\cos(\alpha)}{\alpha^2 - \sin^2(\alpha)}\right]$$

where V is the velocity of the squeegee, r is the distance from the tip of the squeegee, and α is the squeegee angle. The shear rates increase as r decreases. At distances of 0.1, 0.01, and 0.001 inches, the shear rates are 2.2 sec^{-1}, 22 sec^{-1}, and 220 sec^{-1}, respectively. According to this model, the shear rates are relatively low, and it is only at very small distances from the squeegee that the shear rates become significant. Studies need to be carried out relating the actual screening behavior of a solder to the various shear rates of the screening process. To date, most of the studies of this nature have been carried out with conductive inks.

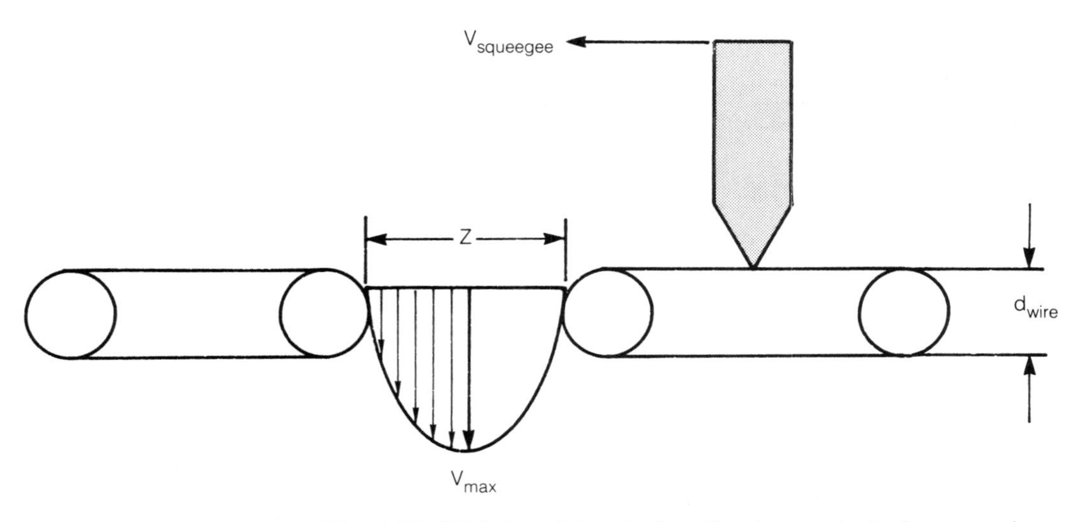

Fig. 4-30. Model used to calculate the shear rate during screening.

Effect of Metal Content on Viscosity

The metal content can have a profound effect on the viscosity and rheological properties of the paste. Metal content is usually specified by weight, 85, 88, and 90 percent being typical. Using weight percent is convenient for certain purposes, but it is the *volume* occupied by the powder that determines the ultimate viscosity. The relation between weight percent and volume percent is shown in the graph of FIG. 8-1. For a 63/37 alloy, a solder paste with a metal loading of 89 percent by weight corresponds to about 50 percent metal by volume.

The apparent volume of a sample of solder powder is the sum of the actual volume of metal plus the volume of the voids. If flux is added to this powder, a certain amount goes to filling the voids and any excess flux is used to further separate the particles of powder. In other words, it is the *excess* flux that gives a solder paste its rheological properties. The amount of void space is dependent on the apparent density of the powder. The apparent density is the weight of the powder divided by the volume that the powder occupies:

$$D_{app} = \frac{\text{weight of powder}}{V_{powder} + V_{voids}} = \frac{\text{weight}}{V_{app}}$$

Figure 4-31 illustrates the decrease in excess flux volume as the metal percent increases. Calculations are based on a 60/40 alloy (d = 8.53 grams per cubic centimeter with an apparent density of 4.7 grams per cubic centimeter

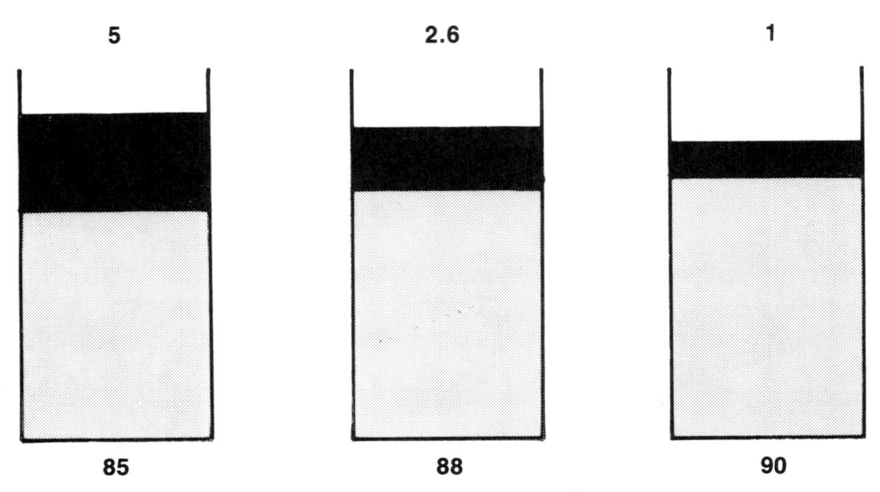

Fig. 4-31. Relative excess flux volume for metal contents of 85, 88, and 90 percent metal by weight. The relative excess flux volumes are in the ratio of 5:2.6:1.

and a flux density of 1 gram per cubic centimeter. For 100 grams of a sample of paste, 88 percent by weight, 88 grams of powder will have a void volume of,

$$V_{apparent} - V_{powder} = \frac{88g}{4.70g/cc} - \frac{88g}{8.53g/cc} = 8.41cc$$

The 12 grams of flux will have a volume of 12 cubic centimeters and 8.41 cubic centimeters of flux will go into filling the void volume, leaving an excess of 3.59 cubic centimeters of flux. Similar calculations were performed for the 85 percent and 90 percent samples, and the excess flux volumes were normalized to obtain the 5:2.6:1 ratio.

It is how the solder particles pack that determines the void volume. This packing is in turn a function of the particle size distribution and shape of the powder. The ratio of the volume of the powder divided by the apparent volume is called the *packing fraction*. The densest known packing for spherical particles of uniform size is obtained for a face-centered-cubic arrangement for which the packing fraction is 0.7405.[17] For a random arrangement of uniform spherical particles, the packing fraction is about 0.636.[18] An increase in the packing fraction can be obtained by mixing particles of various sizes, which allows smaller particles to occupy the interstices provided by the larger particles. The relationship between the packing fractions and particle size distributions is complex. Patton[19] and others[20,21] provide equations and algorithms for calculating maximum packing fractions for multimodal distributions.

In general, the viscosity of a disperse system increases as the volume fraction of the suspended solids increases. If solids are dispersed in a Newtonian fluid, the dispersion still tends to be Newtonian for volume fractions of up to approximately 0.25. At volume fractions above 0.25, non-Newtonian effects (i.e., shear thinning, thixotropy, and yield point) come into play. Figure 4-32 is typical of the effect of an increasing volume fraction on viscosity, usually plotted as relative viscosity, where η_{ref} equals viscosity of suspension divided by the viscosity of the suspending fluid.

The first attempt to account for a dispersed phase on the viscosity was carried out by Einstein.[22] For an infinitely dilute suspension of uniform spheres in a Newtonian fluid, the relative viscosity is given by

$$\eta_{rel} = 1 + 2.5\phi$$

where ϕ is the volume fraction of the suspended phase. Since the publication of this equation in 1906, numerous equations (about 250) have been proposed

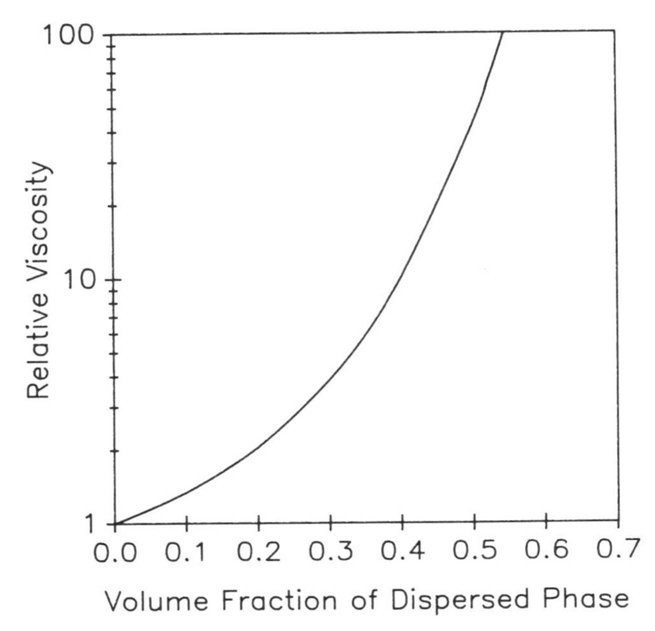

Fig. 4-32. Typical plot of the relative viscosity versus the percent volume of dispersed solids (disperse system of uniform spheres).

that attempt to provide better agreement with experimental data. Mooney took into account the effect of the crowding of particles on each other to obtain

$$\eta_{rel} = EXP\left[\frac{2.5\phi}{1 - k\phi}\right]$$

where k is a self-crowding factor.[23]

The experimental data used to evaluate the accuracy of such equations is usually limited to suspensions of neutrally buoyant particles in Newtonian fluids. The viscosities of the suspensions are obtained at conditions for which the dispersed system still exhibits Newtonian behavior. One of the more successful models is that of Krieger and Dougherty, which has been extended to account for the non-Newtonian behavior of suspensions which occurs at higher volume fractions.[24,25,26] The extended KD equation is

$$\eta_{rel} = \left[1 - \frac{\phi}{\phi_m(\dot{\gamma})}\right]^{-[\eta]\phi_m(\dot{\gamma})}$$

where $[\eta]$ is called the intrinsic viscosity and represents an extrapolated viscosity at zero concentration, ϕ is the usual volume fraction occupied by the solids in the suspension and ϕ_m is the maximum packing fraction for a given distribution of particles. The non-Newtonian behavior is taken into account by considering the maximum packing fraction to be a function of shear rate. While the applicability of these equations to systems such as solder paste is quite limited because of many restrictive assumptions, they do illustrate the complexities of dispersed systems that must be contended with.

While the volume fraction is a determining factor in the resultant rheological properties of such substances, other factors might come into play. The physical and chemical interactions of the solvents and vehicles of the suspending medium might play a role also. For example, systems with smaller particle size can produce suspensions with a higher viscosity than larger particle size.[27] Such effects might be due to absorption of solvent on the surface of the suspended particulate matter, producing an effective larger size and hence a larger effective volume fraction. Such effects, of course, are dependent on the exact nature of the solid and suspending phases.

Finally, note that the microstructure (the relative configuration of the particles with respect to each other) can also influence the rheological properties of a suspension.[28] The final rheological properties of a solder paste are thus strongly dependent on the particle size, shape, and size distribution of the powder and the rheological properties of the flux itself.

Effect of Temperature on Viscosity

The viscosity of a fluid is temperature dependent. In general, the viscosity decreases as temperature increases. It is crucial that when measuring viscosity the temperature be controlled. If viscosity is being used as a method of incoming inspection, the sample must be conditioned to the temperature recommended by the manufacturer of the paste.

The performance characteristics of a paste can also be affected by temperature. Higher temperatures can decrease the viscosity of a paste so that a screening or stenciling operation can suddenly develop poor print quality. Temperature extremes in storage and use of the paste should be controlled. Solder pastes can vary widely in formulation, thus producing different behavior with respect to temperature change. If necessary, detailed information about the effect of temperature should be obtained from the manufacturer. Figure 4-33 shows the relationship of viscosity with temperature for a rosin-based paste with a nominal viscosity of 500,000 cps (Brookfield RVTD, 5 rpm, TF, 2 minutes) and has a change in viscosity of about 12,000 cps/degree F.

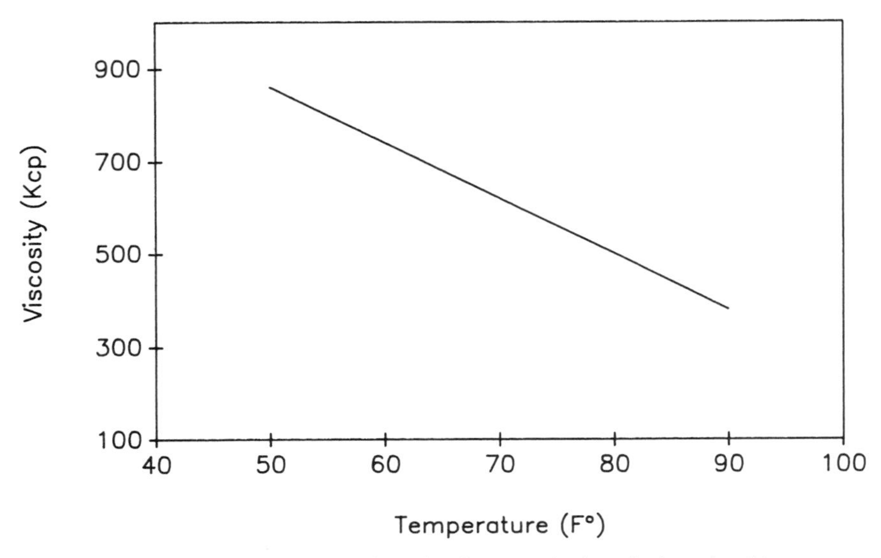

Fig. 4-33. Temperature versus viscosity for a typical rosin-based solder paste.

VISCOELASTICITY

Viscoelasticity Defined

An ideal solid, due to its elasticity, undergoes reversible deformation. The energy of deformation is recovered when the stress is removed and the body returns to its original shape. For ideal elasticity, the strain is proportional to the stress and Hooke's law is obeyed. When subjected to a shearing force τ, a deformation γ occurs and $\tau = G\gamma$. The proportionality constant G is called the *shear modulus*. If an ideal elastic body is "stressed" so that a state of constant strain is maintained, the induced stress will also remain constant.

An ideal liquid is characterized as being viscous. The energy input that causes deformation is dissipated as heat so that when the stress is removed, the deformation does not recover to its original state. For the ideal (Newtonian) fluid, the stress is directly proportional to the rate of strain or shear rate, $\tau = \eta\dot{\gamma}$. However, if an ideal liquid is stressed, the stress will instantaneously decay to zero once the shearing is stopped, shown in FIG. 4-34.

Most substances, especially those composed of polymeric materials, exhibit a rheological behavior that is between an ideal elastic solid and an ideal liquid; such substances are said to be *viscoelastic*. A viscoelastic substance flowing under a constant stress can store some of its energy and partially recover to its original state when the stress is removed. Thus, a constant deformation cannot be sustained under a constant stress. A slow deformation or "creep" over time

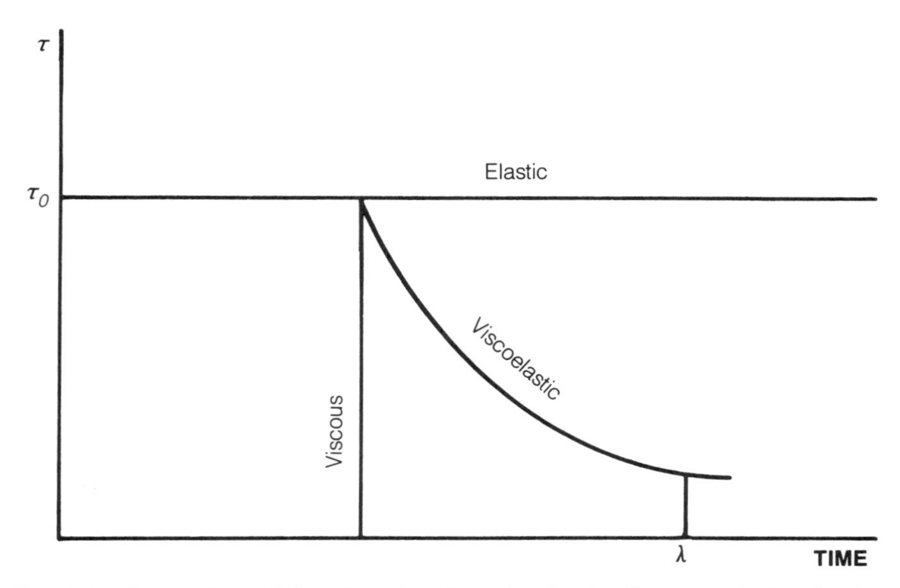

Fig. 4-34. Comparison of the relaxation times for elastic, viscous, and viscoelastic behavior.

takes place, and the stress decays or relaxes, as shown in FIG. 4-34 as an exponential decay ($\tau = \tau_0 e^{-t/\lambda}$). In this case, the relaxation time is defined as the time required for the stress to decay to 1/e of its initial value; hence, λ is the relaxation time.

The property of being viscous or elastic is itself somewhat nebulous, and the decision as to whether a substance is viscous or elastic can depend on the time frame over which the observation is made. For example, a yield point when observed over a long period of time could appear to be a viscous flow. The behavior of a substance can be classified by comparing the experimental time, i.e., the elapsed time of the observation, to the relaxation time. The ratio of the experimental time divided by the relaxation time is called the *Deborah number* and is denoted by D_e. If $D_e << 1$, the substance tends to be viscous for $D_e >> 1$, elasticity dominates, and when $D_e \approx 1$, there is viscoelastic behavior. This is shown in FIG. 4-34, where the relaxation times for the ideal elastic and viscous substances are zero and infinity respectively.

Viscoelastic behavior is typical of polymeric substances, suspensions, and dispersions. Solder paste falls into this category, and viscoelasticity properties play an important role in its performance. Simple shear experiments are not sufficient to completely characterize a paste. Unfortunately, published data relating viscoelastic behavior and solder paste performance is almost nonexistent. It is only in the recent past that moderately priced instrumentation that includes viscoelastic measurements has become available. This is a fertile area for further study and should lead to many new insights in solder paste behavior.

Normal Forces

Viscoelasticity is also associated with another phenomenon that can cause further complications. This was briefly alluded to before in discussing some of the pitfalls of absolute viscometry. When subjected to simple shear, viscoelastic substances tend to generate forces that are normal to the shearing direction. Normal forces can cause flow in directions other than the direction of the imposed shear stress. This is the reason why solder paste can creep out of a cup and rotor or a cone and plate. Although these normal stresses provide further rheological information, the accurate determination of these forces is difficult and requires specialized equipment such as the Rheometrics Fluids Spectrometer or the Weissengerg Rheogoniometer. Only the difference of the normal stresses can be measured. The first and second normal stress differences are defined as

$$\nu_1 = \tau_{xx} - \tau_{yy}$$
$$\nu_2 = \tau_{yy} - \tau_{zz}$$

In most cases, the first normal stress difference is negative and greater in magnitude than the second normal stress difference.

On a molecular basis, the effects of elasticity and normal forces can be thought of as arising from changes in the three-dimensional spatial arrangement of the molecular network when a shear stress is applied. To account for a negative first normal stress difference, i.e. $\tau_{yy} > \tau_{xx}$ an additional source of tension must be accounted for. In polymeric-type substances, the additional tension can be thought of as arising from an elastic energy of deformation. A ''tangled'' molecule that has been ''straightened'' out by the imposed shear flow will have a tendency, due to thermal motions, to ''recoil'' back to some random orientation, producing a force or tension in opposition to the force of deformation.[29,30] This tension arises along the lines of flow and ultimately manifests itself as a force normal to the direction of flow.

Figure 4-35 gives a simplified view of the creeping of a viscoelastic substance up a rotating rod. The stretched molecules on the circular flow lines snap back to their original orientation, producing a contracted flow line, ultimately forcing the fluid inward and up the rotating shaft. The origin of second normal stress differences is not simply accounted for. Figure 4-36 shows a sample of solder paste flux and solder paste for which the first normal stress differences were recorded over varying shear rates. The presence of the solder powder has contributed to elastic nature of the paste.

Mechanical Analysis of Viscoelasticity

This section discusses the basic principles of linear viscoelasticity, that is, it is assumed that the substance obeys the equations of ideal viscous flow, ($\tau = \eta\dot{\gamma}$) and perfect elasticity ($\tau = G\gamma$). This behavior is also assumed valid when the substances undergo "small" deformations.(Full treatments of the subject are in notes 31 and 32.) Experimentally, viscoelasticity is studied using either transient or dynamic measurements. The dynamic method subjects the sample to oscillations, resulting in a sinusoidal strain and the oscillatory forces that result are measured. From the phase and amplitude differences between the input

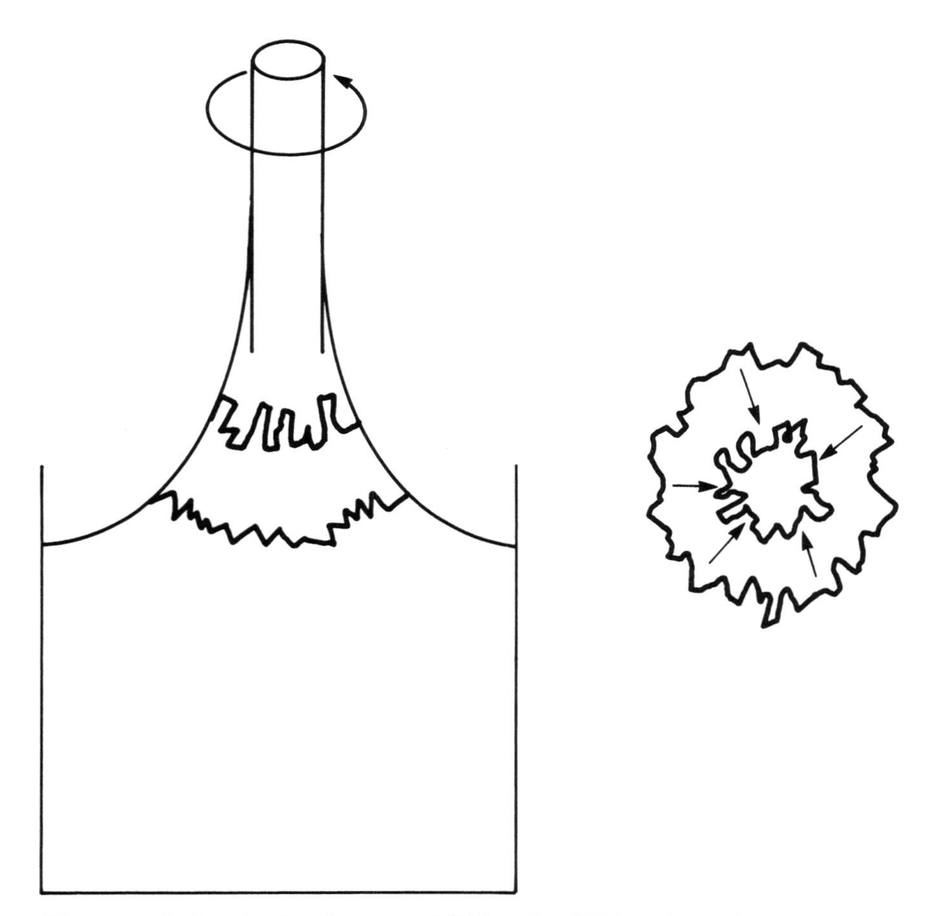

Fig. 4-35. A viscoelastic substance exhibiting the "Weissenberg effect." The fluid tends to climb up the rotating rod as it is being sheared.

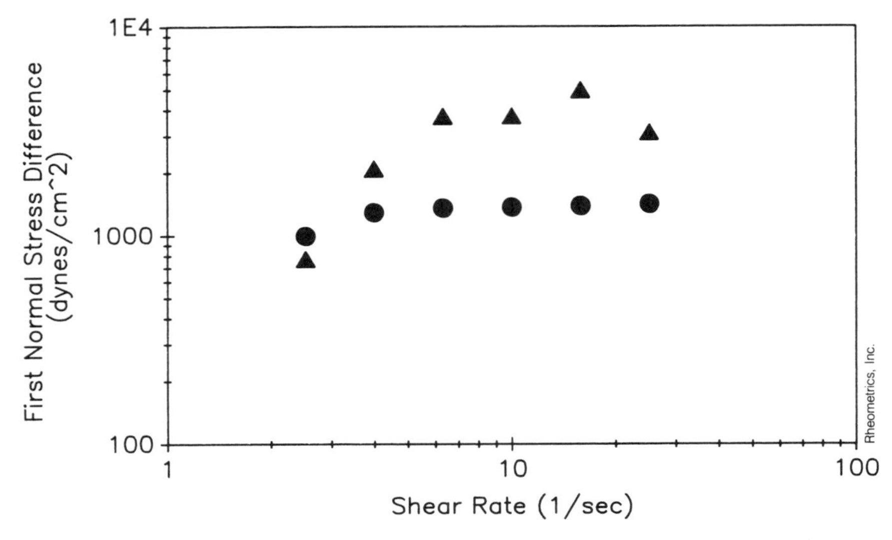

Fig. 4-36. Comparison of first normal stress differences for a solder paste flux (●) and solder paste (▲).

sine wave and the detected wave, the viscous and elastic properties can be determined. In a transient experiment, the substance is subjected to a constant stress, and the deformation is recorded over time. Using an appropriate model, you can deduce viscoelastic characteristics.

It is standard practice when discussing viscoelasticity to use mechanical analogies to represent viscoelastic substances. The ideal elastic substance is represented as a spring, shown in FIG. 4-37. The force applied is directly proportional to the strain. Being an ideal elastic substance, the spring returns to its original position when released. A purely viscous fluid is represented by a dashpot, as shown in FIG. 4-37. The applied force is proportional to the rate of deformation. When released, it will not return to its original position the applied energy is lost. These basic elements can be combined in series or parallel to obtain the Voigt body and Maxwell body (FIG. 4-38). For the Voigt body, the applied stress is shared between the elements, and each element is subjected to the same deformation. The Voigt body is the mechanical analog for a fluid whose behavior is given by

$$\tau = G\gamma + \eta\dot{\gamma}$$

where G is the shear modulus and η is the viscosity. A Maxwell body consists of a spring and dashpot in series. The stress is the same for both elements and

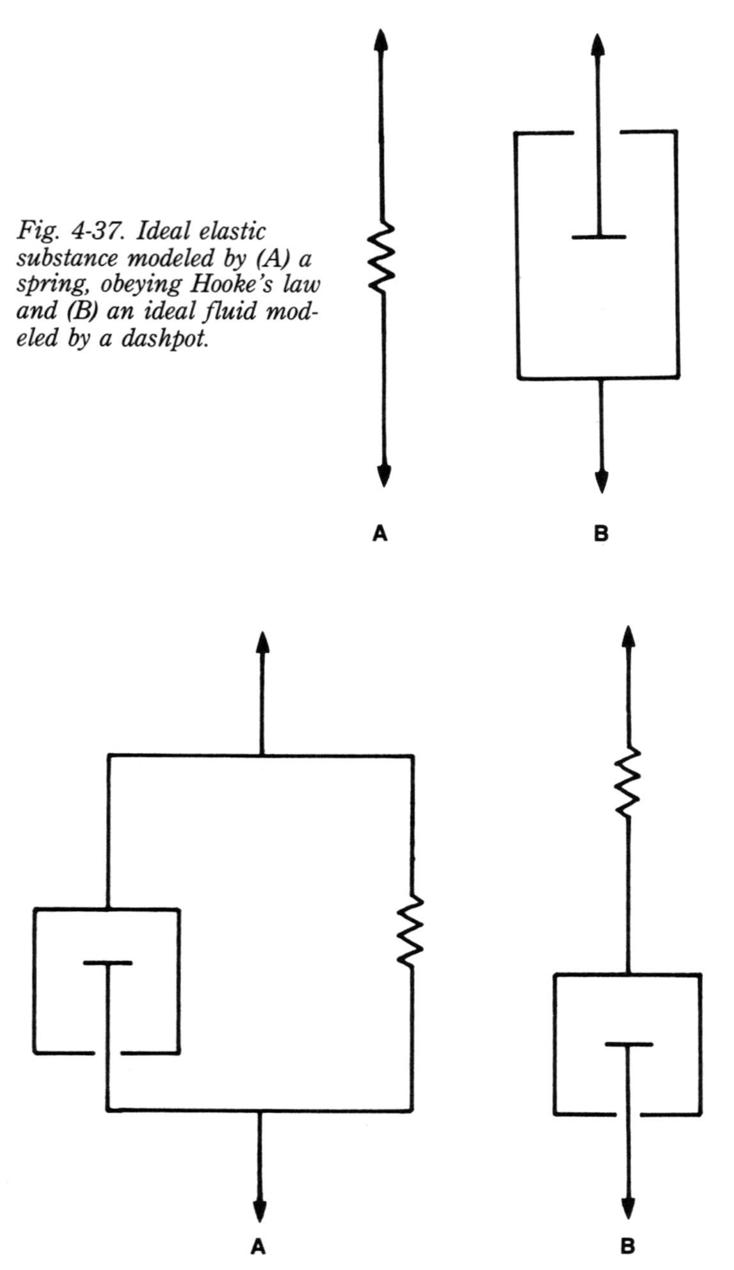

Fig. 4-37. Ideal elastic substance modeled by (A) a spring, obeying Hooke's law and (B) an ideal fluid modeled by a dashpot.

Fig. 4-38. Combination of a spring and dashpot to produce (A) a Voigt body and (B) a Maxwell body.

the total strain is the sum of the strains of each element. The rheological equation for its behavior is given by

$$\dot{\gamma} = \dot{\tau}/G + \tau/\eta$$

The Voigt and Maxwell models are usually too simple to represent real fluids. Various combinations of the Voigt and Maxwell bodies can be combined to give a more realistic model.

Strain Retardation (Creep)

Strain retardation or *creep* is a transient measurement in which a substance is subjected to a constant stress and the resulting deformation is recorded over time. Creep is a convenient method of studying viscoelastic substances that tend to be more of a solid rather than a liquid, such as solder paste. The technique of strain retardation is ideally suited for studying low shear phenomena such as settling in suspensions, structure build-up, and breakdown over time, and leveling (slump) under the influence of gravity. Using a stress-controlled mode, the effects of structure can be studied without actually breaking the structure.

To interpret a creep curve, a general model is used to describe viscoelastic behavior that consists of a combination of Voigt and Maxwell bodies. The creep curve is assumed to be a combination of instantaneous elastic strain, viscous flow, and delayed elastic strain as shown in FIG. 4-39. On the initial application of the stress, there is an immediate deformation due to undamped elastic behavior that is accounted for by the spring J_0. The other springs and dashpots start to move, and the particular rate at which these units move is dependent on the relative strength of the damping effect of the dashpots and the strength of the springs. If the stress being applied is sufficient to break the structure, then steady flow will occur, which is accounted for by the element η_0.

To demonstrate an example of a creep analysis, compare two different solder paste formulations. The sample labeled A has excellent dispensing characteristics. This particular sample provided a consistent flow of paste under pulsing during the dispensing operation. Furthermore, if the dispensing was stopped, the flow of paste could be restored immediately, even if the cartridge had been sitting for many days. On the other hand, sample B did not flow as uniformly, had more of a tendency to clog, and was difficult to start after standing for short periods of time.

A comparison of the creep curves in FIG. 4-40 shows sample A as having a linear response to the imposed stress and undergoes viscous flow according to model B of FIG. 4-40. This means that the internal structure has been broken so that flow readily occurs under the stress loads as provided in dispensing.

Sample B is characterized by a delayed elastic strain and flow does not occur as readily.

Oscillatory Measurements

In an oscillatory measurement, the sample is placed in a measuring head such as a cone and plate or plate and plate. The cone or plate oscillates with a frequency ω rad/sec producing sinusoidal deformations for which the corresponding stresses are determined.

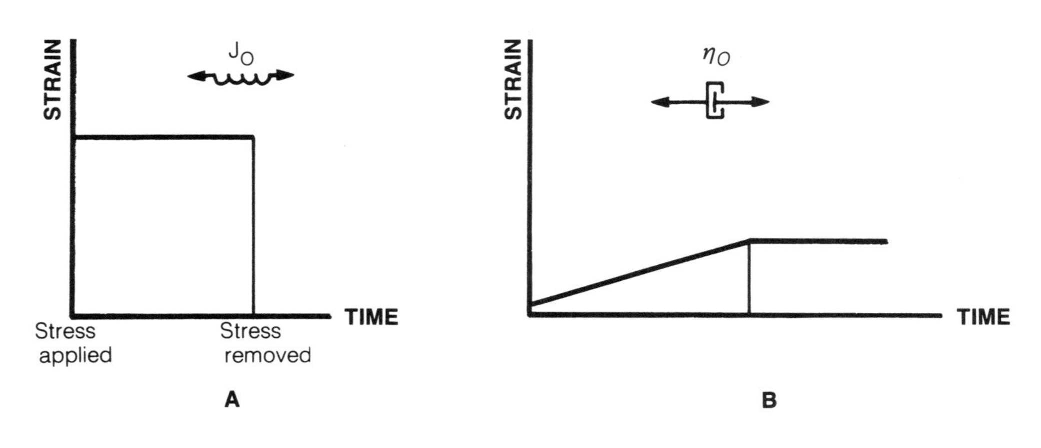

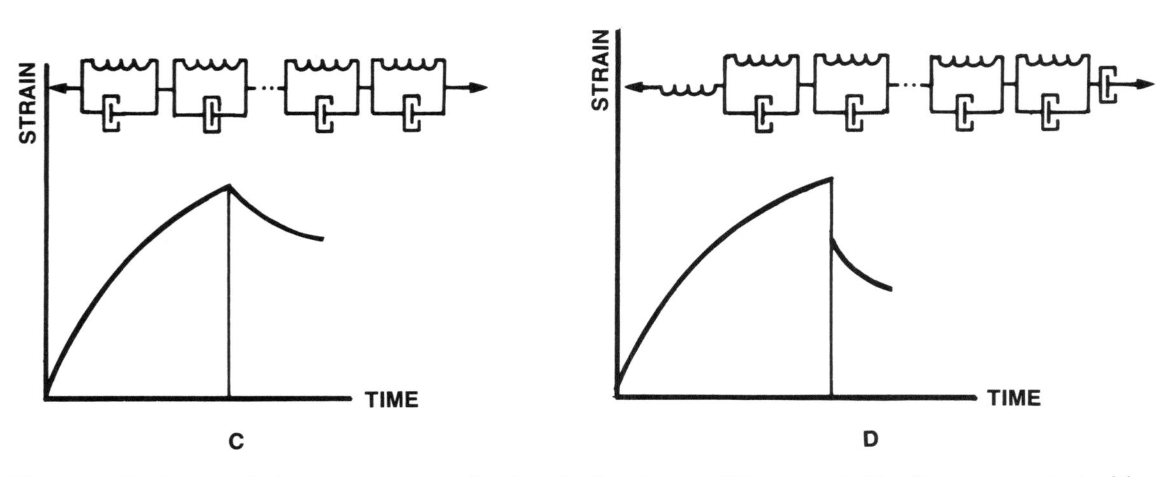

Fig. 4-39. Strain retardation or creep curve of a viscoelastic substance (D) represented by the sum or superposition of instantaneous elasticity, viscous flow, and delayed elastic strain in A, B, and C, respectively. The moduli of the spring and dashpot are denoted by $J = 1/G$ and η.

For a purely viscous substance the stress and strain are out of phase by 90 degrees. The strain is at a maximum (minimum) when the stress is at a maximum (minimum). This is shown by the curves in FIG. 4-41. The maximum rate of change for a sine wave occurs when it passes through zero amplitude and the minimum rate of change occurs when the amplitude is at a maximum. Thus, a viscoelastic substance will have a phase shift between 0 and 90 degrees. A phase shift closer to 0 degrees means that the substance tends to be more elastic than viscous and vice versa. In order to give a simple physical interpretation of the stress associated with a viscoelastic substance, it is convenient to think of this stress as being composed of an elastic component of stress τ' and a viscous component of stress τ''. The elastic and viscous components of stress are combined to form a complex number τ^* called the complex stress, so that

$$\tau^* = \tau' + i\tau''$$

Likewise, the complex strain γ^* is defined as

$$\gamma^* = \gamma' + i\gamma''$$

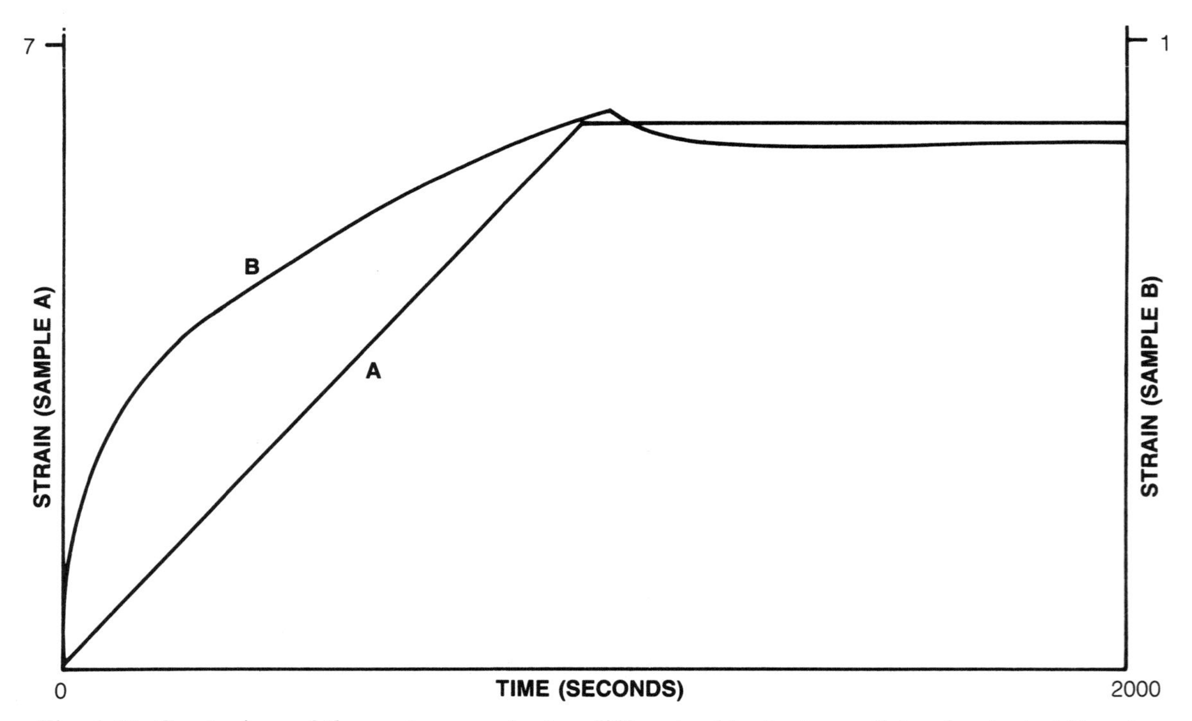

Fig. 4-40. Comparison of the creep curves for two different solder pastes: a dispensing paste (A) and a nondispensing paste (B).

The shear modulus (τ/γ) becomes $G^* = \tau^*/\gamma^*$ and

$$G^* = G' + iG''$$

G' is the storage modulus and is associated with the energy stored in elastic deformation. G'' is the loss modulus and is associated with viscous energy dissipation. This interpretation becomes clear if you consider the Voigt body for which $\tau = G\gamma + \eta\dot{\gamma}$. Since $\gamma = \gamma_0 \sin\omega t$ and $\dot{\gamma} = d\gamma/dt = \omega\gamma_0 \cos\omega t$, this gives

$$\tau = \gamma_0(G\sin\omega t + \eta\omega\cos\omega\tau)$$

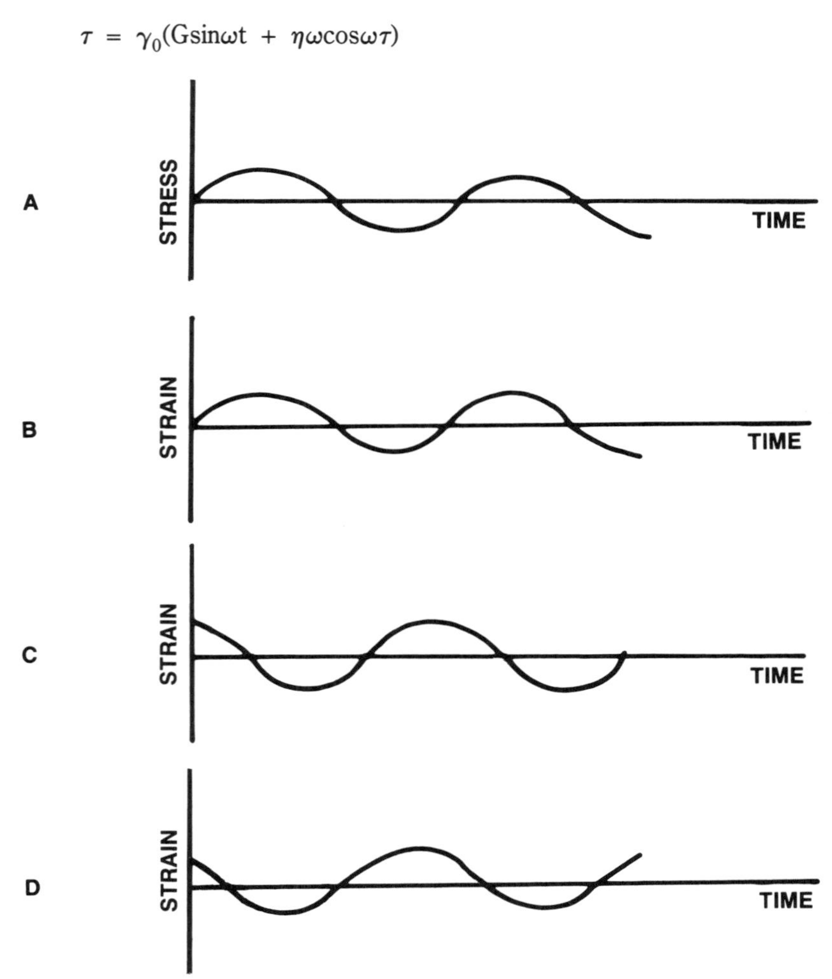

Fig. 4-42. Illustration of the relationship between stress and strain in oscillatory measurements. (A) Sinusoidal stress "input." (B) Strain "output" for an ideal elastic substance. (C) Strain "output" for a purely viscous substance. (D) Strain "output" for a viscoelastic substance.

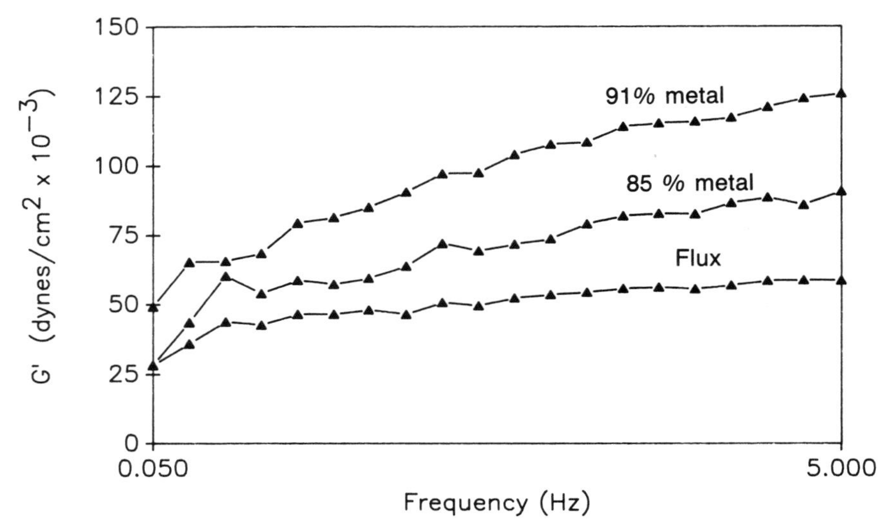

Fig. 4-42. Comparison of the storage moduli for a paste flux and two solder pastes with differing metal contents.

If G is then designated as G′ and $\eta\omega$ becomes G″, then G′ and G″ are associated with the elastic and viscous modulus respectively. Further manipulations show that

$$\tan\delta = G''/G'$$

which is called the *loss tangent*. A complex viscosity is also defined as

$$\eta^* = \eta' + i\eta''$$

It can be shown that $\eta^* = G^*/i\omega$ and $\eta' = G''/\omega$, $\eta'' = G'/\omega$. All these quantities can be determined experimentally and used to characterize a substance.

Hsu carried out a survey of the viscoelastic properties of various thick film materials including solder paste.[33] The pastes were classified according to the relation between the storage and loss modulus:

Type		
Type	I	$G' > G''$
	II	$G' < G''$
	III	$G' > G''$
	IV	$G' <> G''$ (crossover)

As a further example, consider the storage and loss moduli for a solder paste flux and two solder pastes with metal contents of 85 and 91 percent by weight. As shown in FIG. 4-42, the addition of solder powder to the flux results in an increase in the elastic nature of the resulting pastes.

5

Methods of Deposition

Solder paste can be deposited in many different ways, some of which are more practicable than others. For most electronic applications, the choice is restricted to those placement methods that are accurate, reliable, and fast. The options available to the solder paste user range from the most primitive manual application to highly sophisticated machines with computer-controlled operation.

This chapter deals with the principal methods, the major applications for which they are used, the equipment that is commercially available, and the paste formulations most suitable for the type of placement under consideration.

In describing formulations, it is important to note that for the purpose of this discussion, the proprietary chemical compositions of individual solder paste suppliers are of no concern. Only the generic material and the effect of metal loading and viscosity, for example, on the ability of a product to be used for a particular method of deposition are examined.

In selecting such a method, especially when using solder paste for the first time, it is important to consider several factors. The minimum volume of solder paste required to provide a good electrically conductive, structurally sound metallurgical bond must be determined, although this is not always easy. A rosin-based solder paste with 88 percent metal by weight, for example, contains approximately 49 percent metal by volume, but this could vary between 47 to 51 percent, depending on the vendor's metal content tolerance, which is normally held at ± 1 percent.

The application method adopted should provide enough repeatability to ensure that the amounts of solder paste placed on printed wiring board pads or whatever else is to be reflowed are always fairly consistent.

Similarly, the paste itself must be reliable enough to be able to be applied in patterns of equal volume, resulting in uniform reflowed metal deposits. Conditions that could drastically affect the performance of a paste in this regard are product separation (resulting in metal- or flux-rich deposits), the loss of metal due to severe solder-balling, and irregular flow of the material (due to viscosity changes).

Production volume requirements influence the choice of method and equipment. An electronics manufacturer beginning the transition from wave soldering to the new technology involving paste by adding one or two surface-mounted components to the top side of a board is advised not to consider the acquisition of expensive printing equipment unless surface-mount volume is expected to grow very rapidly and a reasonable estimate of that rate of growth can be made. Otherwise, for such an application, a much cheaper alternative would be a dispensing unit, even one requiring solder paste deposition by means of a hand-held syringe.

The other consideration in such a situation would be the sequence of solder bonding. Would the solder paste be applied first, before through-hole components are inserted, or are the boards to be first wave soldered? In the case of the former, all suitable conventional deposition methods could be considered, while in the latter, the paste could only be dispensed in individual dots or stripes (not printed) because of the physical obstacle presented by the wave-soldered component bodies to a printing screen or stencil.

PAINTING

Painting is undertaken with a brush. A typical application for this method is automobile body filling, and coating of metallic surfaces to be joined together in so-called structural soldering, such as in the assembly or repair of automobile gasoline tanks and fittings. Metal loading for such an application could range from 50 to 80 percent, and viscosity is fairly low—in the range 200,000 to 400,000 centipoises.*

For this type of work, a sticky rosin flux with its fairly limited activity should normally be avoided. Depending on the metals to be bonded together, an organic or inorganic acid formulation is recommended, because the residues can then be removed by a hot water rinse, as long as such cleaning is carried out promptly after reflow. Care should be taken when reflowing organic acid fluxes containing water-soluble waxes so these are not held at temperatures in excess of 482

*All viscosity measurements are quoted in centipoises (cps). They are based on readings obtained on a Brookfield Model RVT Viscometer with Helipath Stand, TF Spindle, 10 turns or 2 minutes at speed 5 rpm, at a temperature of 25°C (77°F), in accordance with specification QQ-S-571E.

degrees F (250 degrees C) for more than a few seconds. Otherwise, the flux might char, rendering subsequent residue removal very difficult. Similarly, with the much stronger inorganic acid fluxes that contain zinc chloride as the activator, failure to quickly wash off the residues often results in the formation of salts that are not soluble in water.

Solder paint, which is made from any of the available metal powders, uses much coarser, and more poorly graded particles than for electronics. It also incorporates a much more active flux. The paint can be used for filling the gaps between welded sheets, as in the construction of automobiles. Reflow is done by large gas flames or gas-air torches, and the metal temperature is roughly controlled by directing the flame onto then away from the area being filled.

SPRAYING

Spraying is a method rarely used, although it is a useful one when attempting to deposit solder paste in unusual patterns or into otherwise inaccessible places. A controlled air flow is directed onto the paste as it is dispensed into a spray nozzle assembly.

Such a system has its limitations with regard to the distance the work can be located from the nozzle opening. Also, the use of lead-containing alloys in such an application has to be very carefully controlled for safety reasons.

Depending on what environmental protection measures are feasible, any of the generic flux systems can be used. Metal loading is low, in most instances less than 80 percent, with viscosities less than 100,000 centipoises. Depending on definition requirements, and as a result, the design of the spray nozzle, powder particle sizes vary considerably between the different applications.

DIP COATING

Dip coating involves the immersion of a workpiece into a reservoir of solder paste in order to coat a metallic surface. It also lends itself to applications in the general engineering industry. This type of deposition is also occasionally encountered in electronics for components such as radial- and axial-leaded devices. In this case, the lead is normally the part coated with wet solder paste, which is then attached and reflowed to a component such as a chip capacitor (FIGS. 5-1 and 5-2).

The flux system used can be rosin-based or acid-based (with the latter either organic or inorganic, always assuming that this level of activity can be tolerated). However, rosin might be preferred because the thixotropic nature of this type of formulation generally permits better transfer of paste from the reservoir to the part being immersed. If the rate of dipping is frequent, then thixotropy plays an important role in ensuring that the coating is consistent, and that the paste in the reservoir re-levels before the immersion of the next batch. The position of the reservoir could, of course, be changed between batches to alleviate this problem if production needs permit this.

Metal loading can be as low as 70 percent and as high as 90 percent, depending upon the desired thickness of the reflowed solder and on the required thixotropic and other properties of the paste. The best powder size is generally −325 mesh, as the smaller particles tend to suspend better to ensure a more uniform coating on the lead bonding surfaces or ''nail heads'' than the coarser −200 mesh. The recommended viscosity range is 150,000 to 350,000 centipoises.

ROLLER COATING

Roller coating can provide an accurate means of applying controlled quantities of solder paste. This method is used primarily by manufacturers of electronic components, especially axial-leaded capacitors. The leads are normally loaded into a carrier or boat, usually constructed from graphite, so that their ''nail heads'' protrude. The carrier is then moved either automatically or by hand to a position where a roller, coated with solder paste from a reservoir through which it rotates, transfers a small quantity of paste to the nail heads. The leads are subsequently assembled onto both metallized terminations of the chip capacitors, and the paste is reflowed to form a bond capable of meeting the

Fig. 5-1. Model 600 Axial Lead Attach System with lead- and chip-loading plates, solder paste tray, and reflow oven.

Palomar Systems and Machines, Inc.

Fig. 5-2. Axial capacitors showing "nail-head" leads soldered to chips by means of solder paste.

manufacturer's particular strength and electrical requirements. Special equipment is commercially available for the operation described.

The same flux systems apply as for dip coating, with similar reservations with regard to the organic and inorganic residues as expressed in the section on painting. Metal loading, however, is generally at a higher starting level of 75 percent to a maximum of about 85 percent. Higher metal loadings have been successfully used but generally need to be thinned to such a degree to meet the required viscosity that their rheology can be virtually destroyed, resulting in insufficient paste on the nail heads. Starting paste viscosity for this method of deposition can be as low as 35,000 to a maximum of between 200,000 and 250,000 centipoises.

Powder particle size has been demonstrated to be critical. A preponderance of large particles above, say, 60 microns in a batch of −200 mesh powder, which is otherwise still within recognized specification limits, can cause streaks to appear in the otherwise desired smooth texture of the paste covering the roller surface, resulting in uneven deposition of material on the leads. The −325 mesh powder is normally a good size to recommend for such an application for the same reasons as propounded in the section on dip coating.

PIN TRANSFER

Pin transfer refers to the method by which an array of pins is dipped into a reservoir of solder paste. Small amounts of paste adhering to the ends of the pins are transferred to a series of pads or a similar pattern on a substrate, where they are either reflowed to create ''bumps'' or are used to hold components in a normal surface-mount process prior to reflow. This method of deposition allows very small dots of solder paste to be accurately placed. Rosin-based paste formulations are preferred for this process, because printed wiring or hybrid circuit assemblies are usually involved.

Metal loading does not vary as widely as with the dip coating method already described and ranges from 85 to 90 percent because of the limited quantity of wet paste that can be picked up by the smaller pins. A −325 mesh powder is normally preferred for reasons given in the section on dip coating. The best viscosities for this application are in the 200,000 to 350,000 centipoises range.

PRESSURE DISPENSING

Pressure dispensing categorizes the application of solder paste by means of a number of processes, ranging from a hand-held syringe for placing small amounts of solder paste for development, or prototype work or repair (FIG. 5-3) to highly automated systems (FIGS. 5-4, 5-5, and 5-6). This method of application is particularly suited to placing solder paste onto difficult-to-reach areas, such as a mixed-technology printed wiring board into which through-hole components have already been wave soldered. In this case, the components on the top side of the board would interfere with an operation involving the use of a screen or mask. Care has to be taken to ensure accurate placement of dots or stripes to avoid bridging of leads or pads. Tombstoning of small chip components can occur if the volume of paste deposited on the termination pads varies between them to any great extent.

The simplest function of such a dispenser is as a syringe. The barrel is normally constructed from high-density polyethylene, with either a thumb-operated plunger or compressed-air controlled piston. Paste is displaced through a dispensing tip.

The plungers are commonly constructed from solid wax or in the form of a rubber diaphragm. It is important that these fit snugly, although not tightly, in the syringe to allow freedom of movement. Bear in mind that most syringe barrels have a very slight taper, being widest at the top, where the plungers are inserted.

Such syringes are available in various sizes. The most popular are 10 cc, which holds about 25 grams (about one ounce) and the 30 cc, with a capacity of approximately 115 grams (4 ounces). The capacity depends on the metal content of the paste and the density of the alloy being used. Larger cartridges containing up to some 1,400 grams each are also available for custom-made or commercially available dispensing equipment. These are sometimes also used for dispensing fresh solder paste automatically for screen- and stencil-printing and other deposition methods, thus reducing handling and error. The syringes supplied by solder paste vendors are designed to fit the adapters of the different dispensing equipment manufacturers, and it is always a good idea to let the paste supplier know for which type of dispensing machine the product is to be used. The cartridges likewise are of a standard type to fit into the metal sleeves provided for this kind of dispenser.

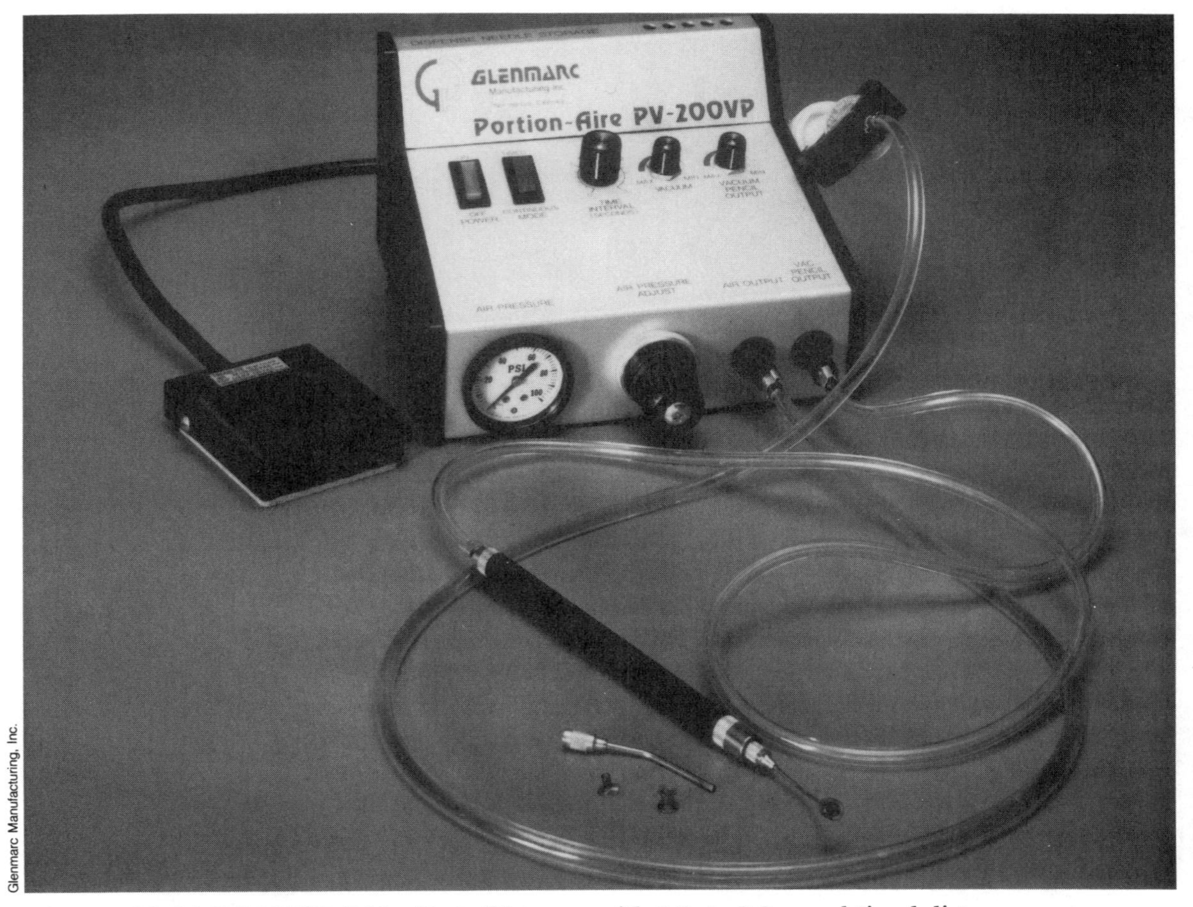

Glenmarc Manufacturing, Inc.

Fig. 5-3. Model PV-200VP Solder Paste Dispenser with 0.1- to 3.0-second timed dispense range.

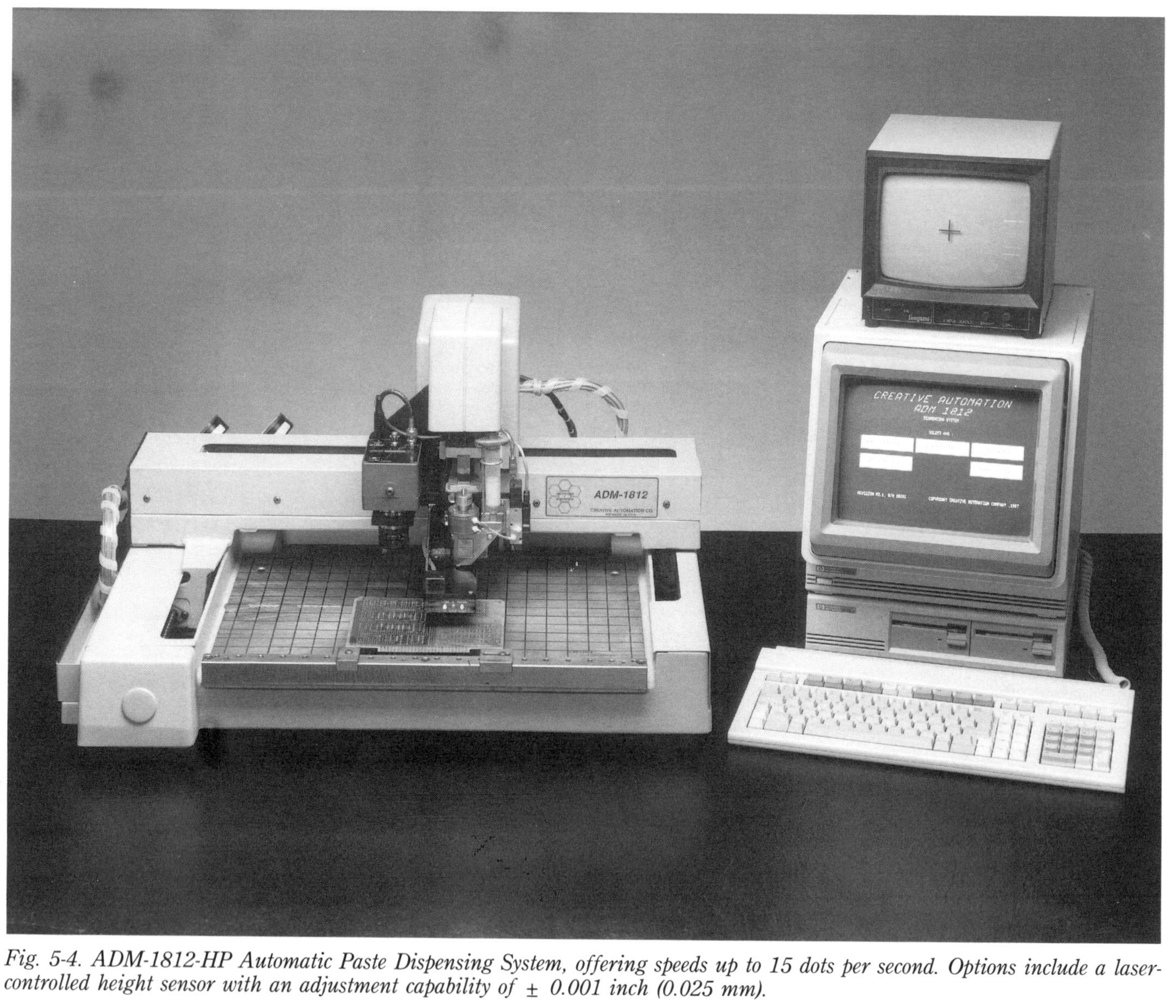

Creative Automation Company

Fig. 5-4. ADM-1812-HP Automatic Paste Dispensing System, offering speeds up to 15 dots per second. Options include a laser-controlled height sensor with an adjustment capability of ± 0.001 inch (0.025 mm).

By combining up to as many as 100 syringes and using these with an optical inspection system and computer-controlled X-Y-Z axis deposition, the accurate application of paste can be coordinated with an automated component pick-and-place operation and a compatible reflow method, such as laser. Multi-orifice manifold assemblies have been built enabling simultaneous dispensing of paste dots

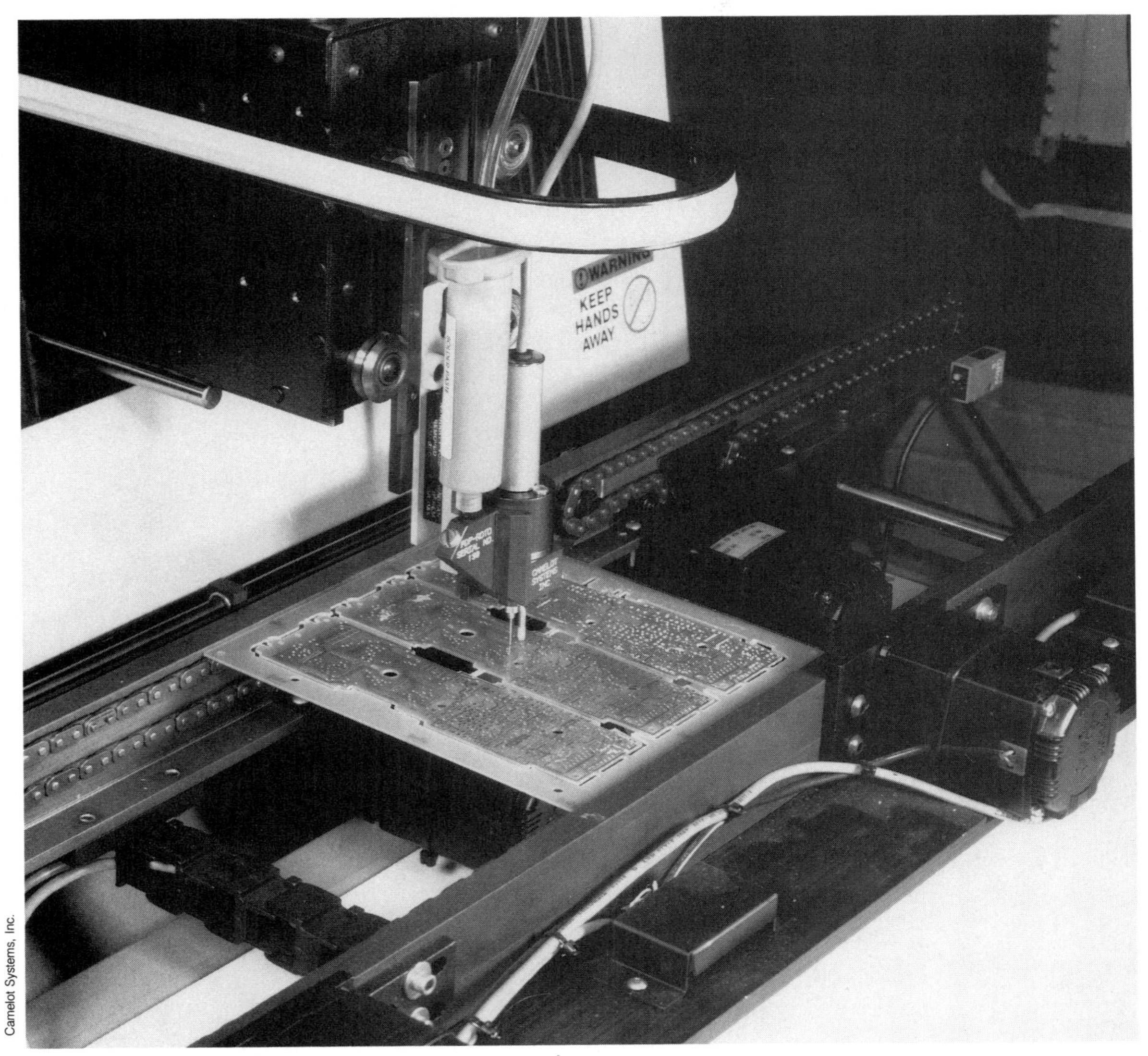

Camelot Systems, Inc.

Fig. 5-5. CAM/ALOT Model 1212 Automated Solder Paste Dispensing System. This is equipped with a rotary positive displacement pump that can dispense up to nine different dot sizes ranging from 0.012 to 0.050 inch (0.300 to 1.27 mm) in diameter.

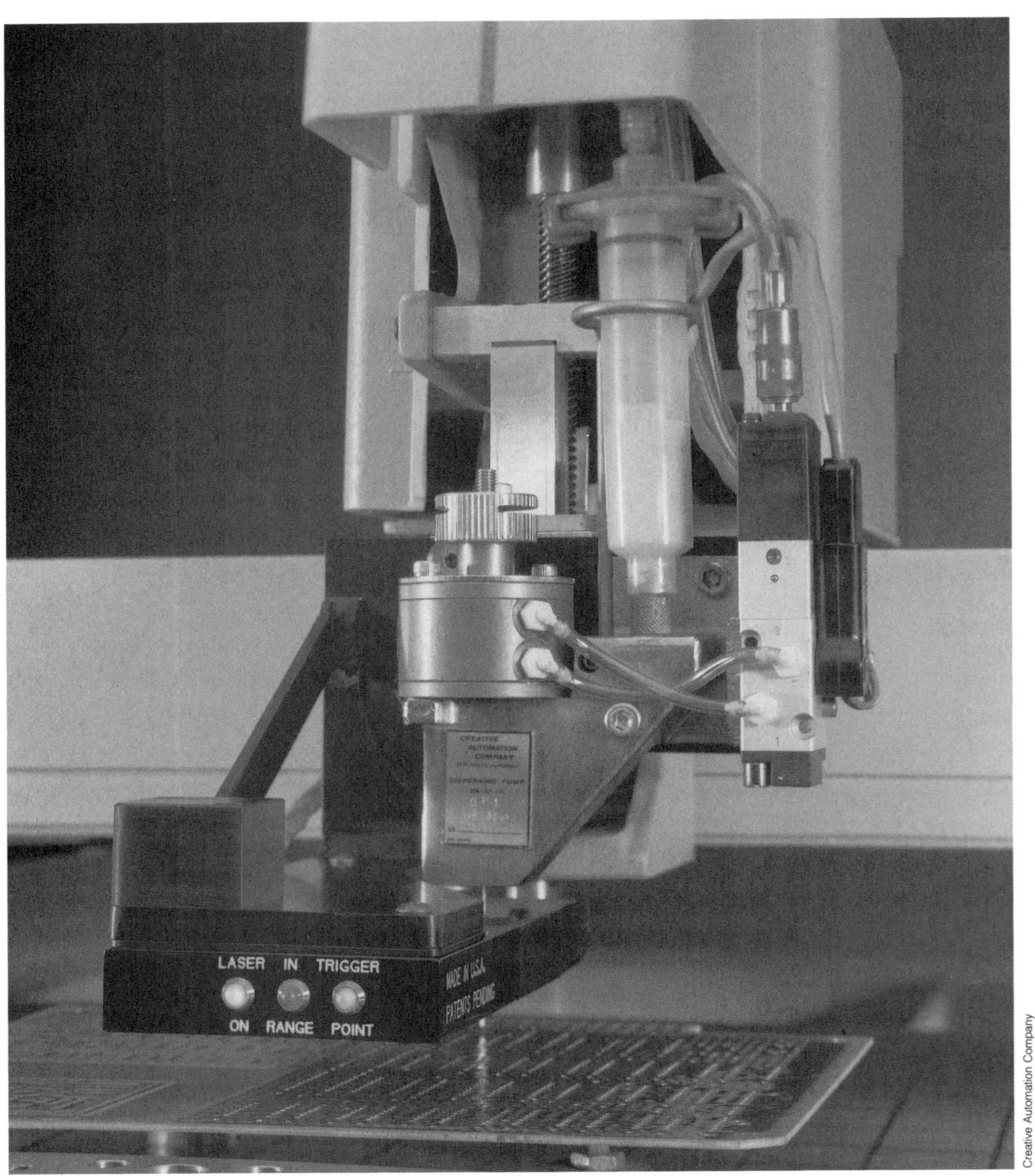

Creative Automation Company

Fig. 5-6. LaserGuide Height Controller ensures paste-height uniformity across warped printed wiring boards.

or stripes for multi-lead surface-mount components. A multi-orifice system must be carefully designed to achieve the desired flow into each orifice. Paste is supplied to these systems from either a reservoir or one or more cartridges. Some dispensing units are even equipped with heating or cooling facilities for viscosity control of the product. Solder paste manufacturers sometimes view such accessories skeptically, because excessive heating could have an effect on the product rheology.

Depending on which equipment is purchased, dispensing is controlled by either air pressure setting or by volumetric displacement to regulate the quantity of paste applied. Both systems rely on the dispensability of the paste to perform their respective functions efficiently.

Traditionally, the needles used for the syringes have been constructed from metal, with an orifice diameter ranging from 0.006 to 0.063 inch (0.152 to 1.600 mm). For solder paste application, however, the minimum practical size is about 0.016 (0.406 mm), provided the powder particle size is carefully controlled. Selection of particle size should normally be based on a ratio of 1:8 to 1:10 in relation to the orifice diameter. For a 0.016 inch (0.406 mm) needle, therefore, a particle size of 0.0016 to 0.002 inch (0.041 to 0.051 mm or 41 to 51 microns) should be selected. This represents a −325 mesh powder with a nominal maximum diameter of 0.0018 inch (45 microns).

If possible, the needle length should be kept to a maximum of about 0.250 inch (6.35 mm) to prevent clogging, but many users find that they can dispense successfully through a length of up to approximately one inch (25.4 mm). Because of the possibility of forming burrs, it is usually not a good idea to cut needles to length, and bending the dispensing needle even slightly to suit the design of the work or the convenience of the operator is unwise, as the flow of paste is further constricted at the angle of the bend. Tapered plastic tips are also available that allow much freer flow of solder paste. Their openings vary between 0.016 and 0.047 inch (0.406 and 1.194 mm). The tapered tips are some 50 percent more expensive than the metal variety, but they are recommended for orifice diameters of less than 0.032 inch (0.813 mm).

Do not assume, however, that the correct nominal powder particle size is all that is required to successfully dispense solder paste. As already noted, the accepted proportion of oversize (−200 to +325) particles in a standard −325 mesh powder is one percent. If a preponderance of that one percent falls within the upper range of the −200 to +325 size, i.e. close to the 75 microns maximum, it is clear that such particles could cause clogging. For work involving needle orifice diameters of less than 0.020 inch (0.508 mm), much better particle size control is required by the paste manufacturer, as well as careful mixing procedures in production. These practices ensure good dispersion of the powder and above all absence of air bubbles in the product, which can lead to severe separation problems.

Metal loading of more than 88 percent is definitely not recommended because the resultant higher density of the paste, combined with low viscosity requirements, will promote separation of the metal and flux, leading to clogging and thus severely reducing flow characteristics.

A loading range of 85 to 88 percent gives most consistent performance. Viscosity plays an even more important role: too thin a paste causes excessive flow, unless the dispensing apparatus is equipped with a vacuum pull-back capability to halt the passage of the material out of the needle. The hand-held unit illustrated (FIG. 5-3) also has an independently controlled vacuum system, with a vacuum pencil for handling small parts. Excessively thick paste might be very difficult, if not impossible, to dispense at the air pressures for which the machine has been designed. Under normal circumstances, the dispenser air pressures for paste range between 10 and 30 psi. It is recommended that viscosities between 350,000 and 450,000 cps be used for needle diameters of 0.062 inch (1.57 mm) and smaller.

Apart from excessive powder particle size, the worst problem the dispensing user normally encounters relates to separation of the paste ingredients in the syringe or cartridge. Where paste is supplied in a jar or similar container and separation occurs, the lid can be removed and the product carefully stirred back to its original homogeneity. With a sealed syringe or cartridge, this corrective action is no longer possible.

The relatively low starting viscosities of dispensable pastes and the shear force to which they are subjected during dispensing make thixotropy of the product quite critical; for precision work, as in electronics, it is important that once deposited, the material should recover its original viscosity almost instantaneously to prevent undue spreading or slumping. Similarly, the paste must break off cleanly at the needle termination once the air pressure has been released and not "string" or form long "tails" on the dispensed dots or stripes. These tails can fall over to create bridges with other pads, circuitry, or isolated solder balls if they drop onto the non-wetting board laminate or substrate ceramic material.

Pressure-dispensing continues to have a valuable role to play in the use of solder paste in many industrial applications. It still provides a very inexpensive means of low throughput placement, with equipment cost normally a fraction of that required for printing equipment. Because of the limitations of high-volume production and solder paste metal-loading, viscosity, and wet-paste thickness constraints, its share of the surface-mount technology market will continue to decline. In the meantime, the solder paste manufacturers continue to seek ways to improve the performance of their products, especially in light of the most recently expressed interest by a leading computer company in a material capable of being dispensed through a needle orifice of only 0.006 inch (0.152 mm)!

SCREEN PRINTING

Screen printing has been used in electronics for well over 40 years, during many of which it was the most popular method of depositing solder paste, especially in the thick-film industry. It began with small, hand-operated machines, which have now developed into highly automated equipment. Printing is achieved by placing a given quantity of solder paste onto a fine woven mesh screen. A squeegee forces it through the openings onto a substrate beneath to reproduce in the form of solder paste the open pattern on the screen. Thicknesses of so-called ''wet'' solder paste for surface-mount applications normally vary between 0.003 and 0.012 inch (0.076 and 0.304 mm), depending on the solder volume required, as well as distances between pad and lead centers. In the hybrid circuit industry, a range of 0.003 to 0.006 inch (0.076 and 0.152 mm) is more usual.

The screens themselves are produced from a variety of materials. For many thin, smoother, non-abrasive materials (such as printing inks), man-made fibers (such as nylon and polyester), or even expensive silk have been employed. ''Silk screening'' was first introduced as a screen material in the textile industry at the turn of the century. For solder paste, stainless steel has long been the primary choice. It wears much better and is much less susceptible to stretch and sagging than the other materials.

Screens are available in different mesh sizes (openings between the wires), which is also expressed as number of wires per linear inch. The wires themselves can also vary in diameter. (See TABLE 5-1.) The screen is stretched tightly to a pre-determined tension within a frame, to which it is bonded, with an epoxy adhesive. The tension is measured in terms of thousandths of an inch deflection at the center of the screen per one pound of pressure applied. The frame itself can be constructed from plastic, wood, or metal, according to the application, but for good-quality electronics work, it is normally of cast or tubular aluminum.

The area of the screen to be imaged normally represents a maximum of about 60 percent of the total area inside the frame. This image area is coated with an emulsion that has been exposed to high-intensity ultraviolet light through the open areas in the artwork, commonly called the ''phototool.'' This film of emulsion is normally not more than 0.008 inch (0.203 mm), but can be as much as 0.020 inch (0.508 mm), in thickness for solder paste applications. It is designed to block out the areas of the screen where solder paste transfer is not desired, as well as increase the thickness of the screen and, therefore, the paste deposit. Theoretically, the thickness (T) of paste deposited through a screen

Table 5-1. Stainless Steel Mesh Dimensions

Mesh Count per inch	Wire Diameter Inches	Wire Diameter Microns	Aperture Size Inches	Aperture Size Microns	Open Area* (percent)
60	0.0045	114	0.0122	310	53–58
80	0.0020	51	0.0105	267	70
80	0.0037	94	0.0088	224	48–50
80	0.0050	127	0.0070	178	31–32
105	0.0030	76	0.0065	165	47–48
120	0.0026	66	0.0057	145	47–48
150	0.0026	66	0.0041	104	37–38
165	0.0020	51	0.0041	104	43–45
180	0.0018	46	0.0038	97	45–46
200	0.0016	41	0.0034	86	46–48
200	0.0021	53	0.0029	74	34–38

*Varies between manufacturers

is equal to the mesh wire diameter (W) plus the thickness of the emulsion (E) times the open area (O) available for the mesh count being used:

$$T = (W + E) \times O$$

As a guide, the higher the open area percentage, the better printing uniformity that can be achieved. Some manufacturers have begun using screens with the mesh filaments etched away from within the open patterns to increase the printed volume of paste. However, this is a fairly expensive and time-consuming practice and has not been widely adopted for surface-mount technology. Another drawback is that the screen can be weakened if too many openings are etched out. One valid reason for doing this, however, is that etched openings can be utilized for selectively printing thicker layers of solder paste where these are needed to provide better structural strength.

Emulsions are described as either *direct*, where a liquid is deposited onto the screen mesh and allowed to dry and is then sensitized and re-dried; or *indirect*, when the emulsion is in the form of a solid pre-sensitized film with a paper backing.

The indirect emulsion is held against the screen wires, moistened with water, and when the emulsion coating has dried and firmly adhered to the screen, the release paper is peeled away.

After the exposure process, the emulsion develops by means of a water spray. The portion of the emulsion covering the intended open areas of the screen that have been masked from the ultraviolet light (and has therefore not hardened) simply rinse away.

Typically, the emulsion is a combination of polyvinyl acetate and polyvinyl alcohol with some type of photosensitizer. There are many different forms of emulsion, but those formulated for printing inks, solder masks, and thick-film inks normally are not suitable for solder pastes. Solder pastes require much greater thicknesses and resistance to wear. Typical tolerances for emulsion thickness are ± 0.0002 inch (0.005 mm) for 0.003-inch (0.076 mm) thickness or less, and 0.0005 inch (0.013 mm) for emulsions thicker than 0.003 inch (0.076 mm). If you intend to make your own screens, you should ask your emulsion supplier to recommend suitable materials; otherwise the screen vendor can generally be relied upon to provide the correct grade ready-for-use or for imaging by the user. Shelf life for the imaged screens is normally six months when kept at about 70 degrees F (21 degrees C) at reasonable humidity levels (50 to 60 percent), but the emulsions can become brittle and unusable even sooner.

The squeegee (FIG. 5-7) is a solid piece of material, usually rectangular in shape, ranging in length from 3 to 26 inches (7.6 to 66.0 cm), or up to a maximum of about 2 inches (5.1 cm) more than the total width of the printing area. It

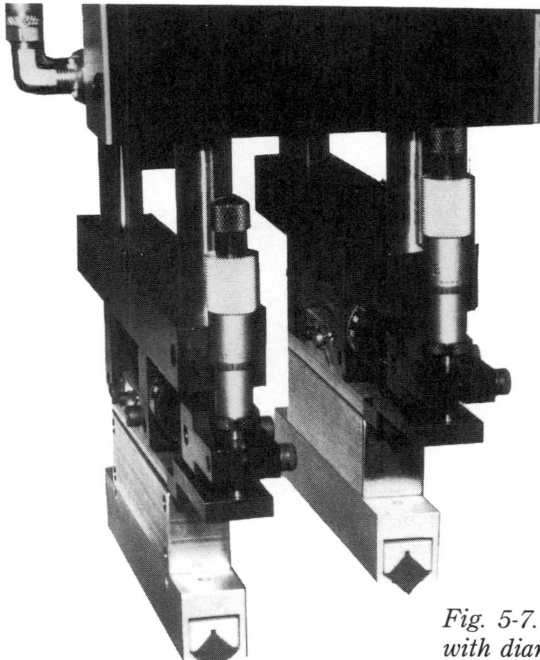

Fig. 5-7. Squeegee holders with diamond-shaped squeegee protruding at 45-degree angle of contact with screen or stencil.

Affiliated Manufacturers, Inc.

is usually made from a synthetic material with good resistance to the solvents used in solder paste, such as polyurethane. The hardness of the material is expressed in durometer and varies to suit the type of printing to be achieved. The higher the durometer, the harder the material.

The squeegee is fitted into a holder and angled at 45 degrees to the screen for most work but is often angled at 60 degrees for multilayer printed wiring boards. It is attached to a carriage on the printing machine. The squeegee lowers to a preset mechanical height setting, or pneumatic pressure is used to hold the edge of the squeegee against the surface of the screen. The squeegee is driven horizontally either manually or via a hydraulic system (speeds in the latter are as high as 20 inches per second). It pushes before it a pre-deposited line of solder paste that is usually about 0.50 to 0.75 inch (12.7 to 19.0 mm) high. When the paste reaches the open areas on the screen, it is forced through onto the substrate beneath.

The squeegee edge must be sharp and free of nicks and any imperfections to keep pressure on the solder paste to a minimum and ensure smooth and quick transfer through the screen mesh. It is essential this edge be maintained in good condition to ensure repeatably successful printing definition and long screen life. The squeegee edge must also be straight so the solder paste is distributed evenly across the whole substrate.

Squeegee pressures and speed of travel can affect print thickness quite dramatically, and the correct hardness is also important in this regard. A fairly soft squeegee of 60 or 70 on a durometer is normally adequate for screen printing solder paste. At low speed and normal pressure, this is able to deflect sufficiently into the screen openings to "scoop out" paste, resulting in a thinner deposit on the substrate. As its speed of travel increases, the squeegee tends more to "skate" across the screen surface, pushing the paste into the openings, but without the excavating effect, providing thicker paste deposits.[1] Squeegee wear also progressively increases print thickness, all other parameters being the same, because the flattening of the edge reduces its ability to penetrate the mesh openings.

Screen mesh orientation should be at an angle of 45 degrees in relation to the direction of travel of the squeegee to reduce screen deformation and improve print definition that might otherwise be impaired by the blocking effect of the wires.

In screen printing, the underside of the screen is almost always set at a height of up to a maximum of about 0.060 inch above the surface of the substrate. The height depends on a number of factors, including screen and squeegee size and screen tension. Inadequate tension (sagging) that results from wear and excessive stretching causes poor print definition (such as ragged edges). The space between the two surfaces is often called "snap-off," "off-contact," "deflection," or "breakaway" distance and has a considerable influence on the quality of the printed paste definition. Paste-filled mesh screens have a propensity to adhere to the substrate, which can badly affect printing definition. However, this is much less prevalent with solder paste than the thinner thick-film inks. Increased snap-off is often required to force the screen to pull back upon release of pressure from the squeegee.

Unfortunately, there are no hard-and-fast rules governing the optimum snap-off distance for a particular board design because of all the variables involved in the process (such as the different types of printers available), but the equipment vendors, and sometimes also the solder paste suppliers, can provide advice in this regard. One formula proposed is that the snap-off distance be 1/250th of the inside length of the shortest side(s) of the printer frame, but this is an arbitrary value that must be coordinated with such parameters as mesh tension and squeegee pressure and speed. More often than not, however, optimal parameters are best established by initial trial and error. Bear in mind that the greater the snap-off distance, the more pressure that must be exerted by the squeegee at a given speed to force the mesh down onto the substrate, and the faster will be the wear on both the screen and squeegee. In addition, the higher pressures involved could be sufficient to cause the squeegee edge to fold, resulting in excessively thick paste deposits.

As the squeegee moves across the screen and its pressure is removed, the mesh begins to peel away behind it until it is once again completely clear of the substrate. For unidirectional, single-pass printing, an adjustable flood bar is normally attached to the squeegee head for returning the paste to its original position. This merely wipes the paste back across the screen to partially re-fill the open areas without exerting the pressure necessary to push the paste through. In alternating double-pass printing, a return flood-bar is not used. The main benefit of flooding, apart from renewing the supply of paste to the squeegee, is that it prevents the solder paste on the screen from excessively drying-out.

The properties of the solder paste itself can be significantly affected by the screening operation, in particular by the pressure and speed of travel of the squeegee. These pressure and speed factors cause the paste to be sheared to varying degrees, depending upon the original character of the product. The original viscosity of the paste lowers, sometimes dramatically, as it travels across the screen and is changed even more so when it is pushed through the mesh openings. Since solder pastes are designed to be thixotropic materials, their viscosity can recover from these mechanical stresses fairly rapidly and completely. However, depending on the inherent stability of the formulation being used, this recovery can be partial, resulting in undesirable slump after deposition. The whole complex subject of viscosity is dealt with comprehensively in Chapter 4, but starting viscosities are important if the screen printing operation is to be carried out effectively.

A number of factors should be considered when selecting a printer. These obviously include price and the application itself and whether the design of the board and components is really suited to screen-printing.

Projected production rates will influence the choice of capacity of the printer, as well as whether or not automatic load and ejection systems are needed (FIG. 5-8). Consideration must also be given to subsequent component mounting operations. There is little point in installing a high-speed printer capable of large volume, if components are being mounted by hand or by relatively slow pick-and-place equipment.

Printers normally fall into one of three categories: manual (FIG. 5-9), semi-automatic, which generally indicates automatic printing capability but manual loading and unloading of the substrates, and automatic units, which provide automatic feeding and off-loading. Maximum board dimensions dictate the screen size and print area, and even the most humble hand-operated machine should

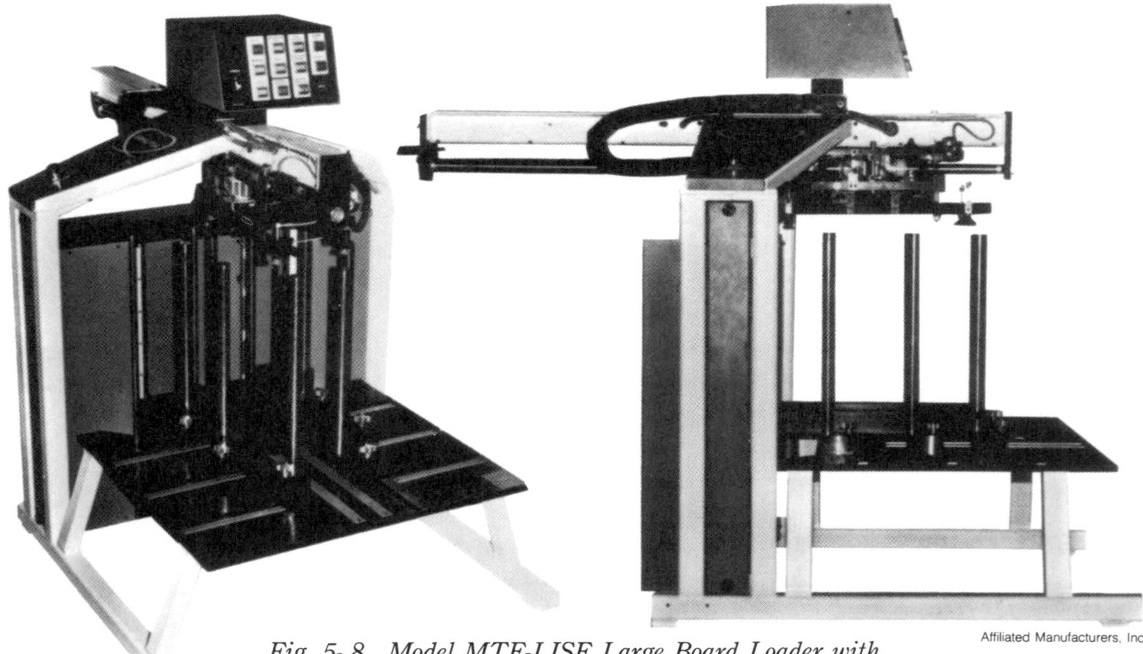

Affiliated Manufacturers, Inc

Fig. 5-8. Model MTF-LISE Large Board Loader with optional Stack Elevator. Provides automatic handling of boards up to 24 by 24 inches (559 by 559 mm).

Fig. 5-9 . Model 11 Manual Screen Printer, designed for small runs of boards or substrates up to 6 by 8 inches (152 by 203 mm).

Aremco Products, Inc.

Fig. 5-10. Accu-Coat Model 3245 Screen Printer, with Vidalign 129 alignment system. Monitor at left provides 10x magnification of screened Mylar image.

be equipped with micrometer adjustment capability for "X-Y" screen image alignment with the substrate. The more expensive printers are fitted with "Z" motion positioning for snap-off distance regulation as well as rotational theta positioning from the center of the screen. All except the more basic units are or can be supplied with a vacuum pump for the purpose of holding warped substrates onto the platen, ensuring a flat surface for the screen and squeegee.

It is often sufficient to use tooling pins on the printer platen and location holes in the laminate for positioning printed wiring boards, as long as print registration repeatability is within 0.0005 inch (0.013 mm). If there is any doubt about the accuracy of the drilling, however, then for densely populated circuits, especially those with fine-pitch pads, the use of assisted image alignment is recommended. For example, place a sheet of clear acetate over the printed circuit or substrate, and then print onto this check alignment. A more sophisticated (and expensive) method is vision-assisted alignment. This consists of a television camera, special lighting facilities, and a television monitor that provides an image of about 10x magnification (FIGS. 5-10 and 5-11).

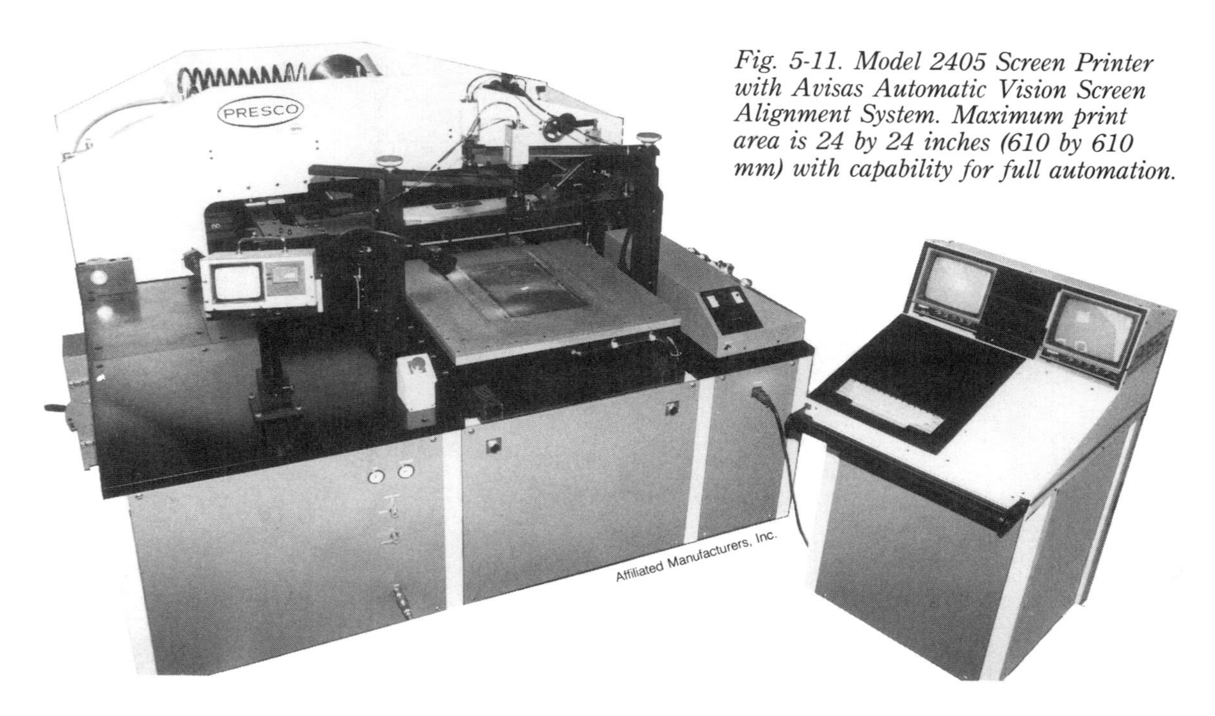

Fig. 5-11. Model 2405 Screen Printer with Avisas Automatic Vision Screen Alignment System. Maximum print area is 24 by 24 inches (610 by 610 mm) with capability for full automation.

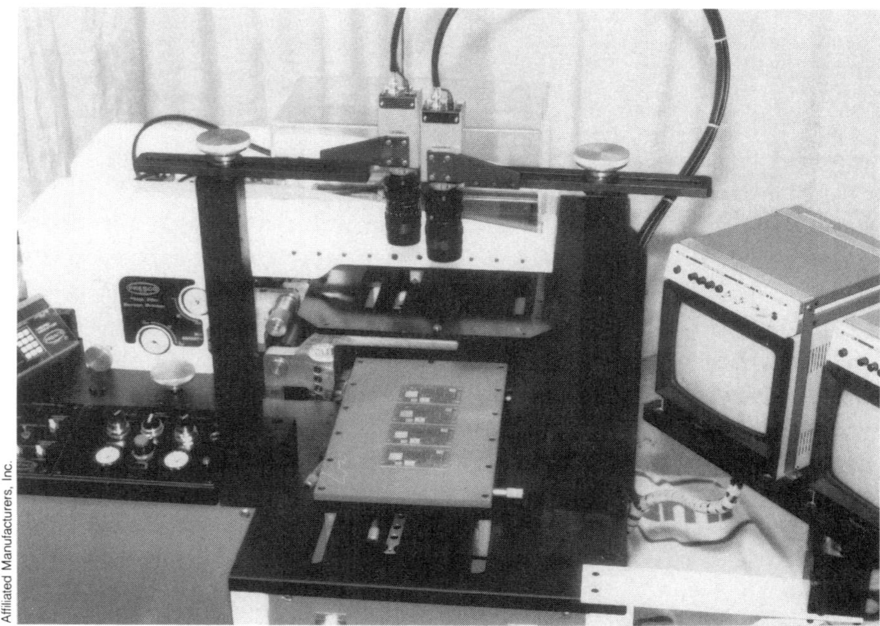

Fig. 5-12. Vision alignment system, incorporating a dual camera arrangement, is designed specifically for registration of solder paste printing.

Once correct registration has been established, some slight adjustment to squeegee pressure might be necessary to compensate for the thickness of acetate film used in the test. Vision-assisted alignment has the advantage of lending itself to automatic substrate-to-substrate registration in a computerized process, using fiducials or other patterns on the board or substrate for positioning purposes (FIG. 5-12).

There are frequently problems with registration or resolution of the screen image that have nothing to do with the actual printing process. These can include poor quality of the original artwork and phototool, the resolution characteristics of the emulsion and its thickness uniformity across the substrate, and the condition of the screen material.

It is vital that the print head be rigid and that there is no undue vibration during operation to cause printing registration errors. It is important also that the screen frames be easily installed and removed and that the underside of the screen be readily accessible for cleaning purposes. Depending on the type of paste being used, paste build-up on the bottom side of the screen might require fairly frequent cleaning to avoid smearing on the substrate and attendant bridging and solder ball problems.

Squeegees and screens should be thoroughly cleaned after each production run, and screens must never be left unattended long enough for the solder paste to dry in the mesh openings, as it will then become very difficult to remove. Cleaning can be undertaken in a number of ways: by immersion in a static bath of solvent that ranges in cleaning power from xylene to 1,1,1 trichloroethane, with or without the persuasive action of a hand-held brush; by immersion and spray, using similar solvents; or in a vapor-degreaser that utilizes one of the chlorofluorocarbons, with or without ultrasonic energy. Care should be exercised when using very flammable materials. It is important that the solvent selected be compatible with the emulsion on the screen, because some solvents, especially the more polar types, frequently attack the emulsion. Excessive cleaning will shorten the life of the screen.

Metal loading and powder particle size and shape are important considerations in selection of a paste for screen printing, as both can affect the thixotropic properties of the product as well as its viscosity stability and therefore its ability to be printed. The size of the particles, of course, must be consistent with the openings through which they are to pass. As a rule of thumb, a ratio of approximately (1:2.5 or 3) should be applied. This means that in an 80-mesh screen, a typical opening of 0.0088 inch (0.224 mm) would require a solder paste particle size of no more than 0.0029 inch (0.074 mm or 74 microns). This is the nominal maximum size of the powder known as 200 mesh. It is, of course, essential that a vendor of powder or paste is maintaining a fairly tight tolerance with regard to particle size distribution. This subject is dealt with in more detail in Chapter 1, as is also particle shape.

Chapter 4 discusses the effects of temperature on rheology and apparent viscosity of solder paste. In screen printing, temperature variations can be quite considerable. The passage of the squeegee and its contact with the solder paste, as well as the friction created between it and the screen, can cause localized increases in temperature, as can heat generated by the equipment hydraulics. In order to ensure consistent screen-printing quality, these temperature changes have to be minimized, and equally important, the surrounding environment should be fairly closely controlled. Temperatures in excess of 90 degrees F (32 degrees C) can cause significant viscosity reduction as an immediate effect and frequently severe viscosity instability later. High environmental moisture levels (50 to 60 percent relative humidity is the maximum level, with 20 to 45 percent preferred) can also be harmful, especially if the paste contains a very hygroscopic solvent system. Moisture normally permanently reduces viscosity to a possibly unacceptable level from the standpoint of slumping, bridging, and, as the result of spattering during the reflow operation, solder-balling.

It is also advisable to set up the screen printing operation away from any major air circulation, for example near fans, and air conditioning and heating ducts, as this turbulence can cause excessive loss of solvent from the paste, causing a rapid increase in viscosity and printing difficulties. This could also lead to loss of tackiness required for subsequent component mounting.

For most electronic applications involving screen printing, the most suitable solder paste would have a metal content by weight of 88 percent ± 1 percent, assuming the alloy and flux activity level are already known. The paste would also have a -200 mesh powder particle size, assuming the use of an 80-mesh-count screen and pads with a center-to-center pitch greater than 0.025 inch (0.635 mm). The paste would have a viscosity range of 450,000 to 700,000 cps, which for ordering purposes would mean a nominal range of 500,000 to 630,000 cps. Most vendors insist on a 10 percent viscosity tolerance for this particular range. For conventional screen printing, higher powder percentages often introduce problems of screen clogging and excessively fast drying on the screen. Paste manufacturers are forced to use much lower viscosity flux in the formulation to meet the range of viscosities required. The increased solvent could exacerbate separation and promote undesirable smearing, slumping, and spreading.

Where screens are being used with larger, etched openings, the printing process is more forgiving, and metal percentages up to 90 percent are theoretically acceptable, with maximum viscosities in the 800,000 to 850,000 cps range.

Where finer pitch pads are involved, then the use of a paste incorporating the finer −325 mesh powder is recommended to provide the better definition needed to avoid bridging. The larger surface area can promote solderability problems (possibly necessitating the use of a more active flux system) viscosity instability, and if the vendor's powder quality is not what it should be, solder-balling (due to the presence of too many fine particles less than 20 microns in diameter). On the other hand, the −325 mesh material can bring substantial benefits in terms of improved printing and tack retention after deposition on the substrate.

In recent months, with increasing concern about miniaturization of circuitry, there has been discussion of aspect ratio. This refers to the relationship between image (or screen) opening and total thickness of the screen, including the emulsion. In the case of mesh screens, this has been defined as 1:1,[2] which means that if the smallest opening in the screen is, say, 0.010 by 0.012 inch (0.254 by 0.308 mm), then the maximum permissible screen plus emulsion thickness for that particular board design would be the dimension of the shortest side, or 0.010 inch (0.254 mm). This is an important point to remember in setting up a new process, as the total screen thickness determines the thickness of solder paste applied. In some cases, because of restrictions imposed by the available open area of the mesh being considered and too low a paste metal content, the reflowed solder thickness achieved might be insufficient. This might be either for electrical or mechanical reasons or for the purpose of providing enough space beneath the component body for flux residue removal.

The thickness of the reflowed solder is, in the final analysis, supplied by the alloy powder in the paste. Assuming that none of the metal is lost through deposition problems or solder-balling, a solder paste with a nominal metal percentage by weight of 88 percent should provide a reflowed thickness equal to between approximately 65 and 70 percent of the original "wet" deposit. Because of the weight-to-volume relationship of the metal in solder paste, users should rarely consider a percentage of much less than 88 percent for screen-printing applications in conventional surface-mount work. A low metal percentage will cause a substantial increase in the volume of flux present, resulting in thinner—perhaps too thin—reflowed deposits that could cause reduced solder joint integrity. Lower metal percentages are, however, permissible for hybrid circuit solder reflow applications where sharp pad definition and slumping or spreading of the paste are not such critical issues. In many cases, a metal content of between 85 and 88 percent is used.

The other major area of interest for solder paste deposited by screen printing methods is printed circuit fabrication. Paste has been used to provide solder selectively on different areas of the boards after the normal print-and-etch and tin-lead electro-plating operations. In such cases, the fabricators have required a thickness greater than the 0.0003-to-0.0007-inch (0.008 to 0.018 mm) thickness typical of electroplating on certain pads that are designed for attachment of particular surface-mount devices. A thickness is desired that is more uniform than a hot-air solder leveling process that in combination with the solder coating on the component leads, would provide sufficient metal for good joint integrity. Solder paste can be printed and reflowed on such boards to provide a repeatably controlled thickness of between 0.001 and 0.002 inch (0.025 and 0.051 mm). This is achieved by using a paste containing from 85 to 88 percent metal with a −325 mesh powder and nominal viscosity between 500,000 and 600,000 cps. Best screen size is 165 mesh, but with care, an 80-mesh screen with 0.002-inch (0.051 mm) diameter filament and maximum 0.001-inch (0.025 mm) emulsion thickness will also work well. Bearing in mind the formula already discussed, the nominal wet paste deposit thickness in this case is approximately 0.0015 to 0.0025 inch (0.038 to 0.064 mm), depending on the effect of the size of the crimps of the warp and weft wires in the screen. This provides a reflowed thickness of 0.0008 to 0.0017 (0.021 to 0.043 mm), according to paste metal content. The reflowed thickness on each pad could be reduced further by under printing, in other words, by using screen openings slightly smaller than the pads, so that the solder paste upon reflow would flow out to cover the whole pad surface. In this way, reflowed thickness could be reduced further by up to about 30 percent. Other applications have required the successful reflowing of thicker paste deposits onto pads up to a thickness of 0.007 inch (0.178 mm). Where it was desired to ensure that sufficient solder was present to provide additional strength or a particularly high stand-off when the component was mounted.

Screen printing is a much more efficient method of applying solder paste on a flat substrate than pressure dispensing. It has less stringent paste characteristics, and the projected production volume usually dictates whether the higher cost of printing equipment is justified. It is a method of deposition requiring careful process controls, and screen-printer operators have justifiably been described as artists in their own right.

Screen printing also has its drawbacks with regard to constraints in the type of solder paste that can be used as well as the limitations and the relatively short working life of the screens themselves. The metal powder in solder paste is very abrasive to both emulsions and screens, which requires frequent replacement and causes production down-time.[3]

Such considerations have prompted many manufacturers in the printed wiring board surface-mount technology industry to change to stencils as a means of applying solder paste.

STENCIL PRINTING

In this method of printing solder paste, the wire mesh is replaced by a chemically etched mask, fabricated from stainless steel, brass, nickel or, beryllium copper (FIG. 5-13). Molybdenum is sometimes (albeit rarely) employed as the base material. No single metal has proved empirically to perform better than others. Stainless steel offers the advantage of increased hardness but the disadvantage of greater difficulty in etching. Brass and beryllium copper, sometimes nickel-plated, are the most commonly used, because they are readily available in the thicknesses and dimensions required for stencils and have the added benefit of being easy to etch.

In fabrication of the stencils, a photo-resist ink is typically applied to one or both sides of the metal sheet, which is then exposed to ultraviolet light through the phototool and developed in a solvent. A metal etch is then sprayed onto the exposed areas of the plate, again onto one or both sides. In the latter case,

Automated Production Concepts, Inc.

Fig. 5-13. Swiss-made Cosy manual stencil printer with vacuum hold-down for surface-mount prototype work.

the etching depth would normally be 50 percent into each side. The maximum aperture size should normally be not more than 1.2 times the stencil thickness. After completion of etching, the remaining resist is stripped away.

Care should be exercised when ordering stencils that reasonably tight tolerances are held in respect to the actual openings in relation to the original artwork. A sideways etching effect during manufacture of the stencil can increase the actual opening in a 0.010 inch (0.254 mm) stencil by up to 15 percent from the dimensions originally provided for in the artwork. Borneman and Rennaker reported that 0.002 to 0.003 inch (0.051 to 0.076 mm) in excess of what was called for can cause bridging and component misalignment with a corresponding decline in yield, especially for plastic-leaded chip carriers (PLCCs).[4] The same precautions need to be taken with surface-to-center tolerances, which vary as a function of this type of etching procedure (FIG. 5-14) . These could affect the volume of solder paste by more than 5 percent, depending on the quality of the stencil maker.

Stencils are available in thicknesses generally ranging between 0.001 and 0.030 inch (0.025 and 0.762 mm) in increments of 0.001 inch (0.025 mm), depending on the type of paste being used, circuitry design, and stencil openings. Thicknesses as low as 0.001 inch (0.025 mm) are also available. The stencil can either be mounted directly onto the frame, which must be rigid so that it remains flat under tension, or affixed by means of an epoxy adhesive to a stretched mesh border of either stainless steel or polyester which is more common today (FIG. 5-15). This provides the user with the flexibility of using the stencil in either an ''on-contact'' mode (with a snap-off of 0.005 inch or 0.127 mm or less) or ''off-contact,'' with a peel-back similar to that available with screen printing and all the advantages it offers. As a general rule, however, snap-off of more than 0.015 inch (0.381 mm) is not needed because the much greater mass of metal normally available in stencil form is sufficient to prevent it from sticking to the

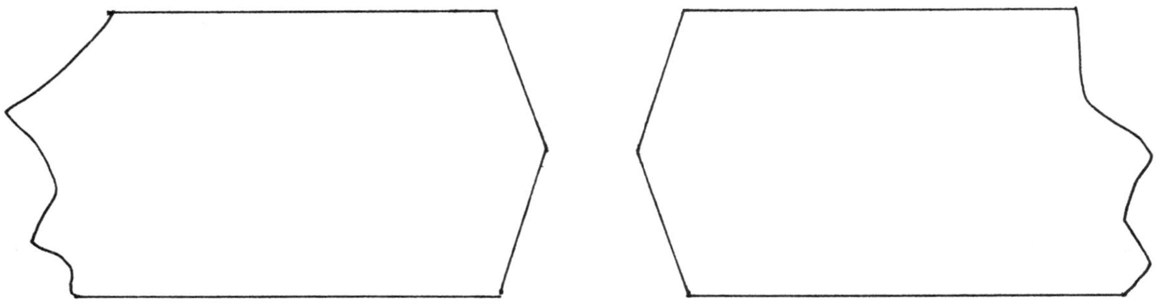

Fig. 5-14. Cutaway view of stencil etched from both sides, showing the surface-to-center dimensional differences that can occur.

board, and it pulls away cleanly after depositing the solder paste even in an on-contact process as soon as the squeegee pressure is relaxed.

It is recommended, when installing the stencil, to ensure that a minimum of 3 to 3½ inches (7.6 to 8.9 cm) and a maximum of about 5 inches (12.7 cm) of border are allowed for between the edge of the stencil and the inside dimensions of the frame. The squeegee length should not exceed the width of the stencil and should not be permitted to make contact with the mesh surround to prevent damage. The aspect ratio for stencils is generally the same as for screens at 1:1 aperture size (shortest side) to mask thickness, although figures of up to 1.6:1 have been quoted in different technical papers and presentations.

Fig. 5-15. Examples of different metal stencils with mesh border.

Squeegee hardnesses at 80 or 90 durometer are almost always higher for stencils than those for screens, depending on board or substrate design and, sometimes, user preference. The initial cost of a stencil is considerably higher than that for a screen, but its useful life is far longer. The latter consideration, in addition to the generally improved results obtainable from stencil printing, have convinced many companies to make the change to stencils.

Stencils provide improved and more consistent printing, and alignment is facilitated because visual inspection of the etched holes is much easier than with the openings in a mesh screen.

Other benefits that affect printing quality include the fact that less squeegee pressure is required and no flood bar is normally needed. On the other hand, stencils do have a disadvantage when compared with screens because screens have an inherent flexibility that can be used for printing onto surfaces that are fairly uneven (due perhaps to selective solder bumping). The rigidity of a stencil in such circumstances can lead to serious pad-to-pad inconsistencies in the thickness of paste deposits. Similar problems can arise with resists: depending on the type being used, heights can vary between less than 0.001 inch (0.025 mm) for wet types to 0.002 to 0.004 inch (0.050 to 0.102 mm) for dry films. The gasketing effect achievable with the emulsion-coated screen enables such unevenness of the board or substrate surface to be overcome.[5]

Stencils cost between two and four times more than a screen but generally last from 20 to 25 times longer. Stencil holes do not easily clog, and it is therefore possible to use solder pastes with much higher metal content than is practicable or wise with a screen. Approximately 91 percent by weight is feasible with the 63 Sn/37 Pb alloy and others in that melting point range, while 92, and possibly 93 percent could be achieved with some of the higher density lead-rich compositions. Any of the conventional powder particle sizes apply, and viscosities of as high as 1,500,000 cps are being successfully stenciled for densely populated boards or for fine-pitch leaded devices, which cannot tolerate any slumping or spreading of the paste deposit. The minimum viscosity recommended for stencil application is 700,000 cps.

With the transition to 0.025-, 0.020-, and even 0.018- and 0.016-inch devices (0.635, 0.508, 0.457, and 0.406 mm respectively), all of which have already been successfully reflowed with solder paste, manufacturers now need to reflow boards that contain both 0.050- and 0.025-inch (1.270 and 0.635 mm) lead pitch components. In many cases, the volume of paste applied for the 0.050 inch (1.270 mm) version is excessive for the smaller parts, causing bridging problems. If the overall paste thickness is then reduced to suit the 0.025 inch (0.635 mm) centers, then the 0.050 inch (1.270 mm) devices could experience "opens" or "starved" solder joints with insufficient metal volume to assure good electrical and mechanical integrity. To answer this problem, stencil makers have developed differentially etched masks, that permit the user to simultaneously print paste in the correct thickness for both types of device. For the 0.025 inch (0.635 mm) centers, the pattern is etched down or recessed to a thickness less than that of the remainder of the stencil pattern. These stencils are fabricated from stainless steel, which provides much more accurate etch definition and greater strength than brass.

Another method of achieving the same result is to bond stencils together in stepped fashion to provide the different total stencil thicknesses required. For this application, a softer squeegee (70 durometer) is recommended, as this then possesses the flexibility to penetrate the extra depth to force paste through the fine-pitch openings. Squeegee pressure should be kept low to prevent too much ''scooping-out'' of the paste in the 0.050-inch (1.270 mm) areas. Pastes with the same high metal content and viscosities should be equally suitable for this type of work. A simpler solution is often to simply under-print the paste, using apertures in the stencil deliberately made smaller than the pads.

Screens will continue to be widely used for the assembly of hybrid circuits because of the size and variety of substrate designs and the small volumes so often involved, as well as the relatively low viscosities needed for printing.

Stencils, however, can be expected in future to dominate the printed wiring board surface-mount industry, as pads and spacings become ever smaller, involving more critical deposition and reflow parameters. In most cases, users of solder paste will be converting to maximum metal loadings, minimum powder particle sizes, and high viscosities. The solder paste manufacturers have a challenging future before them.

Note: The viscosity measurements in this chapter are quoted in centipoises (cps) and are based on readings obtained on a Brookfield model RVT Viscometer with Helipath Stand, TF Spindle, 10 turns or 2 minutes at speed 5 rpm, and at a temperature of 25°C (77°F) in accordance with specification QQ-S-571E.

6

Reflow Procedures

Various methods of solder reflow have been tried and tested in surface-mount technology during the several years since its introduction to printed circuit assemblers. These methods are based frequently on what has traditionally been used in the hybrid circuit industry.

Reflow is said to take place when solder is heated to a temperature equal to or greater than that of its melting point. In the case of solder paste, this refers to its metal constituent. Heating is achieved by the transfer of sufficient energy to the solder paste, and this is achieved in a number of different ways, either separately, or in combination. Much has already been written on the physics of heat creation, and there are many publications with excellent material on this subject. This chapter is confined to a description of the ways in which solder paste can be reflowed and the various advantages and disadvantages of each. No attempt is made here to review reflow methods such as those accomplished by means of induction and resistance heating because these are rarely, if ever, found in a surface-mount production environment.

PREHEATING AND CURING

Preheating is always an important step during almost any reflow process. Its purpose is four-fold:

- to release solvent from the flux;
- to promote activation of the flux, thus preparing both the powder surface and board or substrate metallization for reflow;

- to render the paste more tenacious in its tackiness, enabling it to firmly hold even relatively large and heavy components;
- to thermally condition the whole assembly for the reflow stage, thus lessening the danger of damage due to thermal shock.

Different preheating parameters must be established in accordance with the method of reflow, the design of the assembly, any production requirements in terms of conveyor speeds etc., and the type of solder paste selected.

In general, however, there are certain useful guidelines that can be followed. The solvent system in a solder paste can be a fairly complex one, often depending on who the manufacturer happens to be. Solder pastes have remained much of a mystery from a formulation point of view, although most of the manufacturers have become more open with regard to the way their products are made (see also Chapter 3) and the general nature, if not the actual names, of the ingredients used.

The powder is obviously straightforward, and as long as the alloy requirements are complied with, reflow temperature is all that need concern the user in terms of heat application. The finer particle sizes, with a potentially larger surface area exposed to oxidation, might require slightly more time for the fluxing action to be completed, and any significant quantity of oxides could call for a slight extension of reflow time and/or temperature. This, however, should have no effect on preheating.

The three major ingredients of concern in a rosin-based formulation used for preheating purposes (whether in the RMA or RA activation category) are (1) the solvent(s), (2) the rosin, and (3) the flux activator(s).

The major solvent normally has a high boiling point, ranging between 190 and 235 degrees C (374 and 455 degrees F), but the more volatile part of the flux solvent system can be progressively driven off from a starting temperature of about 50 degrees C (122 degrees F). In normal circumstances, most solder paste manufacturers recommend an initial preheat or ''curing'' temperature of 85 to 90 degrees C (185 to 194 degrees F), which is usually adequate for the purpose of preventing the excessive spreading and slumping problems explained in Chapter 8. As long as the correct formulation is being used, this preheat temperature should also eliminate solder-balling and outgassing. The time of exposure to this temperature depends on the particular production process but usually varies between 1 and 4 minutes.

One of the reasons the 85 to 90 degrees C (185 to 194 degrees F) range is used as a preheat yardstick is that rosin softens and melts at the same temperatures, so vaporization of solvent and release of the activators from the liquid rosin can take place simultaneously.

The next temperature level of significance is where the activator system begins to act as a truly efficient fluxing agent. Some oxide removal does take place at room temperature, as a simple test by depositing solder paste on an oxidized copper coupon will demonstrate. A temperature level of about 120 to 130 degrees C (248 to 266 degrees F), however, is required before real fluxing power is observed, and this continues through the reflow stage to ensure proper bonding of clean metal surfaces.

Once the flux activation stage has been reached, it is common practice to then quickly raise the temperature to that of reflow, if, in fact, this has not taken place earlier.

The duration at a specific temperature or range of temperatures is a function of the best heating profile that can be established for an individual assembly design, and many different alternatives can be followed to optimize a process. Suggestions in this regard are in the following sections on the different reflow methods used for solder paste. See FIGS. 6-1 through 6-5.

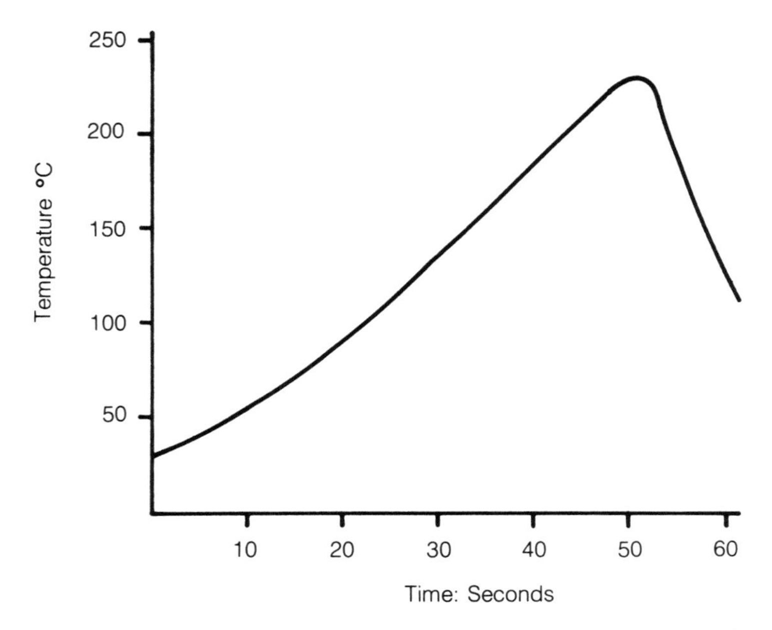

Fig. 6-1. Temperature profile for rapid reflow of solder paste by lamp infrared. Such a process often results in excessive slump or flux spread, causing bridging and solder-balling problems.

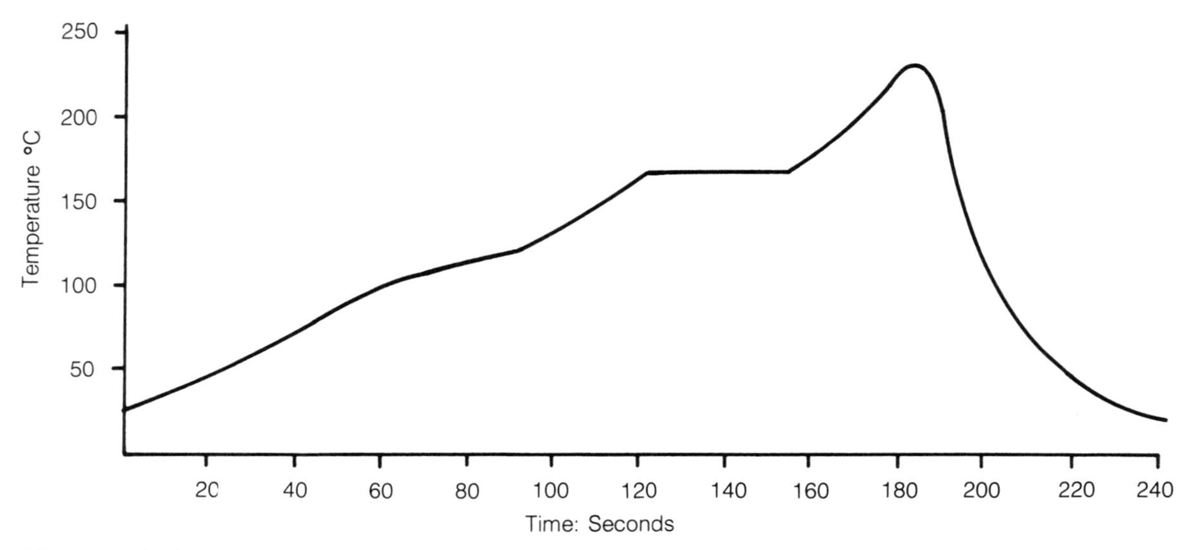

Fig. 6-2. Typical temperature profile for lamp infrared reflow.

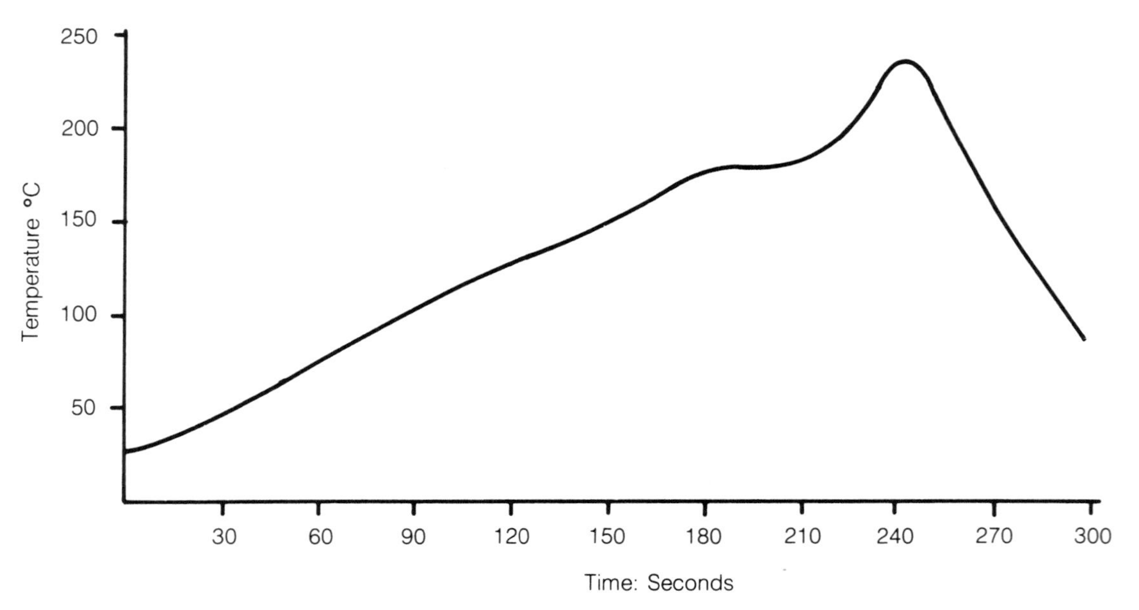

Fig. 6-3. Typical temperature profile for panel infrared reflow.

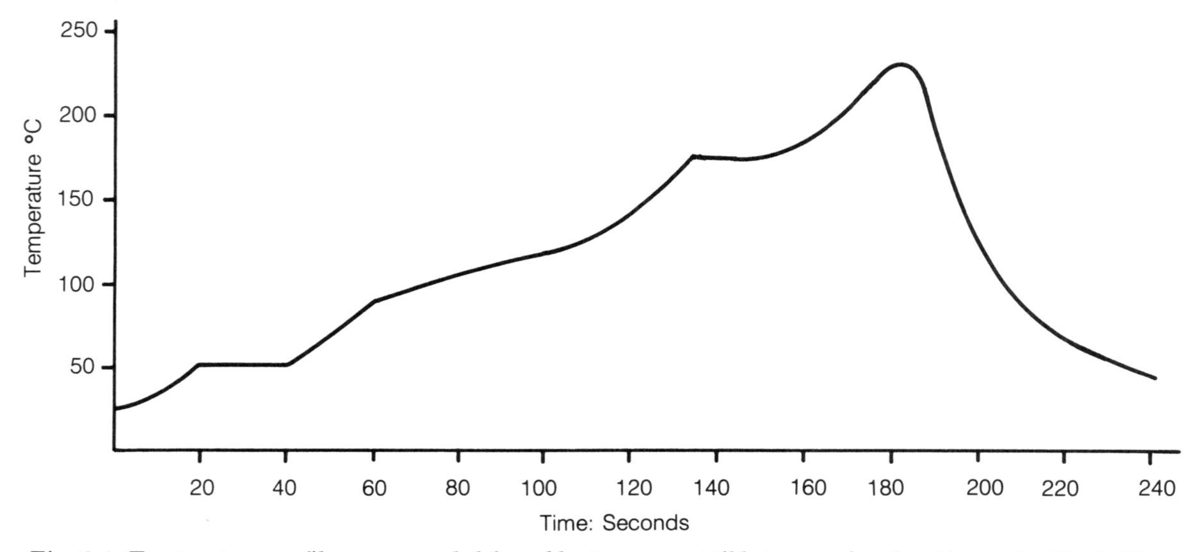

Fig. 6-4. Temperature profile recommended for solder paste susceptible to excessive slumping and solder-balling. The preheat stage at 50 degrees C (122 degrees F) frequently helps to prevent or reduce flux spread.

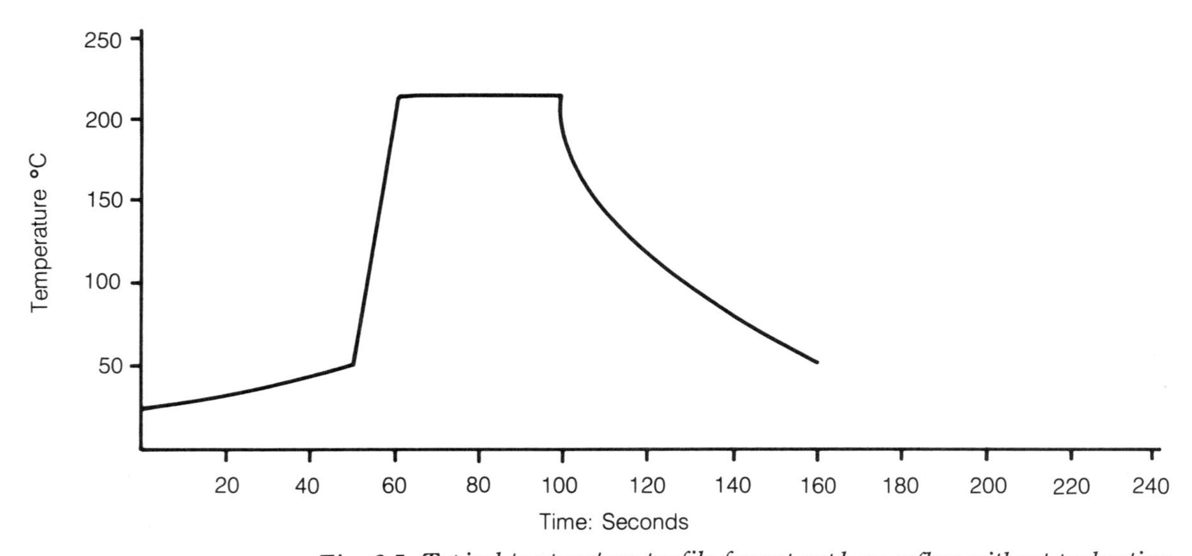

Fig. 6-5. Typical temperature profile for vapor-phase reflow without preheating.

SOLDERING IRONS

Soldering irons, used for many years in electronics, really have no place in surface-mount production. They can, however, fulfill a useful role in prototype operations and in repair and rework situations.

Heat is conducted to the work by means of the heated soldering iron tip. It is not within the scope of this chapter to discuss hand-soldering of this kind in any detail. Suffice it to say that for solder paste reflow purposes, the irons must be designed to produce the correct tip temperature, the profile of which must be determined rather carefully to avoid damage to delicate components. The tips themselves have a configuration somewhat different from that applicable for conventional hand- or wave-soldered assemblies. This is because of the closer spaces they need to penetrate and to provide easier handling for the operator.

HOT GAS REFLOW

Hot gas reflow of solder paste has long been used in what is often termed structural bonding, for example the assembly of gas feed lines, gas tanks, and radiators in the automotive industry. The operation can be automated by installing a series of gas jets in a conveyorized system, either in-line or as a carousel. Individual gas jets or even a simple array in the form of a perforated pipe can be utilized.

In such applications, the paste is normally dispensed as a dot or bead onto the location where it is to be reflowed. For this type of work, cost is an important consideration, and the paste supplied normally reflects this, with a rather coarse powder quality and relatively low metal and high flux content. The dispensing needles or manifolds therefore must have fairly large openings to permit its flow onto the workpiece.

In structural soldering, the gas temperature is usually substantially higher (in excess of 100 degrees C, or 180 degrees F) than the alloy melting point in order to overcome the heatsink properties of the metal part being bonded. This means that the paste will tend to spread or run very rapidly, not only because of the effect on the flux of the high temperature, but also as a result of the pressure of the gas jet. Rarely is any form of preheating employed in this type of operation.

For these reasons, the use of pressurized hot gas reflow of this type for electronic assemblies is not recommended because even with a good-quality solder paste and some form of preheating, it would be extremely difficult to restrict excessive flow of the flux, and problems of bridging and solder-balling could result.

The concept is similar to that for manually operated heatguns, a method by which a fine jet of air or gas is heated to a suitable temperature by being forced past a heating element and is blown through a nozzle or orifice to reflow solder paste. One advantage of this process is that it can be used to apply the heat quite accurately to a specific location without affecting the surrounding area, always provided that the air flow can be accurately directed. It can thus also be used to repair solder joints.

A disadvantage is that high temperatures are required to compensate for energy lost from the heated air flow to the atmosphere through which it passes. Although such hot gas reflow has occasionally been employed in surface-mount technology, this has usually involved a very small number of components soldered manually. In some cases, an inert gas such as nitrogen has been used to exclude oxygen from the air from the location being heated.

If gas-jet size and temperature are closely controlled, this reflow method can be used for remelting solder joints to remove and replace defective components without disturbing neighboring devices.

Hot gas ovens have been used to pass high-temperature air or gas over hybrid circuit and surface-mount assemblies. However, since the transfer of heat is a slow process, the method does not really lend itself to a high rate of production. Large furnaces are usually needed, and these in turn create high energy costs. Since high temperatures might have to be used to increase production rates, there is always the danger of overheating substrates or components.

An efficient furnace can, however, provide good uniformity of heating, but because of limitations regarding oven size and problems that can be caused by the effects of movement of both solder paste and components by high air or gas flow, the hot gas system has very definite limitations in today's solder paste reflow applications.

Fig. 6-6. Conduction reflow. Falcon Five Hybrid Soldering System with optional hood for nitrogen or forming gas.

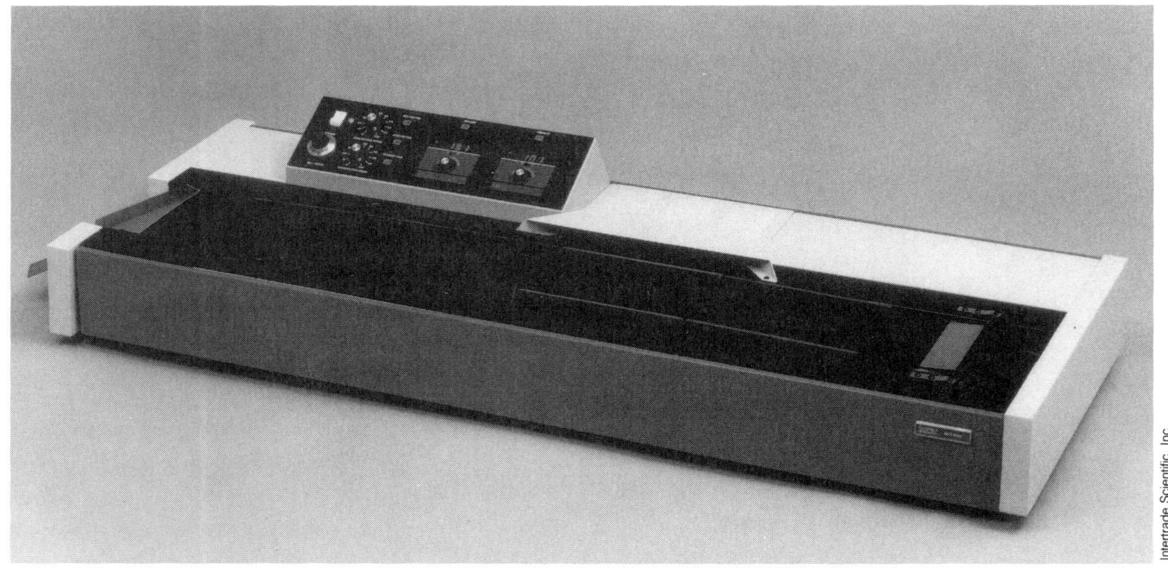

Intertrade Scientific, Inc.

Fig. 6-7. Conduction reflow. Model 6820 Belt-type Hot Platen Reflow System with hood for use with inert atmosphere.

HOT PLATES

In the case of a hot plate, heat is conducted from a series of hot plates through a metallic or polytetrafluorethylene (PTFE) belt into an assembly (FIGS. 6-6 and 6-7). Only single-sided boards or substrates can be reflowed, and larger sizes may be difficult, if not impossible, to process due to flatness considerations. The air gaps created usually lead to variable heating rates, with the attendant dangers of either over-compensation of heater settings or insufficient temperatures being reached by the solder paste in different areas on the substrate to enable it to reflow.

Such conductive belt systems have for many years been successfully employed in the hybrid industry for reflowing solder paste onto ceramic substrates. The high thermal conductivity and relatively small size of these enable the process to be performed very efficiently. Unfortunately, the nature and dimensions of the printed wiring board materials required for conventional surface mounting generally preclude conveyorized hot-plate soldering as a reflow method for the current technology.

VAPOR-PHASE OR CONDENSATION REFLOW

When printed wiring board surface-mount technology began its dramatic development at the beginning of this decade, the vapor-phase or condensation method of reflow soldering was the only one capable of providing the process repeatability and consistent reliability of results. The manufacturers of infrared and other conventional equipment that were so long involved in the thick-film industry seemed at first almost overwhelmed by the prospect of reflowing solder paste on fiberglass/epoxy with its poor thermal conductivity and very different thermal mass characteristics, compared with the aluminum oxide and beryllium oxide materials they had previously been accustomed to.

The one manufacturer of the vapor-phase equipment and their supplier of the processing liquid were left to enjoy a virtual monopoly of the solder paste reflow market in the United States for some two years. Those halcyon days have long since come to an end, mainly as the result of the introduction of reliable infrared furnace designs. The process remains popular, however, and significant improvements have been achieved meanwhile with regard to both equipment design, as well as the cost and efficiency of the process, enabling both the original and more recently established manufacturers of these systems to continue to compete for business in the rapidly expanding world markets.

Development of the vapor-phase technique began at the Engineering Research Center of Western Electric Company, Inc. in Princeton, New Jersey in 1971, and it was the subject of a first publicly presented technical paper at the Nepcon West exhibition in Anaheim, California in February, 1974.[1]

The process utilized the latent heat given up by the vapor of a boiling liquid as it condenses on a cooler surface, which in the modern surface-mount process would be a printed wiring assembly with components held to the circuitry by solder paste. The vapor, with its greater molecular energy over that of the static fluid, was used as the heat transfer medium. The process was very similar to that used in a vapor degreasing machine, where the action of condensation of the cleaning solvent vapor upon an cooler object dissolves and rinses away flux residues and other contaminants.

Impetus for the Western Electric work came originally from problems in reflowing flux-coated preforms for bonding gold-plated wire-wrap pins into plated-through holes of multilayer boards and achieving satisfactory solder fillets on both sides through capillary action. There were inconsistent results from other reflow methods, such as infrared, due to lack of temperature uniformity that caused wide temperature variations across the boards. Hence, the Western Electric engineers undertook tests with a fluorinated polyoxypropylene fluid with a boiling point of 224 degrees C (436 degrees F) that was manufactured by the DuPont Company and originally given the nomenclature E5. The results were very successful, with reject rates of less than 0.1 percent in a total of more than 1 million solder joints, compared with up to 5 percent for other reflow methods. Note that neither dip- nor wave-soldering procedures could be considered because the gold pins themselves had to remain free of solder to facilitate the subsequent wire-wrapping operation. Following these tests, the first production machine was developed that was similar in concept to a batch vapor degreasing machine, with two sets of condensing coils above the vapor level to reduce fluid loss. It was installed at the Western Electric facility at the Merrimack Valley Works in North Andover, Massachusetts.

The major initial concern at this time was loss of the very expensive fluid, mainly due to vapor escaping into the open atmosphere above the condensing coils as well as (to a lesser extent) drag-out of condensed liquid by the printed circuit assemblies. The loss of vapor experienced could well have been at least partially due to the "aerosoling" effect described by Ruckriegel and to disturbance of the sharply defined interface existing between the surface of the vapor and the air above.[2]

It was not very long before those same Western Electric engineers solved the problem by introducing a secondary and inexpensive solvent vapor with a much lower boiling point upon the primary fluid vapor to effectively prevent its passage into the atmosphere. It was also necessary that this new material be compatible with the primary fluid. Additional condensing coils were installed to prevent excessive loss of the secondary vapor—the refrigerant trichloro-trifluoroethane (R-113)—with a boiling point of 47 degrees C (117 degrees F), whose cost was said to be 1/50 of that of the primary fluid. This meant that any primary fluid vapor surviving passage past the first set of condensing coils would condense with the secondary vapor on its condensation coils and drop back to the sump, where density and their different boiling points would ensure the re-separation of the two fluids. They realized that some of the secondary fluid would be lost to the atmosphere, but its rate of diffusion was considerably less than that of the primary fluid because of the much smaller difference between its temperature and that of the air above. The secondary vapor also reduced the sharp differential between the temperature of the primary vapor and that of the atmosphere, thereby helping to prevent aerosoling. Later, the equipment manufacturers recognized that fluid losses were usually proportional to the size of the open area above the vapor zone; they therefore tapered down this opening, substantially reducing usage of the fluorocarbon.

The boiling point of the liquid subsequently developed and made commercially available in 1975 was 215 degrees C (419 degrees F), or approximately the desired 30 degrees C (54 degrees F) above the melting points of the major alloys used for surface-mount reflow—63Sn/37Pb, 60Sn/40Pb, and 62Sn/36Pb/2Ag. The major benefits of such a system were immediately obvious: no part of the assembly could reach a temperature of more than the liquid's boiling point, which was also the temperature of the vapor, thus eliminating any overheating of components; and since the reflow operation was undertaken at normal atmospheric pressure, the workpiece would be heated extremely uniformly to 215 degrees C (419 degrees F), provided that exposure time to the vapor was sufficient. Experience since has shown that this ranges from approximately 10 to a maximum of between 45 and 60 seconds, depending upon configuration of the assembly.

This fairly short reflow time at a low reflow temperature is beneficial to the user, because the formation of intermetallics such as copper/tin (primarily Cu_6Sn_5) and gold/tin ($AuSn_2$ and $AuSn_4$) is slowed, and the solubility of any silver from board or substrate metallization or on component terminations in tin-lead is greatly reduced. Board geometries are no longer such a concern, because even assemblies with high heatsink areas can now be easily reflowed without the danger of overheating other sections of the boards. Uniformity is achieved primarily because condensation of the vapor occurs on all surfaces whose temperature is below that of the boiling point of the fluid. The efficiency of the heating method meant that a lower reflow temperature than those for other processes could be used. Another advantage of the process was that the density of the saturated vapor would effectively displace air and moisture from the board surfaces, preventing the formation of oxides during the reflow operation.

The announcement of the new process created much interest, despite initial excessively high operating costs, and the first commercially available reflow units in 1975 were basically a scaled-down version of the large upright batch system built by Western Electric, each one's size dictated by the thermal mass of the panels being processed. The equipment utilized a fluid produced by the Minnesota Mining and Manufacturing Company (3M), which had a boiling point of 215 degrees C or 419 degrees F. This product was called Fluorinert, with the designation FC-70.

The fluid, which is now produced under different designations according to its boiling point, is chemically a fluorocarbon. It is colorless, odorless, and non-flammable, and all the different boiling point versions of any particular manufacturer are fully miscible with each other. It contains no hydrogen or chlorine and does not pose any threat to the earth's ozone layer; it is not related to chlorofluorocarbons (CFCs). It is nonpolar and, contrary to a misconception sometimes encountered, has no solvent cleaning action. Its water solubility is very low, being measured as a few parts per million.

The first reflow units to be generally marketed in 1975 comprised a chamber to hold the fluid, fabricated from stainless steel, and containing electrical immersion heating elements. It employed a dc-motor-driven, SCR-speed-controlled elevator platform that carried the assemblies down into the primary fluid vapor where they were held for a predetermined immersion period before being transported back up through the secondary vapor area and condensing coils for unloading and inspection. Automatic timing later also became available to control elevator speed, dwell position, and dwell times in both the primary and secondary vapor zones. Capacity was limited to only about 6 pounds per load (2.7 kg) for boards up to 14 by 16 inches (356 by 406 mm) with typical cycle length of the order of 3 to 5 minutes.

Originally, connector pin reflow was still the major application for the new process, and to some extent also the reflow of termination leads (also by means of a preform) to a hybrid integrated circuit. Only much later was it realized how suitable this could be also for solder pastes. Operating costs initially were claimed to be some $3 to $5 per hour, which compared favorably with those experienced with other conventional forms of printed wiring board soldering.

From the batch reflow units with their load constraints (FIGS. 6-8 and 6-9), a rapid transition was made to in-line systems that enabled manufacturers to expand their reflow operations (FIGS. 6-10 and 6-11). One of the first applications for this improved process method was the reflow of tin-lead-plated printed wiring boards, but it excited most interest in the burgeoning surface-mount industry and played an important role in promoting the use of solder paste as a bonding medium for components on printed wiring boards. One of the first disadvantages of in-line vapor-phase reflow was that a secondary vapor zone was no longer practical.

Considerable improvements have been made in recent years with regard to equipment design and reliability and—just as important—operating costs, which have been estimated by some critics to be as high as $10 to $15 per hour for fluid alone in the in-line machines. Diffusion is still an important factor in fluid loss, as is the entrapment of condensed fluid on printed wiring boards.

Much improved techniques in vapor condensation, recirculation, and recovery have resulted in claims by the equipment manufacturers of fluid cost reductions to between $3 and $6 per hour. Especially important in lowering such costs have been tighter control of vapor height, better adjustment of fluid heating rates to meet process load requirements, and correct ventilation procedures preventing excessive vapor loss.

Computer-controlled systems are generally preferred to avoid operator error (which can cause significant additional expense if fluid loss is involved) and to better automatically adjust parameters to deal with different thermal loads processed. The size of the vapor zone and the amount of heat applied to the fluid usually governs the energy available for reflow.

Safety considerations have played an important role in discussions about condensation reflow since it was realized at a fairly early stage that harmful vapors

Fig. 6-8. Model 1214 Batch Vapor-Phase Reflow System. Handles assemblies up to 12 by 14 inches (305 by 356 mm) and total load up to 4 pounds (1.8 kilograms).

Fig. 6-9. Model M10A Batch Vapor-Phase Reflow unit for laboratory or prototype application. Shown with Auto-Mate elevator.

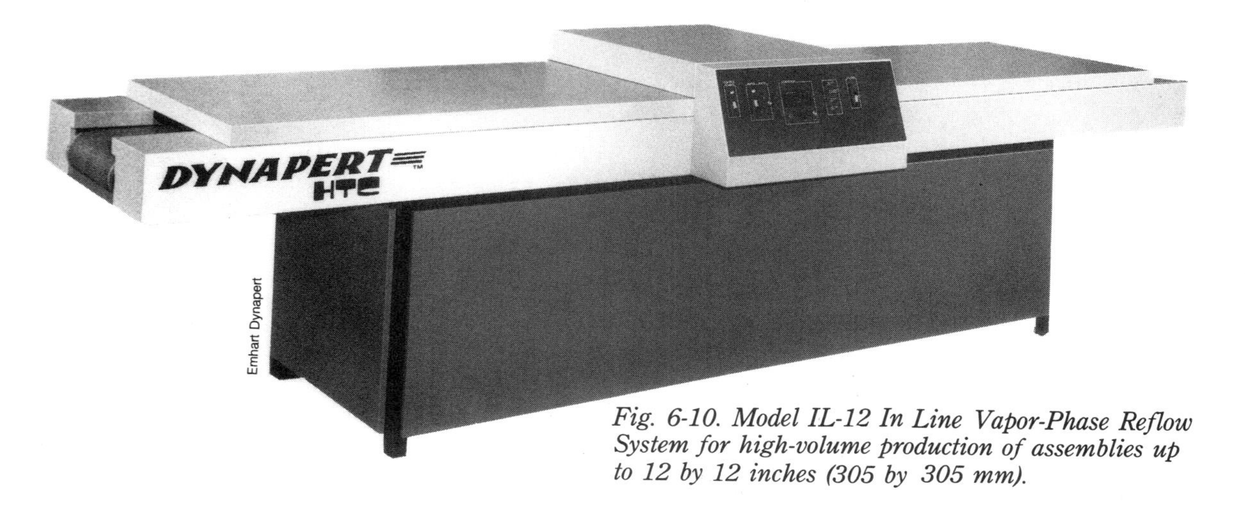

Fig. 6-10. Model IL-12 In Line Vapor-Phase Reflow System for high-volume production of assemblies up to 12 by 12 inches (305 by 305 mm).

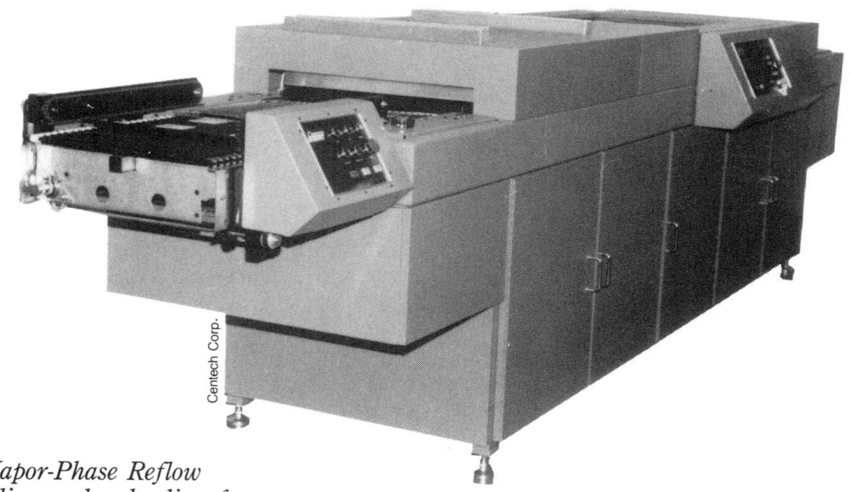

Fig. 6-11. Model CLUL Vapor-Phase Reflow System with automatic loading and unloading for assemblies up to 18 by 24 inches (457 by 610 mm).

and potentially corrosive acids could be created during the process. Thermal decomposition of the reflow fluid can result in the creation of vapors containing perfluoroisobutylene (PFIB), which is extremely toxic and reacts with moisture to form hydrofluoric acid (HF). Such decomposition would usually be caused by flux used during the reflow process and not absorbed by the fluid being deposited upon the surfaces of the immersion heating element surfaces. This tends to harden and char, providing an insulating effect and prompting the elements to overheat. Such a residue is normally readily visible.

Because of possible vapor toxicity, therefore, good ventilation is always very desirable if not essential when vapor-phase equipment is used, as well as preventive maintenance in keeping the heating elements clear of flux residue. This is usually accomplished by filtering the excess flux from the sump.

The hydrofluoric acid already mentioned, and any hydrochloric acid formed by thermal degradation of the R-113 secondary fluid, should best be removed by means of a water scrubber. Failure to eliminate these acids often results in corrosion and even cracking of the secondary vapor cooling coils.

The 3M Company now supplies FC-5311 (formerly known as Flutec PP11 and marketed by ISC Limited, England) as a more stable alternative to FC-70. Air Products and Chemicals, Inc. of Allentown, Pennsylvania and the Italian company, Montedison, manufacture a similar range of fluids, which are sold under the names of Multifluor and Galden, respectively. The latter are marketed by Alpha Metals of Jersey City, NJ. The vapor-phase process is a safe one from the viewpoint of not presenting any smoke or fire hazards.

The rapid form of heating utilized can cause severe volatilization of the solvents in the solder paste flux system. The temperatures can vary from 15 to 50 degrees C (27 to 90 degrees F) per second, compared with 2 to 6 degrees C (4 to 13 degrees F) for other reflow methods such as infrared. This leads to spattering with its attendant solder-balling, blow-holes, voids, component misalignment, and tombstoning of chip components. Excessive movement of components is sometimes attributed to the inclination of in-line conveyor systems, especially if the solder paste has been improperly cured. Most, if not all, of these pitfalls can be avoided by effective pre-heating. The duration of this can range from 5 to 20 minutes, depending on the size and configuration of the assembly.

Thermal shock to boards or components also appears more prevalent with vapor-phase than with other mass-production reflow methods, which can likewise be eliminated by judicious temperature profiling. The vapor-phase method of transferring energy is a very efficient one, and it will clearly remain an important, although no longer dominant, method of reflow in electronics in general and for surface-mount technology in particular.

INFRARED

Infrared has been used in electronics for many years, not only for solder paste, but also thick-film ink compositions and for fusing the tin-lead electroplate on printed wiring boards.

As the result of significant improvements by the equipment manufacturers, the use of infrared radiation for the reflow of solder paste has been growing steadily during the past five years. This method has now taken over from vapor-phase the lion's share of the market for reflow equipment in surface-mount technology involving printed wiring boards.

In this method of heating, infrared energy is carried by electromagnetic waves that travel at the speed of light. The waves are classified according to their frequency and length in the electromagnetic spectrum. Infrared falls within a wavelength range of 0.72 to 1,000 microns or 1 millimeter, falling in the spectrum between visible light and Hertzian waves. However, for the purpose of reflow soldering, the practical bandwidth is from 0.72 to about 2.5 microns (near infrared) and 2.5 to 5.0 microns (middle infrared). Full spectrum infrared, occasionally used in reflow applications, would combine both near and far infrared wavelengths. The energy transmitted by the wave is converted into heat when it contacts and is absorbed by an object.

Infrared differs from light in that it has a lower frequency and longer wavelength, both of which are important factors in making infrared suitable for reflow soldering. The amount of energy absorbed depends also upon the thickness of the material.

Different materials absorb to a greater or lesser degree the energy given off by the infrared radiation, depending on the length of the wave. Thus, metals best absorb near infrared, while organics such as the chemical constituents of solder pastes and printed circuit board materials such as epoxy/fiberglass and polyimide/fiberglass absorb middle infrared wavelengths better. All this means that the latter heats up faster than the metal powder in the solder paste when subjected to infrared wavelengths in the middle infrared range. Selection of a particular reflow equipment system therefore depends upon the nature of the materials being reflowed, the thickness and thermal mass of these materials, and also the type of soldering medium being used, i.e., solder paste, plain, flux-coated, or flux-filled preforms, etc.

All such materials have an emissivity factor, which can be defined as follows:

Reflective the object does not permit any absorption of infrared energy to heat it;

Transparent infrared waves can pass through the object without giving up any energy in the form of heat;

Semi-transparent enough of the infrared energy is absorbed and retained inside the object to create heat;

Opaque the energy is resisted at the surface of the object, does not penetrate it, and is therefore converted to heat at the surface.

The wavelengths used in infrared for reflow soldering, particularly near-infrared, normally ensure that little energy is transferred to a gas such as nitrogen, forming gas, or for that matter, to air, blanketing the assembly. Such gases are transparent to near-infrared. This means that the energy passes through to be either partially or completely reflected or absorbed by the assembly to be reflowed, which is how materials will react to infrared, depending upon the wavelength being used.

An instrument such as a spectrophotometer can be very helpful in determining minimum and maximum absorption rates of a material, reproduced as a series of peaks and valleys in a spectrograph. This is essentially the same method employed to monitor the chemical characteristics of, for example, soldering fluxes, including those used in pastes. Since the absorptivity rates of the different ingredients varies according to wavelength, changes from a pre-established standard or "fingerprint" readily become apparent and quickly identify the presence of undesirable chemicals, especially different or new activators that might later cause corrosion or electrical problems.

There are two types of infrared radiation sources used in the industry. Each emits energy at a different wavelength, and new solder paste users are frequently at a loss to decide between the systems, assuming that infrared is the reflow method preferred.

The first type of infrared source uses near infrared with short wavelengths that, according to Wien's Displacement Law, provide the highest amount of energy.[3] Near infrared is also partly color selective, so absorptivity of the energy by a black component body, for example, is greater than by a white ceramic substrate, thus requiring careful setting of reflow parameters.

The results of some tests, however, show that the color-selectivity issue has been greatly overstated, and the only real disadvantage of lamps is that of shadowing of solder joints by components.[4, 5] The system employs up to more than 50 non-focused tungsten filaments or nickel-chromium quartz lamps to emit the infrared energy on wavelengths in the range 0.7 to 2.2 microns from both the top and bottom of the heating chambers or zones (FIGS. 6-12 and 6-13). The tungsten filament lamps are filled with an inert gas such as argon to prevent degradation of the filament by oxidation. The average operating life of such lamps is estimated to be from approximately 5,000 to 20,000 hours, (depending on the filament temperature). The tungsten filament temperature is reported to be in excess of 2,200 degrees C, or nearly 4,000 degrees F, while that of the nickel-chromium is of the order of 1,100 degrees C, or just over 2,000 degrees F.[5] Such systems normally operate with a controlled atmosphere, capable of being flowed at a fairly high volume; this has greatly alleviated the component shadowing problem.

The atmosphere for these near-infrared systems is often air but can also be nitrogen or forming gas (up to 12/15 percent hydrogen, balance nitrogen). Rarely, if ever, with solder pastes would a pure hydrogen atmosphere be used.

Inert atmospheres containing not more than about 150 parts per million of oxygen can be advantageous in improving solder joint quality by eliminating the formation of additional oxides on the metal surfaces as well as in reducing the tendency of the fluxes to char, especially at the elevated reflow temperatures required for higher-melting-point alloys. They can also be helpful in reducing discoloration of board materials, although this cannot be avoided if they are exposed to excessive temperatures. The presence in the atmosphere of only 5 percent hydrogen can facilitate solder wetting and has been shown to inhibit excessive flux spreading, an important cause of solder-balling. This can also aid in preventing excessive component movement (which sometimes tend to slide about on flux residues), and in reducing the increased area of board containing flux residue,

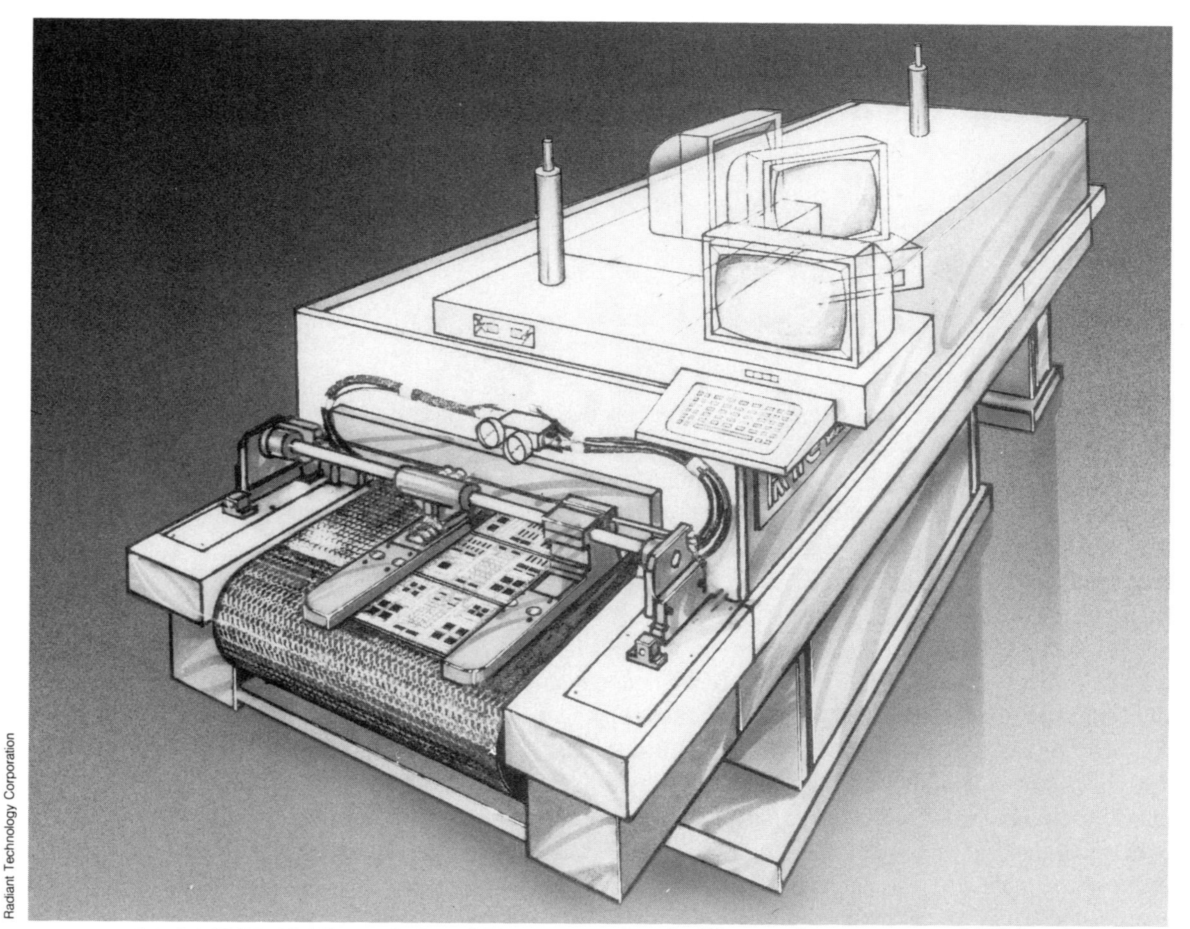

Radiant Technology Corporation

Fig. 6-12. Model SMD-624 Lamp Infrared Furnace, equipped with new edge conveyor and PC-based computer controller with swing-out feature.

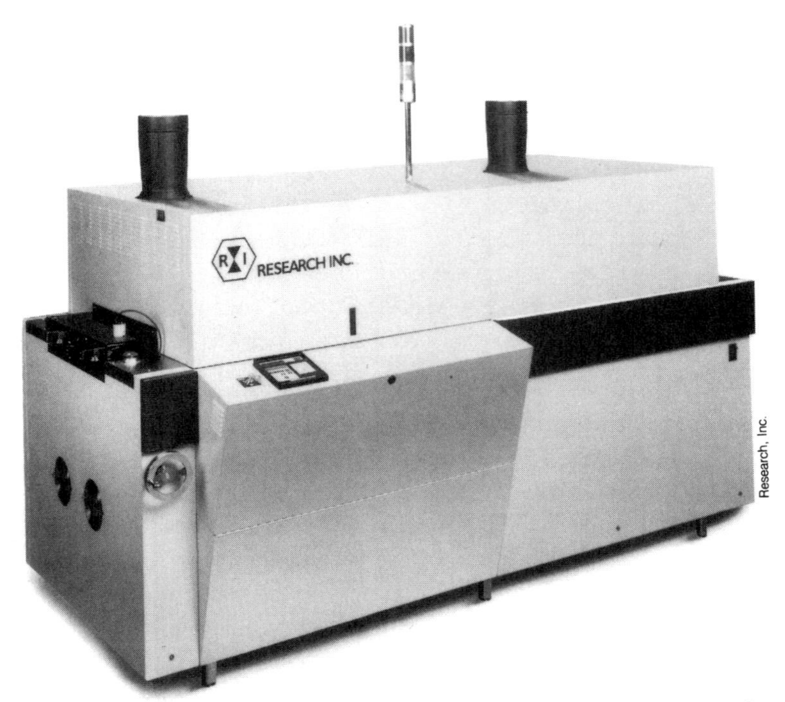

Fig. 6-13. Model 4470 Lamp Infrared Reflow System with edge conveyor for two-sided processing.

which has to be removed after reflow, especially underneath components.[4] The use of such gases does, of course, increase processing costs.

The energy can be applied to either or both surfaces of the assembly being reflowed, and systems are available with several heating zones to enable the user to establish very precise temperature profiles. Temperature control is maintained by variation of voltage applied to the lamp filament by means of SCRs in conjunction with a microprocessor.

In the most basic case, let's assume the use of a 63 Sn/37 Pb alloy solder paste. The assembly is loaded onto a variable-speed stainless steel open mesh conveyor belt up to 24 inches in width. It enters a short first zone that is invariably set at a high energy level for assembly preheating, with a temperature increase rate of between 2 and 7 degrees C (4 to 13 degrees F) per second. Most of the volatiles contained in the solder paste are removed at this stage to prevent spattering and excessive flux spread later (which could cause solder-balling). A too-rapid increase in temperature can, however, cause the flux solvents to boil, causing component movement or even tombstoning of ceramic chip components. What is most desirable, as has been stated before, is that when the solder paste leaves the preheating area, it is no longer wet but is almost dry and very tacky, able to hold even the largest or heaviest components firmly in place on the pads. The heating rate is determined by the assembly geometry and any concerns with respect to temperature-sensitive devices. The duration of preheating is typically between 15 seconds and 1 minute for lamp infrared.

As in other forms of soldering, care should be taken not to overheat the solder paste in the preheating zone. This could cause deterioration of the flux activators and even excessive hardening of the flux. In the worst case, the result could be poor wettability during reflow and difficulty in subsequently removing flux residue in the cleaning stage.

The second section, sometimes called a "holding zone," is designed to give the whole assembly time to reach a uniform temperature equilibrium to overcome any heat absorption differences that might have occurred in the first preheating zone. The assembly leaves this holding zone at a temperature between 160 and 170 degrees C (320 and 338 degrees F), or just below the melting point of the alloy. This is an important stage in ensuring success of the reflow operation.

In the third zone, reflow, the temperature is raised rapidly to approximately 215 to 230 degrees C (419 to 446 degrees F), or some 35 to 50 degrees C (63 to 90 degrees F) above the melting point of the alloy. The rapid temperature increase and short dwell time at reflow temperature are important not only to minimize potential damage to board materials and components, the temperature of which should remain well below that of the actual reflow, but also in preventing the formation of excessive intermetallics at the interface between the board metallization and the solder joints. Rapid cool-down of the solder after reflow is most desirable for the formation of fine metallic grain structure of the joints and a clean, shiny appearance. Furthermore, the same comment made earlier with regard to hardened flux residues applies. Temperature is normally controlled within a tolerance of about plus or minus 2 to 3 degrees C (4 to 6 degrees F) across the conveyor belt.

Other systems are available with a much greater degree of sophistication, for example having as many as seven heating zones or a post-reflow cooling section using filtered air.

Opponents of the tungsten and nickel-chromium filament lamp methods claim that their short peak emission wavelengths of 1.15 and 2.11 microns, respectively, can cause problems with regard to delamination of multilayer printed circuit boards if these contain pockets of moisture between the laminate layers. Pre-baking the boards for 20 to 30 minutes at a temperature of about 120 degrees C (248 degrees F) normally resolves this problem, although higher temperatures and/or dwell times might be necessary for certain multilayer types.

The other system in common use, known as Convection/Infrared, employs what is known as secondary emission in the middle-infrared range of about 2 to 4 microns. What this means is that rather than energy being directly transmitted from a source such as a filament, the heat is conducted by embedding a resistive element or emitter in a thermally conductive ceramic, the rear of which is covered by a material with high thermally insulating properties. The emitter is designed to have a rated peak temperature of 800 degrees C (1,472 degrees F). To enhance the efficiency of the emitting side, high emissivity material with electrically insulating properties is attached to this.

Manufacturers of this type of reflow equipment, also called ''Area Source,'' claim that the wavelengths they use provide a more even distribution of heat than is possible with filament-type emission because the near infrared results in different rates of absorption by the various solder paste ingredients, such as powder and flux. They point out that as the printed circuit board material absorbs middle infrared better than near-infrared the material will heat up faster to ensure better heat distribution across the assembly that would be prevented from overheating by its thermal mass. In addition, middle-infrared wavelengths are not color selective up to 600 degrees C (1,112 degrees F), and thus the danger of creating hot spots or of underheating certain areas of the assembly are eliminated. Work involving near-infrared emissivities and different printed wiring board colors seem to indicate very little difference in heating rates at a typical 63Sn/37Pb reflow temperature of about 230 degrees C (446 degrees F). It has been concluded that thermal mass of the assembly is probably a more important consideration with regard to heating rates.

A working-life of up to 8,000 hours is claimed for the emitters because they are protected from oxidation and corrosion, and they require little maintenance.[6]

A typical conveyorized area source furnace also uses a short preheating area, a vented transition zone where the volatiles accumulated from the solder paste during preheating are driven off, and one or more additional zones with top and bottom heating capability before actual reflow takes place (FIGS. 6-14, 6-15, 6-16, and 6-17). The preheating cycle with this type of infrared reflow is normally between 20 seconds and 4 minutes.

One manufacturer uses stainless steel tubes installed in the furnace insulation to carry heated air into the processing area to provide even distribution of heat. Another uses air pulled through the upper panels of the furnace, which is heated and directed onto the assembly. This type of heated atmosphere is estimated to contribute more than 50 percent of the energy imparted to the assembly and is beneficial in helping to volatilize and remove the solder paste solvents and in overcoming any possible "shadowing" of components, especially the larger plastic chip carriers (PLCCs).

The heating profile is somewhat similar to that employed for lamp reflow equipment, and a cooling station or zone can be installed after reflow.

Problems have occasionally been encountered with regard to higher temperatures along the outside edges of the assemblies, but these can often be resolved by designing the board so that larger components with a greater thermal mass are placed close to those edges while the smaller devices are moved closer to the center.

Infrared heating is considerably more efficient than convection or conduction reflow methods because the latter need direct contact between the workpiece and the heated solid, liquid, or gas. The reflow operation can often be accomplished in less than half the time required by a convection furnace, and the fast response of the infrared heating system, especially lamps, enables the user to

Intertrade Scientific, Inc.

Fig. 6-14. Model 6830IR Infrared Reflow System incorporating both panel area source infrared (top) and conduction through hot platens (bottom).

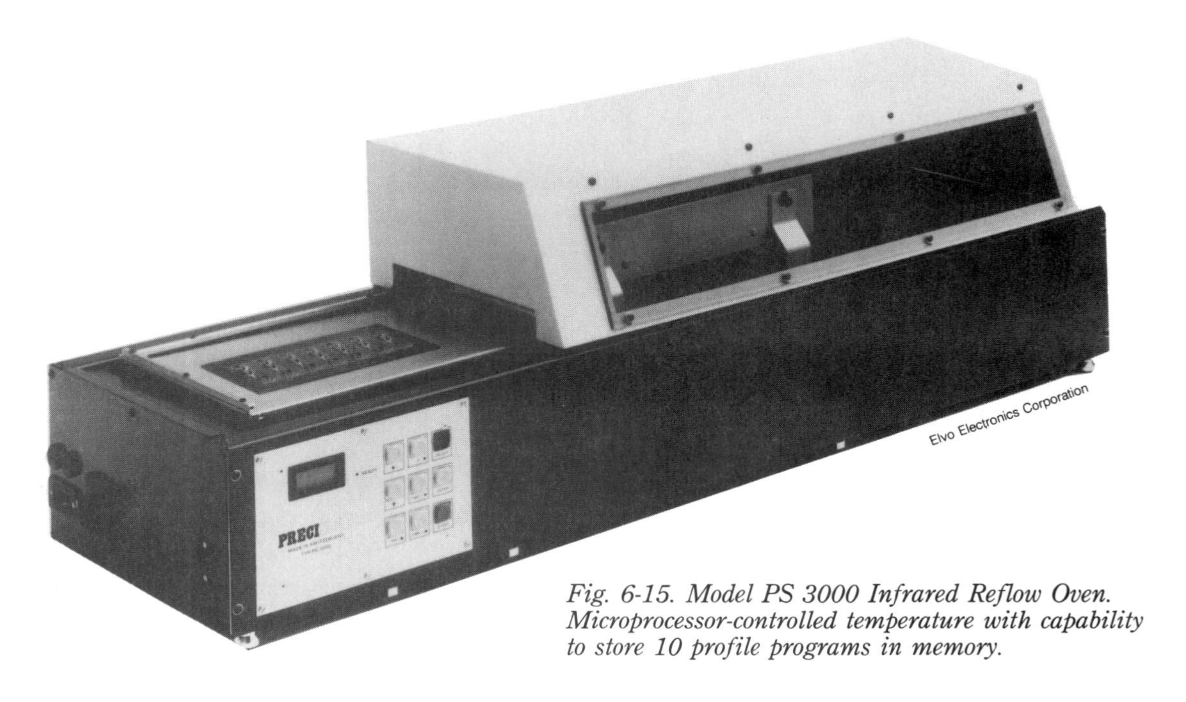

Fig. 6-15. Model PS 3000 Infrared Reflow Oven. Microprocessor-controlled temperature with capability to store 10 profile programs in memory.

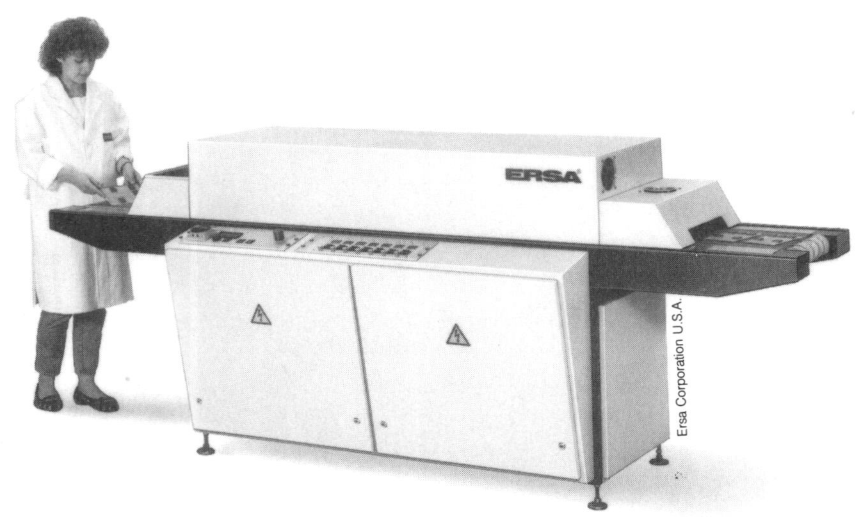

Fig. 6-16. Model ERS 450 Infrared Reflow System combining infrared and convection heating. Microprocessor-controlled temperatures. Middle infrared. Belt width is 18 inches (457 mm).

easily and quickly change temperature profiles to suit different assemblies. In some cases with lamp systems it might even be possible to realize energy savings by switching off power to the infrared furnace between runs.

Computer-controlled systems have been developed to set reflow parameters, each of which can be stored and recalled for use when specific assemblies are being run. Repeatability is thereby much easier to achieve, and such a feature is especially valuable when boards containing very temperature-sensitive components are to be reflowed.

The heat to which parts are subjected is very time and temperature related, and it is essential that the infrared system selected be equipped with accurate controls for setting and monitoring both temperature and conveyor speed. The distance between the energy emitter and the assembly is also a factor to be considered in establishing suitable profiles, and facilities should exist for such adjustments to be carried out easily. Changing this parameter can also help to reduce energy costs. The temperature can be measured by a thermocouple within the emitter, as in the panel system, or within the process chamber, as is usually the case with filament infrared furnaces. Correlation has to be established between these temperatures and that of the assembly, and this can be achieved and continuously monitored by means of thermocouples attached to the assemblies. In some cases, especially with single-sided assemblies, adjustments

Vitronics Energy Systems, Inc.

Fig. 6-17. Model SMD-722 Convection/Infrared Solder Reflow System with microprocessor-controlled heating and adjustable-width conveyor.

might be necessary to provide a difference in energy transmitted by the top and bottom emitters.

Infrared is a cost-effective reflow system to use for both low and high production rates. It requires less power than, for example, a vapor-phase unit with similar production capacity while also offering higher yields in terms of solder joint quality. It has long been the process favored in Japan for surface-mount assembly reflow. It offers the advantage of individually controlled heating zones, while the penetrating effect of the infrared wavelengths helps to ensure very uniform heating. This also reduces stress created by temperature differences between substrate and components, which can result in printed circuit board warpage and cracking of devices such as ceramic capacitors.

The discussion on the merits and disadvantages of the panel and filament infrared systems could continue as long as this method of heating is employed for solder paste reflow. Proponents of area source emitters claim that the middle-infrared wavelengths employed can more efficiently reflow solder paste by the high absorptivity of this energy by the flux, which is then transferred to the powdered alloy.[6] It is equally reasonable to accept the view of the filament lamp supporters that the ability of the metal powder to absorb the near-infrared wavelengths provides just as practical a means of generating the heat necessary to melt the solder. On the other hand, studies have shown the absorptivity of solder paste to be almost 100 percent in both the near- and middle-infrared wavelengths.

In addition, note that the ability of lamp systems to operate at both low and high emission power renders them more versatile in operation than the panel design, which allows low emission only. Lamps are also able to respond much more quickly than panels to temperature changes, and this is advantageous when product is continuously being fed through the system. If panel temperatures have to be increased to compensate for higher throughput and conveyor speeds and if the widths of the assemblies are narrower than those of the heating panels, then there is a danger that the board edges will be seriously overheated. It is equally true that because conventional panels transmit less energy, they cause fewer overheating concerns, and the need for an expensive nitrogen atmosphere rarely arises.

Our view is that excellent results have been achieved from both, and that final selection of equipment will inevitably be influenced by many other factors, including reliability, cost, and servicing facilities.

LASER

Lasers, since the construction of the first operating emitter more than twenty years ago, have been used for reflowing solder paste. However, this method has largely been confined to research and development and prototype work in view of its expense and difficulties in automating such a process. In the surface-mount industry, laser reflow has nevertheless attracted a lot of attention because solder paste lends itself very readily to this form of soldering.

The laser, an acronym for *l*ight *a*mplification by *s*timulated *e*mission of *r*adiation, emits an intense beam of light that, unlike that transmitted by an ordinary electric lamp for example, is unidirectional and does not diffuse or spread out. The laser energy beam is termed *coherent* because it travels on a single frequency. These properties render it fairly easy to focus the beam accurately on a very small area onto which the whole of the beam energy can be concentrated. The potential advantages of such a system become apparent when you consider the degree of increasing component population density being used for printed circuit boards and substrates.

The fact that the laser can be directed so accurately means that only those locations with which the beam is in direct contact can be heated (FIG. 6-18). The surrounding areas are not affected, so if solder paste is being used to provide the bond, there is much less likelihood of a bridge being formed between adjacent pads or leads as can frequently occur with other reflow methods when component leads and metallized pads are heated simultaneously to the same temperature.

The efficiency of lasers when transmitting heat substantially shortens reflow time. Cooling is also very rapid after the energy source is withdrawn because the heat is immediately diffused into the cool surrounding area of the board or substrate. This serves to create a fine metallurgical grain structure in the joint, providing it with a shiny, lustrous appearance.[7] The reduced reflow also helps keep intermetallic growth to a minimum, and Lea has reported that such intermetallic formation from laser reflow should be only 20 to 25 percent of that produced by wave soldering for 3 seconds at an interface temperature of about 230 degrees C (446 degrees F).[7] The copper-tin intermetallic growth created by reflowing a high-percentage tin-lead solder paste onto a board or substrate with copper circuitry is both time and temperature dependent. At best, this intermetallic reduces the natural ductility of the joint. In the case of mismatching thermal coefficients of expansion between, say, the pads and chip capacitor terminations, this reduced ductility could result in the ceramic cracking. At worst, if too thick an intermetallic layer is permitted to form, this could cause extreme brittleness in the solder joint.

Two types of lasers have found application in electronic soldering, and each has its special attributes and disadvantages. The neodymium-doped yttrium-aluminum-garnet source provides a short wavelength of 1.06 microns, which falls within the near-infrared range of the electromagnetic spectrum. This means that it is readily absorbed by metals, including solder. The disadvantage here from an operator's safety point of view is that the Nd:YAG beam will pass through any transparent shielding such as glass or plastic and can cause eye damage.

The carbon dioxide (CO_2) laser uses a wavelength of 10.6 microns—in the far-infrared range—which is reflected by most metals but is very suitable for plastics, printed circuit board materials, and the fluxes and other nonmetallic ingredients of solder paste. It is also completely absorbed by transparent safety shielding such as those already described. The CO_2 laser is regarded as being cheaper with regard both to initial equipment outlay and operating costs.

Careful selection of a solder paste to be used for laser reflow is important. The extremely rapid reflow can cause the flux solvents to boil and even to spatter, resulting in solder-balling or voids. Depending on the way in which the paste is deposited, it is probably wise to choose a formulation with a fairly high viscosity

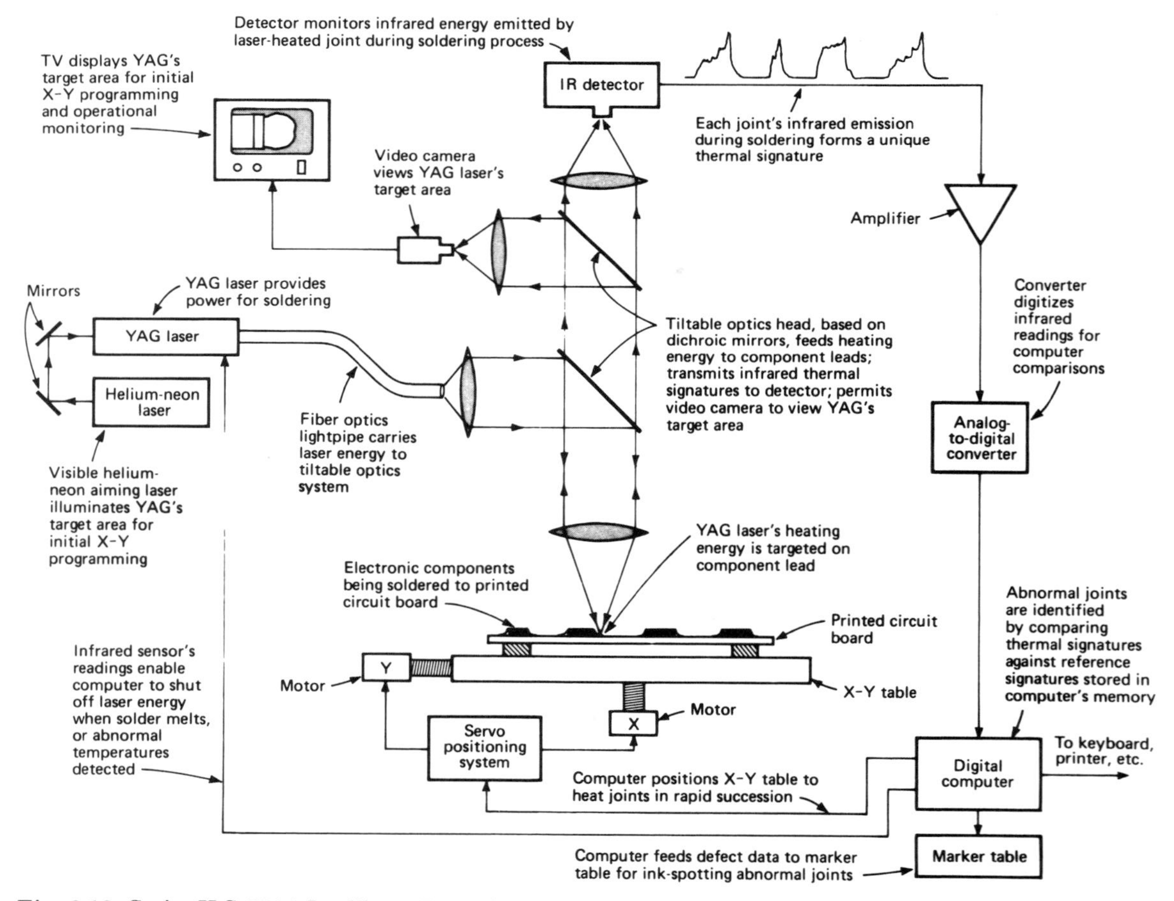

Fig. 6-18. Series ILS-7000 Intelligent Laser Soldering System uses an infrared detector to monitor the heating of each solder joint. Vanzetti Systems, Inc.

(greater than 800,000 cps) so that solvent content is reduced to a reasonably low level. Some solder paste manufacturers have been developing new formulations around high boiling point solvents (greater than 250 degrees C, or 482 degrees F) in an attempt to prevent voids in joints attributed to outgassing. While such solvents would not vaporize during reflow, there is no guarantee that pockets of these solvents would not be formed within the joints, thus also affecting their mechanical integrity. Metal content should largely be determined by the reflowed solder thickness required, but in printed circuit board surface-mount assembly, this is unlikely to be less than 88 percent.

Laser reflow systems are available with computer-controlled positioning capability (FIG. 6-19). The beam position can be varied by either adjusting the mirrors and lenses or by using a step-and-repeat table for the board or substrate.

Unfortunately, however, laser reflow is still slow and costly when compared with other conventional methods. Factors such as amount and duration of the laser power have to be calculated for each area to be reflowed to ensure the provision of adequate heat and time for a reliable solder connection to be made.

Vanzetti Systems, Inc.

Fig. 6-19. Series ILS-7000 Intelligent Laser Soldering System embodies a sophisticated optics head that can be tilted to permit the laser to be directed at component leads that are difficult to access.

7

Residue Removal

This chapter considers the removal of reflowed solder paste residues. So much has already been written and presented on the subject that all that is required in the present instance is to summarize the basic reasons for cleaning and how this might best be achieved. No attempt is made here to cover cleaning processes in any but the surface-mount industry as it is known today, although much of what is recommended for printed wiring board assembly should apply equally well for thick-film work.

The cleaning of surface-mounted electronic assemblies following reflow has probably received more attention in the last year or two than any other aspect of the new technology. This is at least partly due to the realization by many more companies how important a role cleanliness plays in the reliability of the printed circuit board or substrate—a vital component of any electronic apparatus. Primarily, however, it is because of the special challenge posed by new component design, more densely populated boards, and the much smaller spaces both around and underneath the components that the solvent needs to penetrate to achieve a satisfactory degree of removal of both flux residues and other contamination, including solder balls. *Minimum component stand-off*, or the space between the component body and substrate surface, typically varies in surface-mount design from less than 0.001 inch (0.025 mm) to approximately 0.008 inch (0.203 mm). These critical conditions are likely to worsen in the future as device sizes increase and lead, pad, and line spacings are further reduced. Chip sizes of 1 inch square and more, and VSLI and VHSIC devices with 0.025 and 0.020 inch (0.635 and 0.508 mm) lead pitches are already in use, making the cleaning operation more complex than ever. Conductor tracks, heat sinks, and chip components under the larger devices make the conditions for residue removal ever more difficult.

Some of the methods that worked well previously for wave-soldered printed wiring boards have had to be discarded. The cleaning equipment manufacturers have been compelled to adopt new ideas to combine the standards already set by the thick- and thin-film industries, and the requirements of the printed circuit assemblers for high-volume production.

The two major factors that initially had to be considered both related to the solder paste used to form the bond between component terminations and board metallization. Although the paste would be selectively placed where it was actually needed, the quantity of flux deposited in any one spot was likely to be up to 10 times greater than is the case with wave soldering, with what one might expect to be proportionately more potentially corrosive flux residue. In addition, these deposits were to be located primarily on pads where they could do most damage and often in almost inaccessible areas between the leads and under the bodies of components. Furthermore, reflow time was to be substantially longer (5 to 12 times) than the 2 to 5 seconds typically expected from the old method, resulting in increased hardening of the flux residues, making them even more difficult to remove.

TYPE OF CONTAMINATION

The expected contamination from the new solder medium (paste) is chemically similar to that encountered with liquid fluxes. The same types of rosin—water-white and tall oil—apply, with their main composition of abietic acid. The other resins and thickening agents used in the paste systems are inert substances and non-ionic in nature. Some of the thickening agents, however, are not very soluble in the solvents used for cleaning after the reflow process and can form tenacious and unsightly residues. The solvents are very different, with a much higher boiling point and lower evaporation rate than those used in wave-soldering fluxes. The reason for this is that enough solvent has to be retained in the paste throughout the reflow process to provide the tackiness needed to hold components in position until the solder becomes molten. The use of highly volatile solvents, such as those used in typical liquid wave-soldering fluxes, would cause rapid drying of the paste with the result that components would become misplaced or dislodged.

A number of the paste activators bear different names but are in many respects very similar to the halides, organic acids, and amines incorporated in conventional rosin fluxes.

One advantage the new technology does offer is the fact that virtually all the products promoted by the solder paste vendors for thick-film and surface-mount technology are formulated around an RMA flux activity. Water extract resistivities are often well in excess of the required 100,000 ohm-cm limit. On the other hand, the percentage of activator incorporated into a paste system is invariably up to five times more on a weight basis per pound than would be the case with the majority of liquid RMA fluxes.

The major reasons for removal of the flux and other residues nevertheless remain the same: assurance of electrical integrity and prevention of corrosion. Also important, however, is the need for a clean surface to prevent the interference that residues would provide to automatic "bed-of-nails" testing and the application of a final, protective, conformal coating.

Even the RMA fluxes can leave potentially ionic residues to cause current leakage and a reduction in the electrical resistance of the board or substrate material. In addition, small amounts of chloride included in the flux to boost its activity together with the presence of moisture can become corrosive, forming hydrochloric acid from the reaction between metal (especially lead) chlorides, and water to attack the solder joints and conductor traces. Electrolytic corrosion on the assembly can be caused by ionic residues connecting metals of different electromotive force values.

These problems would be exacerbated by the use of stronger RA or water-soluble OA (organic acid) fluxes, but fortunately, the surface-mount industry has largely avoided these. This is perhaps a tribute to the progress the industry has made towards perfecting the reflow operation in respect to both the process itself and the quality of the materials it utilizes. If water-soluble pastes eventually receive general acceptance, and that time is moving closer (albeit slowly), then even more careful attention will have to be paid to residue removal, especially with regard to the tendency of these fluxes to decompose at the higher reflow temperatures (above 250 degrees C or 482 degrees F) and to leave behind tenacious, difficult-to-remove white residues.

CLEANING SOLVENTS

Cleaning solvents have been broadly classified by Manko and others as being either polar or non-polar in nature.[1]

The non-polar type, which include all the chlorinated and fluorinated hydrocarbons approved for use in the electronic industry, as well as a number of alcohols and other materials, solubilize rosin, oil, and grease, for example, but cannot ionize, or in other words, take into solution the ions of the activator ingredients in rosin-based fluxes.

The polar solvents, which comprise many alcohols and water, do have such an ionizing effect, enabling them to easily remove such contaminants as well as potentially dangerous sodium chloride or salt from fingerprints and any harmful residues from the printed circuit fabrication process, such as plating chemicals and etchants. Unfortunately, the polar solvents are either too flammable to be practicable or are not able to dissolve rosin residues, that often encapsulate the activators, and can therefore only deal with exposed polar soil.

The solution has been to combine the qualities of the two chemical cleaning agents into what is then termed a bi-polar solvent, although a few companies have taken a different path by using separate, sequential cleaning stages such as a non-polar chemical followed by a polar wash in water. Important considerations for selection of such a blended solvent include safety from flammability, which precludes most of the alcohols from being used alone, and toxicity, as well as thermal stability. Threshold limit values (TLV) and time-weighted averages (TWA) are safety values for solvents as established by the American Conference of Governmental and Industrial Hygienists (ACGIH) and the Occupational Safety and Health Administration (OSHA). Lea and others have noted that maximum allowable concentrations (MAC) would be a better way to assess the toxicity of a solvent than TLV/TWA, which represents only an average of airborne concentration of substances, without regard to the much higher and more dangerous levels to which operators might sporadically be subjected.[2]

In a cleaning operation involving a vapor degreaser, if the chlorinated and fluorinated solvents are heated to temperatures much above their boiling points, they will decompose, forming hydrochloric and hydrofluoric acid respectively. To forestall the creation of such undesirable by-products, an inhibitor or stabilizing agent (usually some form of amine) is added to the solvent blends. It is important that the pH value of the solvent in the boiling sump be monitored regularly so that it can be adjusted once it has reached the level specified by the suppliers. A pH of 4 is considered the safe minimum level. If it falls below this level, an acid acceptance test should be carried out as recommended by the vendors.

The solvent must be blended using two products with a similar boiling point and vapor pressure. The boiling point of the mixture, which is then termed an *azeotrope*, is usually slightly lower than that of the two individual components. But it is important that these vaporize at the same temperature in order to retain the bipolar characteristics of the blended solvent. Any wide variation in evaporation rates could result in a change in the characteristics of the blend in the boiling sump of the degreaser, which could affect the flammability of the product. Some studies do indicate that the composition of certain chlorinated blends can change during use due to the vaporization of components with the lowest boiling points, with a deleterious effect on cleaning results.[3]

The cleaning solvents now being used for surface-mount use can be classified under five principal headings, and the major advantages and disadvantages of each are discussed.

Apart from the safety aspects, other important factors are cleaning efficiency (determined by solvency power and surface tension), solvent consumption (usually strongly influenced by the product's rate of evaporation), and energy requirements. In addition, consideration has to be given to any investment required in terms of new or improved equipment.

Surface tension—the phenomenon by which molecular forces acting on a liquid compel it to reduce itself to a form with the smallest surface area (a droplet)—should be examined carefully when evaluating solvents, especially for surface mount. It is quoted in dynes per centimeter. Surface tensions vary considerably, both among solvents and the objects to which they are exposed for cleaning purposes. As a rule of thumb, however, the higher the product's surface tension, the less the solvent tends to spread or circulate. In the case of narrow spaces beneath components and between leads, a high surface tension liquid could be rather ineffective as a cleaner. On the other hand, Wang and Seghal have found that low-surface-tension solvents, with their greater propensity to spread especially when combined with low viscosity, have poorer capillary action than high-surface-tension products.[4] Therefore, these solvents with low-surface-tension cannot penetrate the small crevices as well when directed as a high-velocity spray.

Kenyon compared several different cleaning agents by establishing a "wetting index" for each, based on surface tension and other factors.[5] His conclusions have caused some controversy, especially since an operating temperature of 25 degrees C (77 degrees F) was used to measure the performance.

An alternative method of calculation, incorporating two additional physical characteristics, was subsequently offered by Brous and Schneider, termed an "efficiency index".[6] One of these characteristics, whose significance is rarely denied, is the boiling point of the solvent. This is a very important consideration, because solubility of residues is substantially increased by heat, which can also reduce surface tension of the solvent. In almost all cases, the solvency power of a fluid is increased as its temperature rises to its boiling point.

The other consideration introduced—the Kauri Butanol value—is more controversial. This indicates the ability of a solvent to hold non-polar resins, such as rosin, in solution. The higher the number attributed, the greater its effectiveness. Some investigations, however, have shown no direct correlation between the Kauri Butanol index value for a particular solvent and its actual defluxing ability, ostensibly because of the different nature of the Kauri resin carried into solution in butyl alcohol.[2] In neither the Kenyon nor the Brous/Schneider indexes were factors such as the method of applying the solvents or the design of the assemblies considered (see TABLE 7-1).

Hale and Steinacker developed a test to physically assess the efficiency of a solvent that involves the use of perspex or glass slides and a suitable shim of a thickness equal to the narrowest component stand-off on the board or substrate to be reflowed and cleaned.[7] If desired, one slide can be cut into one or more segments to represent the actual size of the more critical larger component bodies. Solder paste is deposited on the bottom slide and the other slide placed over it and the shim. These are then clamped together with a spring clip or similar device. The whole assembly is immersed and reflowed in a vapor-phase fluorocarbon fluid with the appropriate boiling point and then cleaned. The results can very easily be visually inspected, and subjected to ionic contamination measurement, if required. The test is simple, fairly effective, and inexpensive.

Initial cost, however, including that of the necessary equipment, is, ultimately on many occasions the single most important issue in determining choice of a particular solvent. Manufacturers too often disregard the superior performance of one product to settle for something that is considered adequate to achieve what frequently become false cost savings.

More energy, for example, is needed to heat the higher-boiling-point solvents. But is the potentially more effective cleaning ability of such products always taken into account when gas, electricity, or oil costs are considered as

Table 7-1. Properties of Solvents Used in Solder Paste Residue Removal

SOLVENT	Liquid Density at 25°C (g/cm³)	Liquid Viscosity (centipoises)	Surface Tension (dynes/cm)	*Wetting Index	Kauri Butanol Value	Boiling Point °K	**Efficiency Index
Fluorocarbon 113 with Methyl Alcohol, stabilized[1]	1.477	0.700	17.4	121.3	31	320.6	121
Fluorocarbon 112 with n-Propyl Alcohol, stabilized[2]	1.423	1.466	23.11	42.0	71	355.2	106
1,1,1-Trichloroethane, with n-Propyl Alcohol, stabilized[3]		0.885	25.72	55.8	124	346.9	240

$$\text{*Wetting Index} = \frac{(\text{Density})(1000)}{(\text{Viscosity})(\text{Surface Tension})}$$

$$\text{**Efficiency Index} = \frac{(\text{Wetting Index})(\text{KB Value})(\text{Boiling Point °K})}{10^4}$$

[1]Trade Name: Du Pont Freon TMS
[2]Trade Name: Alpha 1003
[3]Trade Name: Alpha 565

Reprinted with permission: Dr. Jack Brous and Alvin F. Schneider, "Cleaning Surface-Mounted Assemblies," *Electronics* (April 1984) part of the process of choosing a cleaning medium? Conversely, the lower boiling solvents might require special chilled-water cooling of a vapor degreaser's condensing coils to prevent excessive evaporation of the product, which is an added cost.

Whichever solvent is selected, especially if it happens to be a chlorinated or fluorinated type, existing or forthcoming regulations concerning pollution have to be borne in mind. You must also be aware of the difficulty and expense of disposal of waste material in accordance with legal requirements. Depending on the location of the company, such logistics items could prove to be very costly. So might be the recovery of contaminated solvent by an outside contractor if located at any distance from the plant.

Choosing a cleaner is further complicated by the universal concern with the depletion of the Earth's protective ozone layer by chlorine released in the stratosphere from chlorofluorocarbons (CFCs) and with health problems associated with chlorinated solvents.

Improper cleaning can result in poor product reliability with consequent costly repairs or replacement and the loss of a manufacturer's reputation. Is such a risk justified?

One factor of relatively new significance can now be expected to exert a major influence on future choices of solvent. Never before has the environmental impact of cleaning chemicals received so much attention. As a result, many more electronic manufacturers are expected to make the transition to aqueous systems unless the solvent makers can develop less harmful alternatives to what is currently being marketed.

Chlorinated

The chlorinated solvents used in electronics are considerably more aggressive as cleaners than the fluorocarbons. Because the chlorinated solvents have much higher boiling points this makes them much more effective as cleaners. As an example, the melting point of rosin is about 85 degrees C or 185 degrees F, so it would seem logical to use a solvent with a boiling point above this so that the otherwise hard reflowed rosin is softened to enable it to be more easily removed.

One of the important requirements of a solvent used in electronics is that it must be compatible with any material it comes into contact with during the cleaning process. The chlorinated products do present problems in that they do attack certain plastics to different degrees, depending upon time of exposure and solvent temperature.

The major solvents in this category that have been employed in surface-mount residue removal are blends of 1,1,1-trichloroethane (sometimes called methyl chloroform) and different alcohols, depending on the manufacturer. The use of perchlorethylene was once fairly widespread among chlorinated cleaner users as a base for bipolar solvents. However, its much lower TWA, higher boiling point, and the fact that at room temperature it is not as effective a solvent as 1,1,1-trichloroethane has caused its popularity to sharply diminish.

If you examine the Kauri-butanol values, the obvious choice as a material for optimum rosin flux removal is methylene chloride, a material with outstanding solvency power. Concerns about its toxicity, lower boiling point, and corrosive nature however, have led to a general decline in its usage in electronics. It is seldom found in surface mount, except occasionally as the constituent in a fluorocarbon blend.

This example illustrates why the potential or existing user should look carefully at all the characteristics of a solvent before making a final choice.

With increasing concern about the effect of solvents on the environment, users of the chlorinated variety can expect in the future to face even more stringent regulations with regard to the use of these products.

Fluorinated

Fluorinated solvents have been used in the electronic industry for more than 25 years. They are generally less toxic than the chlorinated type but have a lower boiling point and higher evaporation rate. One of their attractions has been that unlike the chlorinated versions, they do not attack plastics as readily and therefore have better component compatibility.

The status of these CFCs (chlorofluorocarbons) has changed dramatically in recent months after the collection of evidence that they have contributed significantly to depletion of the protective ozone layer above the earth's surface. It has been suggested that the innate chemical stability of the CFC-113 (trichlorotrifluoroethane) compounds, which was always an advantage in electronic cleaning applications, has contributed to the present accumulation of CFCs in the atmosphere.

Since the signing of the international Montreal Protocol agreement on September 16, 1987 requiring a 50-percent reduction in the production of CFCs by the mid 1990s based on 1986 output, various organizations, including the IPC, have established test programs to evaluate alternatives to CFC cleaning in electronics.

Commercially, the best-known products in the 113 group are the Freons and Genesolvs. The Freons, manufactured by E.I. Du Pont de Nemours & Co., Inc., include types TF (113 only), TES (113 plus ethanol and a stabilizing agent, nitromethane), TMC (113 plus methylene chloride), and, dominating United

States electronics consumption, TMS (113 plus methanol and nitromethane). Allied-Signal, Inc. supplies the Genesolv D Standard, DES, DM, and DMS in competition with Du Pont.

Du Pont's declared goal is to phase out 95 percent of its CFCs by the turn of the century, or sooner, if a suitable replacement can be found before then.[8] In the meantime, two new products with reduced 113 content have been introduced: Freon SMT for electronics applications, and Freon MCA for general cleaning. Allied-Signal has developed a substitute solvent, HCFC-141b (1,1-dichloro-1-fluoroethane), which has a greatly reduced ozone depletion range (ODP).[9]

A fluorocarbon not included in the Montreal Protocol provisions is the compound 112, tetrachlorodifluoroethane. By virtue of its higher boiling temperature it has superior solvency power when compared with CFC-113. (A bipolar version of this is manufactured in the United States by Alpha Metals).

Water

Water by itself cannot solubilize rosin, oil, or grease and is only effective with truly water-soluble flux residues. Its other major drawback is its inability to penetrate tight spaces, such as those to be expected underneath surface-mount components on a board, which can be attributed to the high surface tension of water.

Most of the longer established solder paste manufacturers have developed water-soluble versions in the past, but until now none of these has been suitable for surface-mount technology, primarily because of the highly hygroscopic nature of the flux systems and their consequently unstable viscosities, which made deposition results very erratic. In addition, adequate residue removal with water alone has until recently proved impossible.

Considerable improvements have taken place during the past months in the design of efficient water-washing systems for surface-mount (FIG. 7-1), and high-pressure jets at an angle of between 10 and 40 degrees have been installed to try to overcome the poor penetration properties of water. The angle ensures no loss of kinetic energy otherwise arising from the spray at the perpendicular. In many cases, these modifications have been successful, but the remaining problem still to be resolved is the evacuation of the water residues from those same crevices on the board. In general, a minimum component stand-off of 0.010 inch (0.254 mm) has been recommended for water-cleaning purposes. The use of ultrasonics would undoubtedly assist cleaning, but the industry in general has been reluctant to consider this (for reasons already stated in the next section, "Cleaning Equipment.").

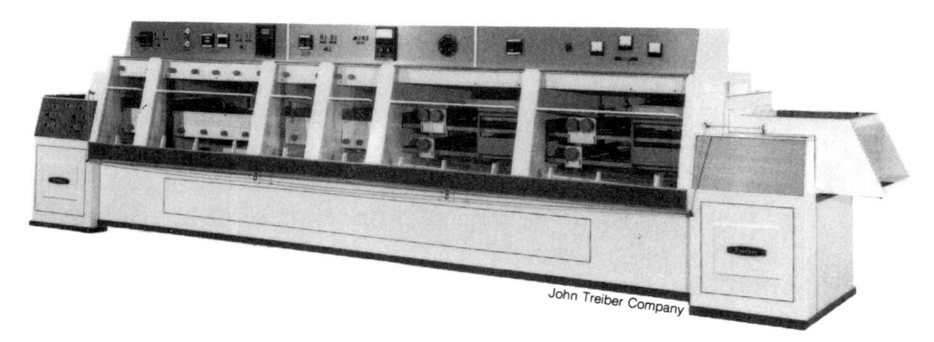

Fig. 7-1. Model TRL-C Aqueous Cleaning System, suitable for water only or water/saponifier processes. Recirculating wash and rinse stages with heated air and infrared drying zone.

Water-soluble organic acid pastes have still not received approval from the military authorities (although the Navy is studying the matter), so the development of new, more reliable water-soluble products has not enjoyed the highest priority among the solder paste manufacturers, still preoccupied with improvement of the rosin-based products. Several companies have developed new water-soluble formulations, some of which exhibit great promise in regard to radically reduced hygroscopicity, improved printing characteristics, and residue removal.[10]

The quality of the water used for cleaning can vary enormously. Ordinary tap water is normally acceptable unless it is very hard, because high levels of heavy metals can precipitate out and cause a scale to build up on equipment. De-ionized water is usually too expensive to use in most surface-mount operations but has been used as a final rinse to contribute greatly to the removal of residual ionic material and to a subsequent reduction in electromigration problems.

Post drying is usually a matter of individual preference, depending upon the quality and the amount of water left on the assemblies after cleaning. It is often sufficient to remove water residue with pressurized air, employing high velocity air knives or even nitrogen, rather than simply allowing the assembly to dry. The advantage of using pressure is that any water droplets containing traces of contamination are blown off the assembly. The air temperature can range between ambient (25 degrees C or 77 degrees F) and 120 degrees C (or 248 degrees F) for best results.

Effluent disposal in the United States is almost always subject to federal and municipal regulations. Pre-treatment is required for the removal of heavy metals such as copper, lead, and zinc and for adjustment of pH.

Saponifiers

Saponifiers have been in use in the electronics industry for about 20 years. They consist of alkaline amines with a pH of between 9 and 12. They can be added to water in concentrations of up to about 10 percent, depending on the quantity of contamination to be removed, for the removal of non-polar materials, such as rosin. They react with the non-polar substance to form a soap that is soluble in water. The major advantage of saponifiers is that they can be used successfully for both rosin and water-soluble flux residues. In the highest concentrations, they have proved quite successful in softening and releasing hard, baked-on flux from high-temperature solder paste reflow processes (greater than 300 degrees C/572 degrees F).

The pH value of the solutions should be regularly monitored and adjusted by means of the titration procedure recommended by the saponifier vendor. Once again, however, because of surface tension considerations (especially during the critical rinsing stages), component stand-off needs to be a minimum of 0.010 inch (0.254 mm).

Large dishwashers can be used, but these have limitations with regard to efficient cleaning of large volumes of boards or substrates. Today, in-line conveyor systems are largely preferred.

Terpenes

The environmental concerns about solvents and the doubts about the efficacy of removing rosin flux paste residues with water-saponifier mixtures have prompted investigations into the use of terpenes. *Terpenes* are materials that are distilled essentially from orange peels, but they exist in virtually all living plants. The major ingredient—approximately 90 percent by weight—is limonene, a terpene hydrocarbon with a surfactant to enhance wetting and rinsing properties. Adherents of this material promote (1) its ability to take into solution both polar and non-polar soils, especially rosins, and (2) the fact that it contains no halides, so it does not have any effect on the ozone layer. It has a long bath life, so waste is therefore minimized. It is also biodegradable and basically nontoxic. Limonene itself presents no problems in regard to current waste disposal regulations, although the same caution should be exercised with regard to heavy metal contamination as described back in the section under ''Water.''

The first limonene-based cleaner to be marketed was the Bioact EC-7 manufactured by Petroferm, Inc. and originally developed for AT&T, who has been conducting tests for more than two years. Results obtained thus far indicate ionic residues at least as low as those achieved with F113 plus methanol. Its cleaning capability is good, even with 0.001 inch (0.025 mm) spacings, while surface insulation resistance values also appear promising. (The product is now being marketed worldwide on an exclusive basis by Alpha Metals, Inc.)

The major obstacle to adoption of this new product with a Cleveland Open Cup flashpoint of 71 degrees C or 160 degrees F is its combustibility and the need for suitable equipment with special cooling to control the temperature of the terpene, which can rise from ambient by 10 degrees C (20 degrees F) or more as the result of heat generated by a cleaning machine's pumping section. The operating temperature of the liquid should normally be held at below 38 degrees C (100 degrees F). The control of the terpene vapors during the cleaning operation and provisions of an air scrubber preventing them from being drawn into a plant's ventilation system are obvious safety precautions. Care must be taken in the disposal of the contents of a cleaner tank.

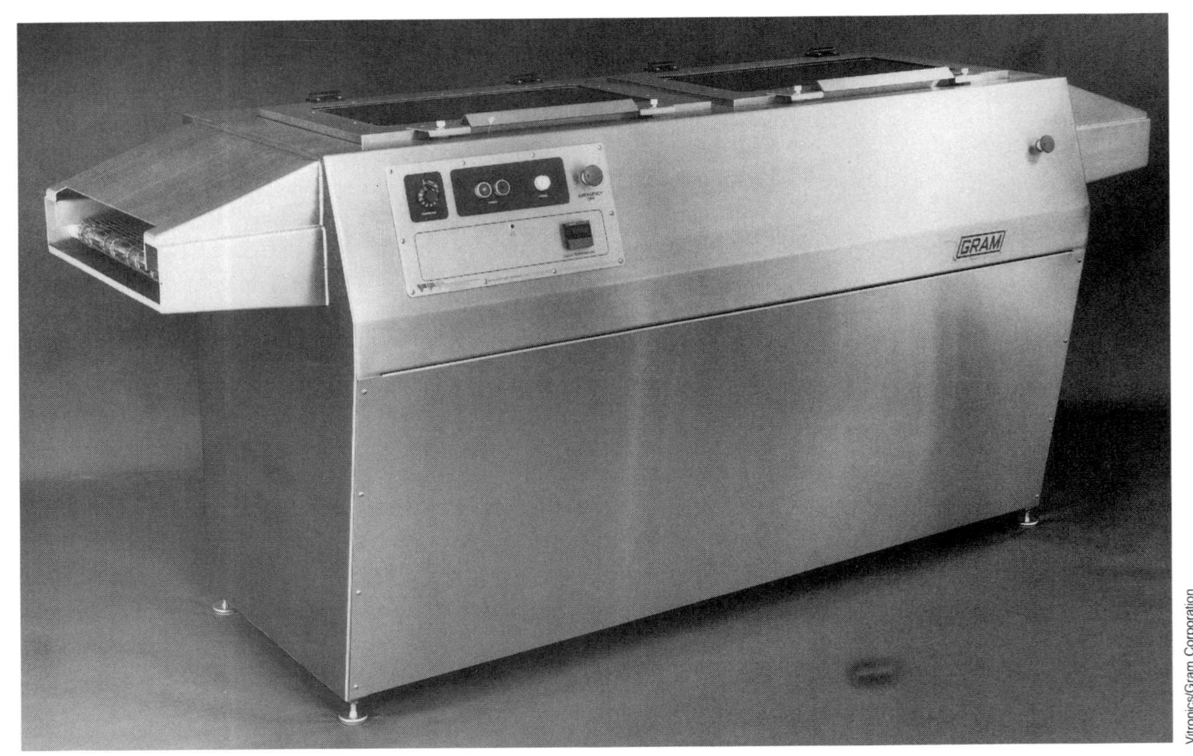

Vitronics/Gram Corporation

Fig. 7-2. Model EC-1850 in-line cleaning system especially designed for use with terpenes. Stainless steel in construction with two high-pressure spray stages.

A less serious but potentially annoying characteristic of the new cleaner is its reported sensitivity to moisture content as low as 1 percent by weight. This can cause rapid rises in viscosity, and at higher water concentrations of more than 5 percent can lead to the formation of a thick gel. All this means the exercise of care during the cleaning cycle to ensure that water from the rinsing sprays does not find its way into the terpene. The fluid itself is normally used as a spray, with a pressure of about 30 to 40 psi, and equipment is under development to eliminate the danger of flammability of the terpene vapor.

The terpene does cause both silicone rubber and polyvinyl chloride (PVC) to swell, and it is generally recommended that equipment be constructed from stainless steel (FIG. 7-2).

Another advantage of this new product are that it requires no distillation and is simply discarded and replaced when spent.

CLEANING EQUIPMENT

Methods of cleaning with chlorinated and fluorinated products include brushing by hand, particularly after re-work, and manual immersion in a cold or heated solvent. For any production mode involving solder paste, however, more sophisticated methods are preferred.

Equipment is a very important feature of successful surface-mount cleaning, and as with all other manufacturing processes in surface-mount technology, significant progress has been made in the design of residue removal systems. Many alternatives are available, up to the optimum combination of in-line vapor condensation, spray, immersion, and ultrasonics.

Vapor now plays a much greater role in removing residues from surface-mounted printed wiring boards than used to be the case for purely wave-soldered assemblies. Then, the boards were simply passed through or across a pumped standing wave of cold or hot (but not boiling) solvent that was invariably a chlorinated blend because of concern about evaporation losses.

The removal of residues in a vapor involves the immersion of an object to be cleaned in the vapor produced by and covering a boiling solvent. Provided that this object is cooler than the boiling point of the solvent, the vapor will condense upon it, dissolving and rinsing away the residues.

The great advantages of vapor degreasing are that the vapor is always clean and free of contamination and the solvent can always be re-used. The vapors of the solvents normally used are much heavier than air and do not tend to be easily lost to the atmosphere.

In the case of batch cleaning, in particular, assemblies should never be immersed into the boiling solvent in the sump because it contains contamination previously washed off boards and substrates. Once re-deposited, they can be very difficult to remove.

Care should always be taken to ensure that flux and other residues do not accumulate in the sump to the point where they might cause an increase in the boiling temperature of the solvent, resulting in the formation of harmful acids. The solvent should be changed once the boiling point exceeds the nominal by more than 3 to 4 degrees C (5 to 7 degrees F).

Small open-top vapor degreasers are very inexpensive and are often fitted with water separators or dryers to remove moisture from the solvent condensate. It is good practice to allow the clean, condensed fluid to be transferred to a second sump, rather than to the contaminated material. This ensures that the boiling sump is continuously fed with clean solvent only.

Flux removal by means of vapor degreasing cannot be undertaken unless the temperature of the part entering the vapor is lower than the boiling point of the solvent and therefore under normal atmospheric pressure conditions, of the vapor also. Cleaning is achieved by the vapor condensing upon the workpiece, dissolving the flux residues, and flowing or dripping off into the machine sump. The best cleaning results in a vapor degreasing operation, all other things being equal, is attained by having the greatest possible temperature differential between the object being immersed and the vapor itself. This results in an increase in the volume of condensate created to perform the cleaning action.

In an in-line system, the boards or substrates are transported on a motor-driven conveyor belt through a series of cleaning zones. Vapor still assures the purity of the solvent, but the major part of the cleaning is accomplished by the use of high-pressure hot liquid sprays employing needle jets. These systems can be linked to a computer, making them much less operator dependent.

Sprays definitely make the cleaning process more efficient, as long as the solvent spray is hot; if it is not, there is the danger that the vapor layer will collapse. From a design point of view, the denser the spray pattern on the spray head, the more consistent the high-velocity solvent impingement onto the part. The angle of contact of the solvent spray onto the part is important in achieving the best flow characteristics. Both pressures as well as solvent emissions can be reduced by increasing solvent volume. Some critics of spray systems claim that these cause static electricity build-up on the board surfaces. Such in-line equipment is generally linked to solvent recovery stills, providing continuous recirculation of clean solvent.

The subject of ultrasonics with either solvents or water has always been a controversial one, and the concept has not proven very popular in the United States, primarily because of concerns about possible damage to components and especially fine wire bonds of semiconductors by the ultrasonic energy. In electronic cleaning, a frequency of between 10 and 100 kHz is used, with 40 kHz probably being the most general. Studies also indicate that ultrasonics might cause certain surface-mount device leads to fracture that have experienced metal fatigue. Ultrasonics are still precluded from assisting the printed wiring board cleaning process under the terms of the new military specification MIL-STD-2000.

The process is based on cavitation—the creation and destruction of tiny bubbles—generated by vibratory forces of transducers at variable frequencies. The transducers release energy upon collapsing or imploding during a compression cycle to dissolve flux residues much more effectively than could be achieved using immersion in a static bath or vapor. Adherents of ultrasonics-assisted cleaning insist this process can achieve results as good as those with high-pressure sprays.

Highly-efficient aqueous cleaning equipment is now available for use with either water alone, or a combination of water/saponifier plus plain or de-ionized water rinsing. The temperature of the washing solution generally ranges between 49 and 82 degrees C (120 and 180 degrees F), with 60 to 66 degrees C (140 to 150 degrees F) the most popular. The water/saponifier mixture, when used with high-pressure, very-accurately-directed spray systems, has performed very well on rosin-based paste flux residues. Results with water alone have shown considerable promise with water-soluble organic acid flux residues. However, in view of the small user data base that is due partially to the reluctance of many manufacturers to evaluate such products, it would be premature at the present time to make any long-term prognosis as to the efficacy of such a system in surface-mount.

A recently introduced method of cleaning uses centrifugal force and the horizontal force called the *Coriolis force* to provide a greater degree of penetration and flow of cleaning solution under the components and in tight crevices.[11] This concept is claimed to be far superior even to high-pressure spray systems because of the energy wasted by the impact of the cleaning fluid on the surfaces of the board and components. The new equipment comprises a closed process chamber where the assembly is spun by a robot. Cleaning solution is then directed at the assembly, where it solubilizes the residues and flies off with any contamination. Rinsing with clean solution and a final dry with hot filtered air or nitrogen complete the process. Drying can also be accomplished with super-heated solvent vapor following the final distillate spray zone to remove all solvent residues.

Remember that the sooner the cleaning process takes place after reflow, the easier the flux residue is to remove. Although the idea of rosin flux polymerization is a misconception, it is true that a structural change of the rosin takes place at reflow temperatures, even those used for the eutectic tin-lead alloy, for example, which results in a hardening of the residue due to isomerization.

Major considerations for the selection of cleaning equipment include:

Projected production volume
Board or substrate size
Desired conveyor speed
Component-solvent compatibility
Solvent
Degree of cleanliness required

Drying is mandatory for boards or substrates to be subjected to in-line testing.

Once these points have been clarified, discussion can take place with the equipment manufacturers as to what will be needed in terms of design and capacity of a suitable system.

CLEANLINESS TESTING

More than 20 years ago, Manko asked the question "how clean is 'clean' in electronics?".[12] There is still no unequivocal answer to that. Indeed, to judge by the variety of testing methods still being conducted or considered, there is still some way to go until the answer is known. There is no doubt, however, that considerable progress in this direction has been made during the past few years.

Among the several methods of cleanliness testing used occasionally are high-performance liquid chromatography (HPLC) and Auger electron spectroscopy, both of which involve the acquisition of expensive equipment, with neither being suitable for use in a production environment. All of these tests are designed to identify the presence of substances on an electronic assembly that could be potentially dangerous from either an electrical or corrosive point of view.

The following discusses the two cleanliness testing methods largely accepted by the electronic industry today, based on existing military and other widely recognized specifications.

Ionics

In 1970, Egan first described the conductivity method he had developed for the detection of residual plating salts, which was subsequently the subject of a published paper.[13] He calculated the increase in conductivity of de-ionized water after plated parts had been immersed and dissolved off with the assistance of an agitator. This method was modified by Hobson and DeNoon of the Naval Avionics Facility in Indianapolis, Indiana using a 75/25 percent isopropyl alcohol/water solution in a spray extraction test to measure changes in resistivity, and this became, and remains to this day, the basis for the procedure described in the 1975 military specification MIL-P-28809, superseded in 1981 by the current MIL-P-28809A. It applies also for MIL-P-5511 OC (1978) and for the more recent Navy specification WS-6536E (1985), being replaced first by MIL-DOD-2000 and now by MIL-STD-2000.

The spray extraction test is simple but time consuming and subject to human error, and new instrumentation has since been developed that accelerates the procedure and significantly reduces operator dependence.

The early equipment comprised the Kenco Omegameter, which measured by means of a probe the change in resistivity of a purified 50/50 percent isopropyl alcohol/water solution agitated in a chamber. Immersion time was fixed, with a maximum of 10 minutes. Nomograms were supplied for conversion of electrical resistivity and other factors to an equivalent value expressed in milligrams per square inch of sodium chloride.

A machine based on dynamic extraction was developed by Dr. J. Brous of Alpha Metals in 1972.[14] In this case, the isopropyl/water mixture was pumped continuously within a closed loop, and after extracting residues from the workpiece in the immersion tank, the solution passed by way of a conductivity meter through an ion exchange column for the removal of ionic contamination before returning to the tank and a new cycle. The measurement method was more thorough than that of the Omegameter but consequently also much slower, and the initial calibration with sodium chloride that was necessary was often a source of irritation.

The new (now also Alpha) Omegameters and Ionographs are very much more sophisticated, with microprocessor controls and user-friendly displays with contamination readings provided in micrograms of sodium chloride per square inch.

The Ionograph 500A (FIG. 7-3) is a general-purpose ionic test instrument for electronics that continues to measure contamination in changes of conductivity. The latest Omegameter, Model 600SMD (FIG. 7-4), features test solution heated to 49 degrees C (120 degrees F), reinforced by submerged spray jets to enable the flux residues underneath surface-mount components to be effectively flushed out for ionic measurement. It calculates ionic content from reductions in resistivity.

Westek, Inc. makes the microprocessor-controlled ICOM 4000 Ionic Contamination Monitoring System that uses the alcohol/water mixture from high-pressure spray nozzles to remove residues from the object after it has been placed in a chamber. The test ends when changes in resistivity have ceased after a rapid series of readings.

Protonique, S.A. of Switzerland manufactures the Contaminometer, which utilizes an agitated alcohol/water solution in a static bath to measure changes in conductivity.

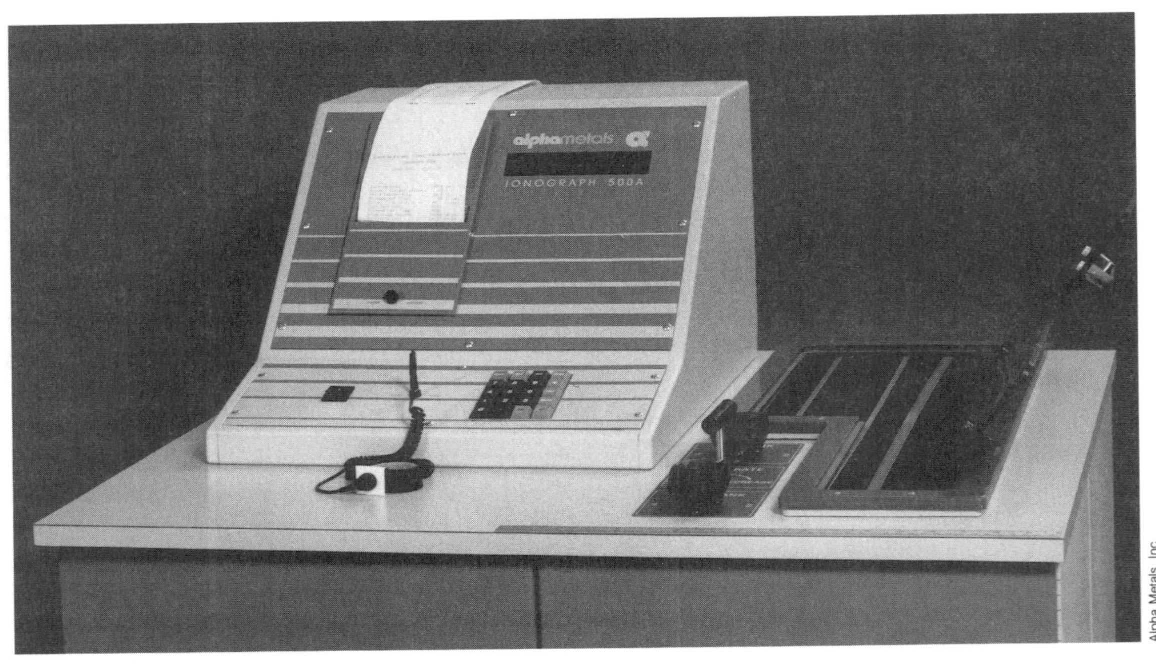

Fig. 7-3. Ionograph Model 500A ionic contamination tester is microprocessor controlled and is especially recommended for high-reliability applications.

Alpha Metals, Inc.

Fig. 7-4. Omegameter Model 600SMD ionic contamination tester is microprocessor controlled and designed for monitoring cleanliness of surface-mount assemblies.

Surface Insulation Resistance

Ionics testing, which uses as its criterion the change in conductivity or resistivity of a solution, has been criticized for the fact that it only measures total contamination across the whole board or substrate area that can actually be dissolved. It does not identify spots where residue-induced localized corrosion might be particularly prevalent as the result of accumulation of contamination. The new test methods developed for determining surface insulation resistance will help to overcome the fears of those not satisfied with completeness of the measurement of ionic contamination.

The surface insulation resistance test measures the effect on the electrical characteristics of a printed wiring board without components—in the presence of moisture and an applied voltage—of contamination remaining on the board or substrate after cleaning. It is of paramount importance for surface-mount technology. This test's major virtue is that it provides a way to functionally test the circuitry and thus usefully complements the ionics testing. The general procedure for this test, which is normally undertaken manually, is described in Chapter 10.

Alpha Metals, Inc. now supplies the Sirometer (FIG. 7-5), a microprocessor-controlled unit specifically designed to carry out this surface insulation resistance test in accordance either with specifications MIL-STD-202 and IPC-TM-650 or Bellcore requirements TR-TSY-000078.

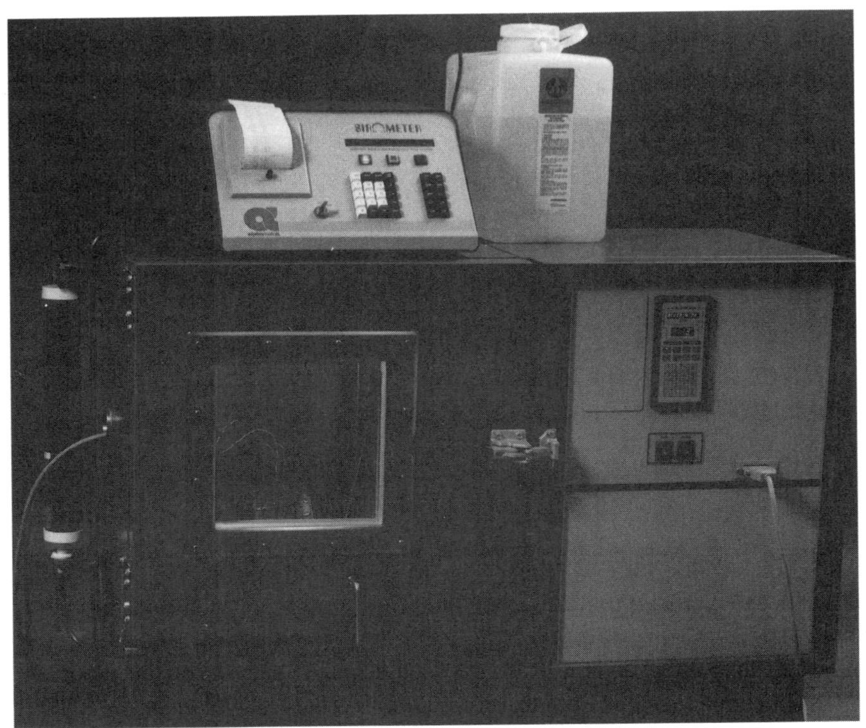

Alpha Metals, Inc.

Fig. 7-5. Sirometer surface insulation resistance tester is microprocessor controlled to accurately detect the presence of both ionic and non-ionic soils during temperature/humidity cycling.

8

Selection Criteria, Assembly Processes, and Other Considerations

This chapter reviews the information contained in the preceding chapters. Of special importance is its ability to offer the reader useful criteria in selecting a solder paste that will meet the needs of a particular application. The most significant of these criteria is summarized here, although it should always be borne in mind that the success of an operation involving solder paste depends not only on the material itself, but on all other factors concerned in its use. A solder paste formulation that works satisfactorily in one assembly method might not always be successfully transferred to another however similar the latter might be with respect to design and process parameters.

The remainder of this chapter contains suggestions with regard to the more mundane subjects of labeling, packaging, and storage. Individually or collectively, all of these topics are of some significance to ensure product quality, reliability, and repeatability.

One cannot emphasize enough how important it is that the customer's needs are accurately transmitted to the vendor. So many problems, large and small, are created in this industry purely as the result of poor communication. Sometimes these occur because solder paste still suffers from neglect by production engineers. Many of these engineers are reaching for loftier status by focusing their attentions on the sophisticated and expensive printing and pick-and-place equipment. More often than not, however, problems arise because no one has thought of writing a material specification, however brief, that at least provides the essential properties of the required solder paste and a basis for incoming quality assurance inspection. In an attempt to bridge that communications gap, a draft solder paste specification is presented that in its basic form should be acceptable to customer and vendor alike.

To end the chapter, we offer some advice on production processes that could be introduced by the utilization of solder paste. Finally, we make a few general suggestions on achieving savings in the use of the product.

SOLDER PASTE FORMULATION

Alloy

For virtually all processes involving glass/epoxy and polyimide printed wiring boards, the normal selection is 63 percent tin/37 percent lead eutectic alloy. Unless there is a pressing technical reason for using 60 percent tin/40 percent lead, it is not advised to do so because of its possibly limited availability as powder and its short, but occasionally significant, plastic range between molten and solid states.

If boards (or hybrid circuits) are metallized with pure gold, the 80 percent gold/20 percent tin eutectic is normally first choice. In some circumstances, however, an indium-lead or even a tin-lead alloy might be feasible, and the solder paste vendor should be able to advise you what would be appropriate.

Occasionally, certain components on the printed wiring board are terminated with silver or a silver composition such as silver-palladium. These might require a non-scavenging solder paste alloy such as 62 percent tin/36 percent lead/2 percent silver.

Hybrid circuit substrates with silver or silver-palladium conductors require a non-silver-scavenging alloy. Selection of this depends on the melting-temperature range desired. The 63 percent tin/37 percent lead alloy should be adequate for substrates with copper conductors but must be compatible with component terminations and, again, temperature constraints.

TABLE 8-1 provides some guidance as to applications for other alloys.

Metal Loading. Metal loading refers to the percentage by weight of the powdered metal contained in the solder paste. Metal loading can greatly affect the properties of a paste, particularly with reference to viscosity, ease of deposition, slumping, tack retention, and shelf life. More than anything else, however, it determines the thickness of the solder remaining after the paste is reflowed.

Due to the great difference in densities between the metal and the flux composition, the percentage by volume of the solder powder is always much less than the percentage by weight. Figure 8-1 exemplifies the relationship between metal content by weight and metal content by volume for the 63 percent

Table 8-1. Major Applications for Different Solder Paste Alloys

Alloy	Application
48Sn/52In	Low-temperature soldering.
50Sn/50In	" " "
52Sn/48In	" " "
58Bi/42Sn	" " "
58Sn/42In	" " "
80In/15Pb/5Ag	" " "
43Sn/43Pb/14Bi	Sequential or step soldering in applications involving copper, nickel, tin, and tin/lead surfaces.
70In/30Pb	Bonding to gold surfaces. For applications requiring this alloy's high ductility.
60In/40Pb	As for 70In/30Pb.
63Sn/37Pb	Bonding to copper, nickel, tin, and tin/lead surfaces when reflow temperatures of 215° to 250°C (419° to 482°F) are required. It is the principal alloy used for printed wiring board surface-mount processes.
60Sn/40	For same metallizations and temperature range as 63Sn/37Pb, but its use is now mainly restricted to non-critical electronic and electrical work.
62Sn/36Pb/2Ag	Bonding to silver and silver/palladium surfaces when reflow temperatures of 215° to 250°C (419° to 482°F) are required. Can be used instead of 63Sn/37Pb or 60Sn/40Pb where its higher bond strength is desirable, but such instances are rare because of fears about solid-state silver migration.
50In/50Pb	As for 70In/30Pb and 60In/40Pb.
96.5Sn/3.5Ag	For applications requiring good bond resistance to thermal cycling; sequential or step soldering especially in thick-film hybrid operations involving silver and silver-palladium metallizations. Used for soldering components exposed to high working temperatures, such as "under-the-hood" automotive parts.
40In/60Pb	As for 70In/30Pb and 60In/40Pb.
95Sn/5Sb	Used as an alternative to 96.5Sn/3.5Ag, when higher tensile strength is needed but is not suitable for thick film and has very limited application in electronics.
95Sn/5Ag	Alternative to 96.5Sn/3.5Ag for higher-melting-point applications.

(Table 8-1. continued)

80Au/20Sn	For soldering to gold and gold alloy metallizations.
19In/81Pb	As 70In/30Pb and 60In/40Pb, but especially as a replacement for 80Au/20Sn with the same melting point.
92.5Pb/5Sn/2.5Ag	For high-temperature thick-film hybrid applications. Has been used also for component assembly involving silver or silver palladium metallization, eg.g. capacitors.
10Sn/88Pb/2Ag	As for 92.5Pb/5Sn/2.5Ag.
92.5Pb/5In/2.5Ag	High-temperature alternative to the indium/lead alloys.
10Sn/90Pb	For component assembly not using silver and silver/palladium metallizations.
95Pb/3Sn/2Ag	Alternative to 92.5Pb/5Sn/2.5Ag with advantage of smaller plastic or pasty range.
97.5Pb1Sn/1.5Ag	Eutectic composition, but because of the higher lead content it is less stable than 95Pb/3Sn/2Ag.
5Sn/95Pb	As for 10Sn/90Pb, but with higher melting point and narrower plastic range.

There is little or no application in electronic soldering for tin/lead alloys other than those listed. In the case of those containing from 15 to 40 percent tin, the plastic range is long, varying between approximately 50° and 100°C (90° and 180°F).

tin/37 percent lead alloy. Thus, a metal loading of 85 percent by weight equals a volume of about 42 percent, while 90 percent by weight represents some 54 percent by volume. Among factors that affect this ratio are the actual metal content of the paste and the density of the alloy, usually expressed in grams per cubic centimeter. This would be 8.39 for 63 percent tin/37 percent lead, but 11.02 for 5 percent tin/95 percent lead. Thus, the same metal loading for these two alloys by weight would result in the higher lead composition occupying much less volume than the other.

Figure 8-2 illustrates the effect of the weight/volume differences on the thickness of solder remaining after both reflow and the removal of the chemical residues of the solder paste. It is apparent, therefore, that metal loading must be given careful consideration when selecting a paste. The reflowed solder thickness must not only be sufficient to provide required electrical conductivity and mechanical strength; it is frequently used to provide adequate spacing or ''stand-off'' between a component body and the board or substrate surface to facilitate cleaning.

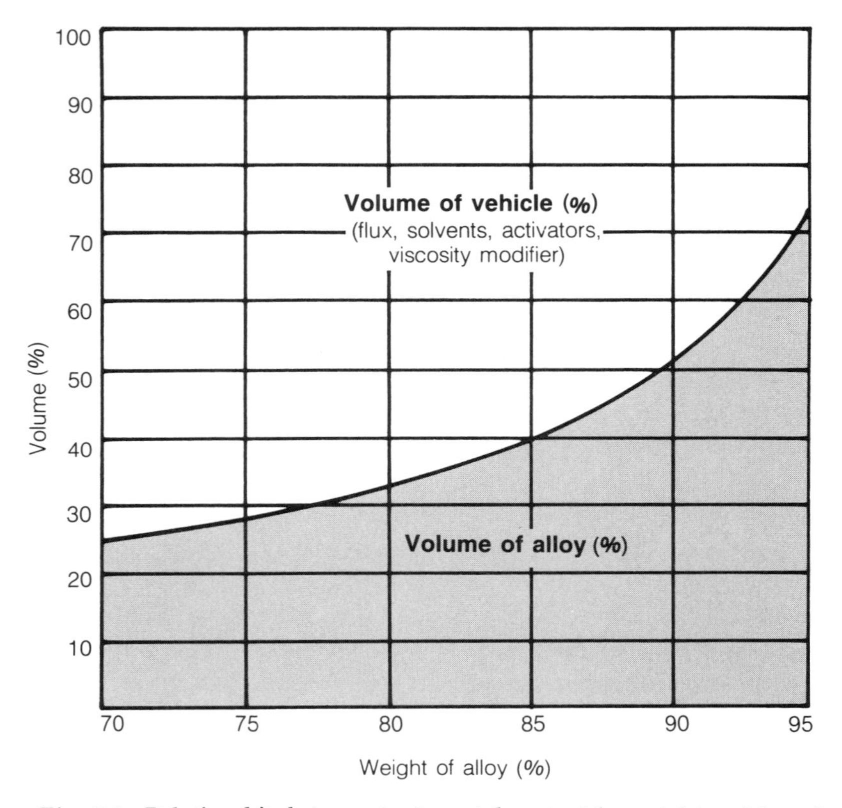

Fig. 8-1. Relationship between paste metal content by weight and by volume.

There are other reasons for caution in specifying metal loading. In preceding chapters, recommendations have been made with regard to minimum and maximum levels to be observed for certain methods of deposition. Too high a weight percentage renders pressure dispensing very difficult, if not impossible, and makes screen printing more onerous than necessary. In combination with a low viscosity, the product is also very susceptible to separation. Too low a percentage promotes excessive spreading, slumping, and bridging. As in so many aspects of solder paste usage, some compromise is called for between the metal loading desired and what can actually be used without inviting processing problems.

TABLE 8-2 provides recommendations as to maximum practicable percentages by weight for the major alloys for conventional deposition methods. Remember that metal weights are normally subject to a ± one percent tolerance.

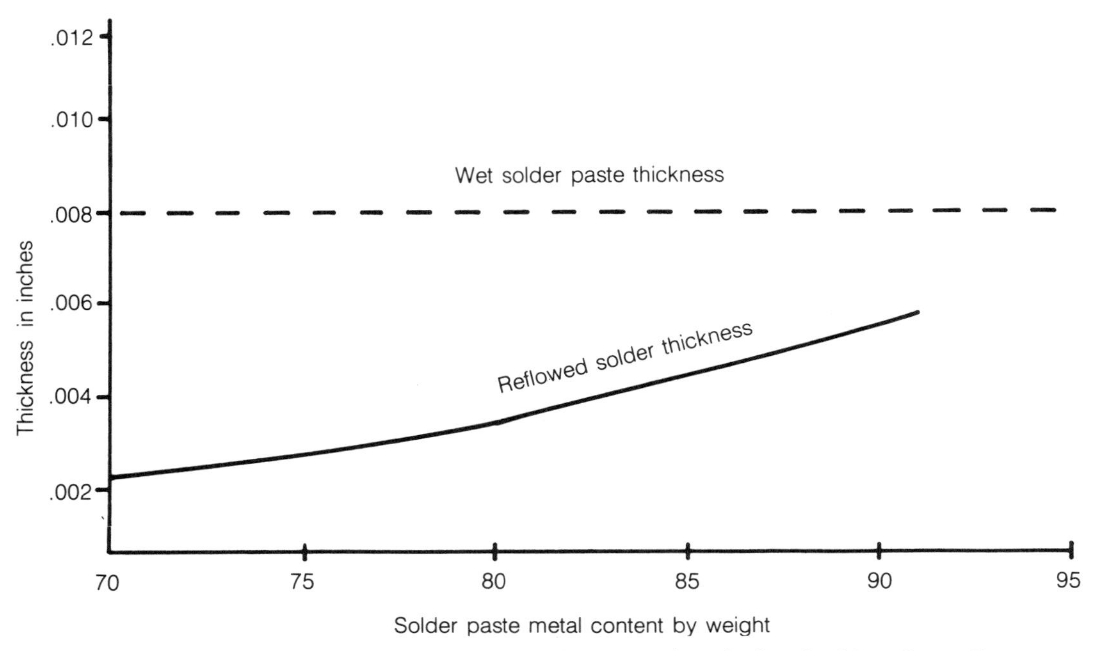

Fig. 8-2. Relationship between thicknesses of wet solder paste deposited and solder after reflow.

Table 8-2. Maximum Recommended Metal Loadings by Weight

	Deposition Method and Loading Percentages			
Alloy	**Dip/Pin Transfer**	**Dispensing**	**Screen**	**Stencil**
58Bi/42Sn	88	85	88	88
43Sn/43Pb/14Bi	88	88	88	90
63Sn/37Pb	90	88	90	91
60Sn/40Pb	90	88	90	91
62Sn/36Pb/2Ag	90	88	90	91
96.5Sn/3.5Ag	90	88	90	91
95Sn/5Sb	88	88	88	88
95Sn/5Ag	90	88	90	90
80Au/20Sn	88	88	90	91
92.5Pb/5Sn/2.5Ag	90	85	88	90
10Sn/88Pb/2Ag	90	85	88	90
10Sn/90Pb	88	85	88	90
95Pb/3Sn/2Ag	88	85	88	90
97.5Pb/1Sn/1.5Ag	90	85	88	88
5Sn/95Pb	90	85	88	90
All indium-containing alloys	88	85	88	88

Theoretically, those alloys containing 88 percent or more lead can be supplied in paste form with a metal loading of 92 or 93 percent. In such cases, however, insufficient flux would be present to ensure good wetting of the powder.

The percentages indicated take this into account. If, for instance, 91 percent loading is required for the 63 percent tin/37 percent lead, this would normally be specified as 90 percent. Any narrowing of solder paste manufacturing tolerances inevitably makes the vendor's task of balancing the composition of the material and its characteristics, such as viscosity and rheology, more difficult. However, some paste producers will agree to a tighter metal loading specification if the customer's application demands this. The advent of fine pitch reflow might render such exceptions to the rule much more common. Some manufacturers claim the ability to produce pastes with higher metal loadings than those listed. Such products can be considered only if they exhibit all the qualities necessary to achieve the desired reflow results.

Particle Size. The selection of particle size is influenced by a number of factors:

- Method of deposition (size of opening in screen or dispensing needle)
- Printed wiring board or substrate pad dimensions
- Slumping
- Tack retention

The relationships between maximum powder particle size (P), dispensing needle orifice diameter (O), and screen opening dimensions (S) were defined in Chapter 5 as

$$P = O \times 8 \text{ and } P = S \times 2.5$$

and for component lead pitches of 0.025 inch (0.635 mm) or less, a maximum particle size of −325 mesh should be utilized.

Particle size can influence slump quite dramatically. Chapter 1 discusses the effect of a powder's apparent density. Density is the volume occupied by a given weight of powder and is normally defined in grams per cubic centimeter. The disadvantages of a very narrow distribution of particle sizes were explained in terms of flux separation and inferior print definition due to slumping when relatively large spaces between individual particles of uniform size become pockets of flux.

There is evidence to suggest that tack retention can be greatly enhanced by a balanced distribution of powder particle diameters and that this is increased further by a reduction in overall size. Figures 8-3 and 8-4 illustrate the reasoning for this hypothesis and also reinforce the need for not attempting to increase metal content beyond 90 percent so that sufficient flux is available to provide adequate tackiness to hold components in place. More study is required of the relationship between the apparent densities of powders in different particle sizes and metal loadings and their effect on solder paste properties.

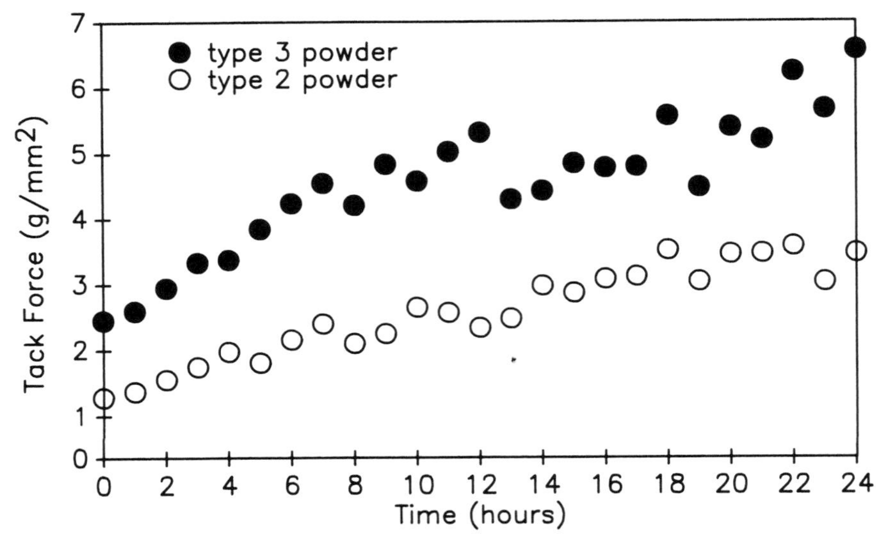

Fig. 8-3. Effect of powder particle size on solder paste tack retention, with the smaller type 3 (−325 mesh) material providing superior tack values. (J. Kevra and D. Mohoric, "Solder Paste: Tack and Cure," Microelectronic Packaging Technology: Materials and Processes, *Proceedings of the 2nd ASM International Electronic Materials and Processing Congress (Philadelphia, April 1989): 321-4.)*

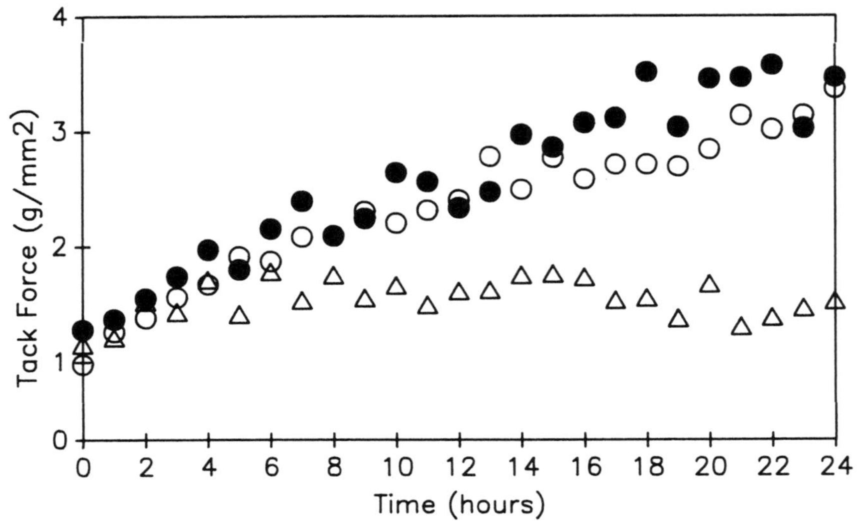

Fig. 8-4. Tests indicate that tack retention of solder paste declines as the metal content is increased beyond the starting level of 87 percent. (J. Kevra and D. Mohoric, "Solder Paste: Tack and Cure," Microelectronic Packaging Technology: Materials and Processes, *Proceedings of the 2nd ASM International Electronic Materials and Processing Congress (Philadelphia, April 1989): 321-4.)*

Flux System

Pure fluxes are discussed at length in Chapters 2 and 3. The solder paste user typically uses the word "flux" to describe everything in the product that is not metallic, and for convenience the same definition is employed in this section.

Apart from powdered metal, solder paste formulations contain fluxing ingredients of varying degrees of activity, thickeners or viscosity modifiers, rheological agents, solvent(s), and the occasional additive for providing special characteristics. It is usually unnecessary for the solder user to know much more than this about the formulation, but it is desirable to be familiar with differences in the systems available for the purpose of paste selection.

Activity Level. For surface-mount technology and hybrid circuit assembly, the RMA mildly-activated rosin flux is the normal choice in almost all applications. Depending on the paste producer, the RMA classification can cover a variety of different activity levels.

Chapter 2 discusses the different flux categories and the restrictions imposed by either government or industrial specifications on the activity level of the RMA flux by copper mirror, halide, and other tests. Provided that such criteria are met, there is no restriction as to the type or quantity of activator that can be incorporated into a paste flux. Taking water extract resistivity as an example, values for solder paste fluxes can typically range from just above the minimum at 110,000 ohm-cm to more than 300,000.

Individual activators used in RMA paste fluxes react differently with different metal surfaces. One type of activator could be eminently suitable for tin/lead coatings (especially for removing the lead oxide on these) but is a poor performer on bare copper. A creative paste producer will therefore utilize the known properties of the different activators to formulate fluxes to suit individual reflow applications; another might incorporate all the different activators into one "universal" flux system. Either approach is acceptable, although in situations where minimal RMA activity is needed, the universal product containing possibly superfluous activators would be at a disadvantage while one formulated for only bare copper traces might not be very effective for the tin/lead coated component leads.

This example demonstrates again the importance of effective communication between customer and supplier. If the solder paste producer is aware of what is to be reflowed as well as any solderability deficiencies, the correct flux activity can be provided from the outset.

Activated RA fluxes are seldom used in surface mount, where board and component solderability is normally good enough to permit RMA. In the hybrid circuit industry, however, substrate metallizations of silver and silver alloys tend to deteriorate rapidly, and RA paste fluxes are occasionally needed. The chloride activators that they incorporate make post-reflow cleaning mandatory, and care must be taken when storing the solder paste, especially in conditions of excessive heat and humidity. The RA solder paste fluxes are very reactive in such circumstances and can cause viscosity instability and surface crusting, depending upon the alloy and/or its particle size.

RSA, or *r*osin *s*uper-*a*ctivated, fluxes have virtually no application in solder pastes except for some structural soldering processes. Their highly corrosive activator systems make their use in electronics unacceptable.

There was initial interest in SA (synthetic-activated) paste formulations for surface mount, but this has faded with the concerns about CFCs. Increasing worries about the effect of solvents upon our environment will render the use of water-soluble organic acid flux systems in surface mount applications much more attractive in the future than has been the case until now.

A number of leading electronic manufacturers in the United States and elsewhere, particularly in the Scandinavian countries, are close to employing WS, or *w*ater-*s*oluble, paste flux formulations, if they aren't already. They might be based on water-soluble resins or organic acids without resins. In the past, the organic acid fluxes suffered from the hygroscopicity of the solvents used, and as a result, especially in conditions of high humidity, it was virtually impossible to control paste viscosity, slump, and spread. In addition, the post-reflow residues were difficult to remove, due partly to the limitations of cleaning equipment then available. These pastes were thus never considered for surface-mount work.

The new organic acid fluxes designed for surface-mount solder pastes incorporate a unique solvent system with a much lower sensitivity to moisture and excellent characteristics with regard to cleanability, from the point of view of both ionic contamination and surface insulation resistance. Continued improvements in the efficiency of available cleaning equipment will help to increase their acceptability in the industry.

Deposition Characteristics. The formulation can influence the quality of deposition by its built-in rheological properties. As has been demonstrated, these are governed by a number of considerations, including powder, but the formulation of the flux can have a significant effect on how well the paste will dispense or print. The paste manufacturer has to strike a balance between the solid ingredients in the product (such as the rosin or resins) and the solvent system in order to meet the rheological needs. These considerations lie beyond the scope of this discussion, because they involve proprietary processes, and the user is normally most concerned with what is visible in terms of dispensability, printability, and the behavior of the wet paste on the pad.

Apart from metal content and particle size, successful deposition relies on the paste's correct viscosity, which can affect the whole process through to

Table 8-3. Recommended Viscosities for Different Deposition Methods

Method	Viscosity Range*
Dip coating	150,000 to 350,000 cps
Roller coating	35,000 to 250,000 cps
Pin transfer	200,000 to 350,000 cps
Pressure dispensing	350,000 to 450,000 cps**
Screen printing	450,000 to 700,000 cps
Stencil printing	700,000 to 1,500,000 cps

*All viscosity measurements are quoted in centipoises (cps). They are based on readings obtained on a Brookfield Model RVT Viscometer, with Helipath Stand, TF Spindle, 10 turns or 2 minutes at Speed 5 rpm, at a temperature of 25°C (77°F), in accordance with specification QQ-S-571E.

**Lower viscosities of 300,000 to 350,000 cps might be required for needle openings of less than 0.025 inch (0.635 mm).

the reflow stage. Table 8-3 restates the criteria recommended in Chapter 5, which are based on empirical data obtained from a large number of solder paste users.

Tack Retention. The introduction of high-output, pick-and-place equipment has reduced the tack retention times originally demanded by solder paste users, which varied from 8 hours to as much as 5 days. Extreme cases still exist.

The time between solder paste deposition and completion of component placement is now typically 1 to 8 hours, depending upon the complexity of the assembly and the speed of the available equipment. For most solder pastes with a metal content by weight of less than 90 percent and a viscosity of less than 800,000 centipoises, a tack retention of up to 8 hours should be readily attainable, provided that the flux formulation and powder meet acceptable standards. The major concern of users is the ability of a paste of high viscosity with 90 percent metal loading to remain tacky long enough.

The effect of powder particle size on tack retention has been explained, but even with ideally distributed −325 mesh particles, the viscosities now needed for fine-pitch devices cause tackiness to be greatly reduced. In the case of a viscosity of 1,300,000 centipoises, the maximum tack retention is frequently only 2 hours. Additives have been developed to extend this to more than 8 hours with virtually no loss of definition or resistance to slump, the attributes that prompted the use of such high-viscosity material in the first place.

Residue Cleanability. The trend towards smaller devices and tighter spacing has increased concern about the ability of cleaning solvents and systems to safely remove solder paste flux residues. Of the rosin/resin formulations currently used routinely in electronics assembly, all leave residues after reflow that are cleanable to a more or less acceptable degree. Care must always be taken to ensure that reflow profiles are not so severe as to bake on the flux, rendering it difficult to dissolve, even in the most potent solvent.

Chapter 3 addresses the most common complaint concerning cleaning that a solder paste manufacturer receives—that of a haze-like white residue left after cleaning around the solder joints and pads. This has usually been attributed to inadequate dissolution of the rosin in the flux due possibly to polymerization and is generally disliked for cosmetic reasons rather than because the residue is harmful from an ionic or electrical standpoint. There are two potential solutions to this problem: one would be to improve the cleaning process; the second to modify the flux formulation. The first is usually achieved by replacing the solvent. The white haze is most frequently associated with one of the relatively mild fluorocarbon solvents. Replacement of the fluorocarbon solvent by a stronger chlorinated cleaner (if compatible with materials and equipment) often eliminates the residue.

The second course of action is to replace the water-white gum rosin in the paste by tall-oil resin, the reflowed residues of which appear to be more soluble in the CFC solvents. Unfortunately, the virtues of tall-oil resin in terms of residue removal are nullified by its deleterious effect on the viscosity stability of paste fluxes, so for this reason its application is very limited.

LABELING

Apart from essential text regarding the safety aspects of the product (with which this section is not concerned), labels should contain sufficient information for identification of the product. Almost as important is the provision of a date by which it should be used. Also acceptable is a date of manufacture from which the user can calculate the expiry of shelf life, which is normally six months.

The label should bear a lot or batch number, to which both the user and manufacturer can refer in case of query or complaint. Depending upon the quantity originally made, the solder paste producer should normally retain a sample of each batch of material supplied for a period of six months or longer. This can serve as a useful checking tool when users experience problems. However, the product under complaint might bear little resemblance to a sample in a controlled environment in the solder paste manufacturer's quality control laboratory.

PACKAGING AND SHIPMENT

Solder paste is supplied mainly in plastic jars sealed with a lined lid. The jars are available in a range of sizes, varying in capacity from about 114 grams

(4 ounces) to 1,500 grams (approximately 53 ounces) or more, depending upon alloy density and metal loading.

For most precision pressure-dispensing applications the paste is supplied in a number of different syringes ranging in volume from 10 cc to 35 cc. They hold from 25 grams, or a little less than one ounce, to about 140 grams, (5 ounces). The solder paste producers do not normally provide needles with the syringes, and they have to be obtained from the dispensing equipment manufacturers.

As described in Chapter 5, cartridges are supplied for other high-volume dispensing methods that are designed to hold between 280 and 1,400 grams (10 to 50 ounces).

The viscosity of solder paste is very susceptible to heat. If exposed for too long to temperatures in excess of 35 degrees C (95 degrees F), it can undergo irreversible changes, resulting in increases ranging from 10,000 to as much as 30,000 centipoises per degree Fahrenheit above 77 degrees F (25 degrees C). The solder paste manufacturer should use an insulating material such as polystyrene for shipping product in jars and dry ice for syringes and cartridges, particularly in the summer months. By maintaining these containers in a cold condition, the flux ingredients are kept immobile and are prevented from seeping into any air pockets created during the manufacturing or filling operations. Separated flux in jars can be stirred back into the paste to restore homogeneity; this is impossible in a sealed syringe or cartridge.

Solder paste users should consider overnight transportation in extremely hot weather, despite the additional expense involved. Even in these circumstances, there is no guarantee that the product will not be left exposed to the sun on an airport ramp or in an unattended vehicle.

WRITING A SOLDER PASTE SPECIFICATION

Having now examined the important properties of a solder paste, a specification can be prepared for the benefit of both the user and the supplier. The following extract, reproduced courtesy of Alpha Metals, Inc., is from a document furnished to their customers for this purpose. See sidebar entitled Solder Paste Specification Outline.

SOLDER PASTE SPECIFICATION OUTLINE

a) Alloy Composition
-List element and percentages
-Reference ASTM grade
-Reference QQ-S-571E alloy type

b) Melting Temperature
The alloy requires a melting temperature of ______
OR
The alloy requires a melting range of ______ solidus and ______ liquidus.

c) Alloy Particle Size
The alloy powder must be classified as ______ grade.
The maximum powder retained on a ______ mesh screen must be ______ percent and no more than ______ percent on a ______ mesh screen.

d) Flux Vehicle
The flux vehicle type must be ______.

e) Concentration of Metal in Paste
The percentage of metal must not vary by more than ______ percent from the established value of ______ percent when determined by the method outlined in ______.

f) Viscosity of Paste
When measured with a ______ viscometer in accordance with the method outlined in ______, the viscosity must be ______ cps ± ______.

g) Visual Graininess
When measured with a Precision Gage and Tool Co. Type CMA 185 Fineness of Grind Gage, the reading must be (80 microns or less for −200 mesh or 50 microns or less for −325 mesh product).

h) Solder Ball Test
After deposition of a standard solder paste pattern of______ that is ______ inches thick on a ______ test coupon, the solder paste must produce a ______. The preparation of the sample and heating method must be as defined in ______.

i) Shelf Life and Storage Conditions
An unopened container of paste must have a minimum storage life of ______ months when stored at a temperature of ______.

j) Packaging
The material must be packaged in any suitable manner that will prevent loss or injury in shipment or in storage. The size of package or container must be as specified on the purchase order.

k) Marking
Each container must be marked as follows:
-Supplier's name
-Lot number
-Alloy type, metal and mesh size
-Vehicle type
-Quantity of material
-Part number

Refer to Chapter 10 for the relevant test procedures.

STORAGE

Jars can be stored at a temperature between 2 and 30 degrees C (35 and 86 degrees F) in relative humidity of 50 to 60 percent.

Sealed syringes and cartridges must always be maintained in a refrigerated temperature between 2 and 7 degrees C (35 and 45 degrees F) for reasons explained earlier in this chapter. As an additional precaution against flux separation, the same syringes and cartridges should be rotated once daily or otherwise as frequently as possible.

Care should be exercised when using paste that has been removed from a refrigerator or cold room. It should be allowed to return to room temperature over a period of 12 to 24 hours before any attempt is made to open the containers. Premature use of the product could result in moisture entrapment in the material formed from condensation by the contact of warm air with the cold paste.

Extremes of temperature should always be avoided when storing solder paste. Heat will affect its homogeneity, viscosity, and rheological properties. Although there is only limited evidence that freezing can permanently damage the paste, it is probably wise to keep the material at temperatures above 0 degrees C or 32 degrees F. This is in view of the experience with liquid machine-soldering fluxes, which have suffered from crystallization of rosin and precipitation of organic acid activators after exposure to freezing conditions.

PROCESS SELECTION

For some years still to come, the surface-mounting of components with solder paste will continue to develop in the electronics industry with a corresponding decline in the number of devices attached by means of wave soldering.

At the present time, although many manufacturing companies are fully committed to reflow soldering, a majority of others are using mixed technology by combining solder paste and machine soldering, while a significant minority have yet to adopt solder paste at all in production.

The three processes described have been classified as follows:

TYPE 1: Surface mounting only, on one or both sides, with no through-hole components.

In this case, board design permitting, solder paste can be employed for both sides of the assembly (TABLE 8-4). Deposition of paste onto the second side requires special fixturing if a printing method is to be used because of the uneven surface created by the components on the first side of the assembly.

It is important that the board is designed so the first side to be reflowed is primarily populated with small components such as small outline transistors (SOTs) or chips. In an in-line system, this then becomes the bottom side during the second reflow stage, and the solder joints already formed will in most situations remelt into a liquid state. The surface tension of the molten solder is usually reliable to hold small, lightweight components in place. Larger, heavier devices will drop off the board.

TABLE 8-5 illustrates a process involving two solder pastes containing different alloys. The first, shown as the eutectic 96.5 percent tin/3.5 percent silver composition, has a melting point of 221 degrees C (430 degrees F), requiring a reflow temperature of about 250 degrees C or 500 degrees F. This is used for the first side. A 63 percent tin/37 percent lead solder paste is then

Table 8-4. Solder Paste Reflow with 63 Sn/37 Pb

Surface Mount only, Single-sided	Surface Mount only, Double-sided
• Print solder paste on first side • Mount components • Preheat assembly • Reflow solder paste • Clean	• Print solder paste on first side • Mount smallest components • Preheat assembly • Reflow solder paste • Turn over assembly • Dispense solder paste • Mount largest components • Preheat assembly • Reflow solder paste • Clean

employed for the second side. By careful control of heating parameters, the solder joints on the high-temperature alloy side will not remelt during reflow of the second side.

It is important that solderability of the second side is not too adversely affected by the reflow of the first, and not all equipment is suitable for such a process. Conveyor belts must hold the assembly level with the reflowed, bottom-side components clear, and in case devices do drop off, some form of catch tray must be installed underneath the conveyor. Both infrared and vapor-phase reflow systems can be used for this process, but the latter requires the use of two different lines or batch units with two separate fluids suitable for the melting points of the two alloys.

Exercise caution when cleaning the first side after reflow. Certain solvents among the chlorinated hydrocarbons have caused degradation of the solderability of printed wiring board circuitry, and tests should be done to determine whether an intermediate cleaning stage is necessary.

Table 8-5. Sequential Solder Paste Reflow with 63 Sn/37 Pb

Surface Mount only, Double-sided

- Print higher melting-point alloy* (e.g. 96.5Sn/3.5Ag) solder paste on first side
- Mount components
- Preheat assembly
- Reflow solder paste
- Turn over assembly
- Dispense 63Sn/37Pb solder paste
- Mount components
- Preheat assembly
- Reflow solder paste
- Clean

*If reflow is by the vapor-phase method, care should be taken to select an alloy with a melting point within the range of boiling points of available fluids. Otherwise, infrared or another suitable reflow method will have to be adopted.

Surface Mount only, Double-sided

- Print 63 Sn/37 Pb solder paste on first side
- Mount components
- Preheat assembly
- Reflow solder paste
- Turn over assembly
- Dispense lower melting-point alloy (e.g. 43 Sn/43 Pb/14 Bi) on second side
- Preheat assembly
- Reflow solder paste
- Clean

TYPE 2: Mixed Technology, with both surface-mounted and through-hole devices.

The lack of availability in surface-mount form of many components has forced companies to compromise between the old and new technologies. Type 2 combines surface-mount devices reflowed on the top side with solder paste, while the leads of through-hole components are wave-soldered from beneath. Table 8-6 shows three different processes that can be achieved with both the conventional 63 percent tin/37 percent lead and higher and lower melting-point alloys. The eutectic tin-lead composition is generally preferred because of its superior wetting characteristics.

Where the through-hole components are to be wave soldered first, then the solder paste can be applied only by some form of dispensing (as described in Chapter 5). Caution is again advised with respect to the possibly deleterious effect of cleaning the first side.

Table 8-6. Sequential Solder Paste Reflow with 63Sn/37 Pb

Mixed Technology, Double-sided

Process 1
- Print 63Sn/37Pb solder paste on first side
- Mount components
- Preheat assembly
- Reflow solder paste
- Insert through-hole components
- Wave solder second side with 63Sn/37Pb or 60Sn/40Pb*
- Clean

*With most double-sided boards, the temperature of the top side during the short wave-soldering operation will typically not exceed a temperature of about 120°C or 248°F, far short of that needed to re-melt the 63Sn/37Pb paste alloy.

Process 2
- Print higher melting-point alloy (e.g. 96.5Sn/3.5Ag)
- Solder paste on first side
- Mount components
- Preheat assembly
- Reflow solder paste
- Insert through-hole components
- Wave solder second side with 63Sn/37Pb or 60Sn/40Pb
- Clean

Process 3
- Insert through-hole components
- Wave solder first side with 63Sn/37Pb or 60Sn/40Pb
- Pressure dispense lower melting-point alloy (e.g. 43Sn/43Pb/14Bi) solder paste on first side
- Mount components
- Preheat assembly
- Reflow solder paste
- Clean

TYPE 3: Bottom-side attachment only with small surface-mounted components that are wave-soldered simultaneously with the leads of conventional through-hole components.

This process does not normally require the use of solder paste. The surface-mount components are placed in position using a small dot of curable epoxy adhesive that holds them firmly during foam, spray, or wave fluxing, preheating, and wave soldering.

HELPFUL HINTS

Good-quality solder paste is not cheap, and it is often remarkable how much of it is wasted. This is beneficial for the supplier, at least in the short term, but it does not encourage new companies with limited resources to adopt the new technology as readily as they should.

The most common criticism of solder paste refers to viscosity changes—invariably upwards. These changes result from any one or several of the reasons already put forward, but unless the paste is new and can be rejected and returned to the vendor, more often than not the unfortunate manufacturer either has to make do with the material or consign it to a scrap solder bin.

Two courses of action can be adopted with excessively high-viscosity solder paste. First, it can be thinned, or second, it can be returned to the factory for reworking for a modest charge (depending on the vendor), which is normally much less than the original cost of the paste.

The viscosity can be reduced by three methods, in descending order of preference.

- Add low-viscosity solder paste. This has the effect of maintaining the exact composition of the product, including the important metal weight percentage. If because of environmental conditions in the workplace rising viscosities are persistent, then it is worthwhile considering the regular purchase of a small reserve quantity of low-viscosity paste. When the nominal viscosity of a paste rises from, say, 800,000 to 1,200,000 centipoises, then the addition of an equal amount of 400,000 centipoises material should reduce the viscosity of the whole to a figure close to the required level. The vendor should be able to provide advice in this respect.
- Add a small quantity of paste flux. The flux comprises everything that is to be found in the paste itself, except the powdered alloy. Exercise care in adding flux so that the metal content does not fall below the minimum required. It is therefore recommended that not more than 0.5

percent paste flux be added for the purpose of in-house viscosity adjustment. The paste flux is normally available fairly cheaply from the paste supplier.

- The third, more potentially damaging alternative is to add pure solvent. This should never be ventured without consulting the solder paste vendor, who will recommend compatible solvents. The thinners normally used for liquid wave-soldering fluxes are not suitable for pastes because of their much greater volatility. They are also, unlike conventional paste solvents, highly flammable. The major difficulty in adding solvents to reduce viscosity is the effect they have on the product's rheological properties and the potential danger of solder-balling that results from outgassing or spattering. For these reasons, they should be employed with great discretion.

One final point with regard to viscosity adjustment: remember that the user can always make changes to bring viscosities down, but no attempt should be made to increase them. There have been instances where heat was applied to drive off solvent in order to raise viscosity levels; the result was disastrous effects on deposition and tack retention.

Finally, take note of a comment now about discarding used or unusable solder paste. The vendor might have a reclamation program to offer whereby such material can be reworked to a condition approaching its original quality. Such modification might require the addition of a substantial percentage of new product to the old, but the total cost is always going to be less than for wholly new paste. In some cases, the condition of the old material precludes such rework, but the supplier is always prepared to accept a sample to determine whether or not reclamation is possible. If it is not, and a sufficiently large quantity is available, then the paste manufacturer might be able to offer a reasonable price for the scrap metal portion of the material.

9

Problems and Troubleshooting Procedures

No book on solder paste would be complete without a review of some of the problems the user can expect to encounter, especially in the early stages of his or her operation.

The paste is customarily blamed for most things that go wrong with solder reflow in a surface-mount process, but very often it is the assembly procedures that are at fault, either because the paste has been improperly used or as the result of an equipment malfunction.

The problems addressed here are the most common. They are classified as pertaining to either the deposition or to the reflow stages and are listed in alphabetical order under each category. There are invariably a number of possible solutions to be offered for each problem; that, unfortunately, is the nature of solder paste. For example, there is no simple, clear-cut answer to a solder-balling problem because there are a number of causes.

DEPOSITION

Bridging

Bridging describes the phenomenon when solder paste links two adjacent conductor lines or pads. Except in severe cases, the paste usually pulls back to one or both lines or pads once reflow takes place because the molten solder is naturally attracted to a metallized surface rather than the glass/epoxy or ceramic substrate material between the two lines or pads. Where this pulling back does not occur, for example when a large quantity of paste is present, any resulting solder bridge will cause electrical shorts. More often than not, however, insufficient metal remains to form a continuous bridge, and the natural surface tension of the molten solder will cause it instead to form into one or more solder balls.

Bridging becomes more serious when pad sizes and spaces become smaller, and even small solder balls are unacceptable. The problem is normally mainly attributable either to the properties of the paste or to the deposition process. Bridging is especially prevalent with pastes of low metal content by weight (less than 88 percent) and viscosities of less than 700,000 centipoises. If the paste is at fault, then possible solutions are a slight increase in metal loading, the replacing −200 mesh powder with the finer −325, and/or raising of the viscosity. Another reason for bridging can include flux separation in the paste, particularly if it is being dispensed from a syringe or cartridge.

It is, however, just as likely that the problem originates at the dispensing station or printer. Heat has a significant influence on paste viscosity and its flow characteristics, and an ambient temperature of more than 27 degrees C (80 degrees F) has been known to make the paste runny. The heat generated by equipment, particularly the hydraulics of a screen printer, can affect paste viscosity. So can that created by the friction of a squeegee on a screen or stencil. The proximity of central-heating ducts can also cause a reduction in paste viscosity.

Another valid reason for bridging is excessive paste deposit thickness. This can be reduced sufficiently sometimes by adjusting screen or stencil snap-off or squeegee pressure or speed. In a dispensing operation, the same goal can be achieved by reducing pressure on the syringe or cartridge piston, by reducing the needle orifice, or by increasing the needle length.

Not to be overlooked is the possibility that excess pressure has been applied on components by an operator or a pick-and-place machine. This can cause paste to be squeezed out away from the pad. In such cases, either the placement pressure can be lowered or the thickness of the paste deposit reduced.

Clogging

Clogging refers to blockages in either a dispensing needle or the openings in a screen or stencil.

Needle clogging is almost always due to oversize powder particles. More rarely it is from undissolved rosin, resin, or thickener in the paste. High viscosities likewise make dispensing much more difficult. Clogging can occasionally occur because needle length is too long or because the needle has become deformed in some way. The design of the syringe or cartridge barrel is also important, as sharply angled shoulders can cause material to accumulate, become tightly packed, and back up behind the opening to the needle. High metal loadings can exacerbate dispensing problems.

In printing, oversize particles and excessively high viscosities can cause difficulties in screens in particular, but sometimes the answer to the problem lies in the tacky nature of the paste because it does not release from the openings. In a stencil, this lack of fluidity might be due to poor etching of the apertures that results in an excessive difference in surface-to-center dimensions.

Crusting

Crusting is normally the product of a reaction between a strong flux activator system, moisture, and powder surface oxides, particularly those of high-lead-containing alloys. Careful storage conditions are required to prevent the occurrence of this, which can seriously affect paste deposition. Complete removal of the surface crusting is essential before the paste is used to avoid creating lumps that can interfere with screen-printing or other applications.

Loss of the crust, if the quantity is not excessive, will not greatly affect the performance of the paste, although inevitably some of the activator will have been exhausted in the reactive process described.

Users of particularly sensitive formulations that are susceptible to crusting should consider packaging the product in syringes or cartridges as a means of excluding moisture, which is an essential ingredient in its formation.

Excessive Paste

The same causes apply here as for bridging.

Insufficient Paste

This is a problem frequently created by the same circumstances as clogging. It can also arise from using the wrong screen wire diameter and emulsion thickness or too thin a stencil. To some extent, adjustment of printing parameters can compensate for insufficient deposits of this kind.

Insufficient Tack Retention

Poor tack retention can be due to several causes. One cause is the loss of solvents due to excessive workplace temperatures and air flow from air conditioning vents, for example, promoting the volatilization process.

Chapter 8 explained that powder particle size and distribution can have a significant influence on tack retention as can metal loading and viscosity. If process changes do not bring about any improvement and the paste formulation cannot be modified, there are certain procedures that can extend the product's tackiness. Apply a low-viscosity flux from a spray bottle onto the pads where longer tack time is essential. Some users, usually without much success, have tried soaking boards in solvent vapor in a sealed chamber.

Slumping

Slumping is caused by virtually the same conditions as bridging.

Smearing

Smearing can result from the same conditions that apply to both bridging and slumping, but usually it is caused by improper printing procedures such as excessive squeegee pressure. Insufficient snap-off and poor screen tension are other reasons for smearing. The poor gasketing provided by a deformed stencil can also be a problem.

REFLOW

Component Movement/Misalignment

Major reasons for component movement are:

- Inaccurate solder paste deposition
- Uneven thicknesses of solder paste
- Improper placement of components
- Non-uniform heat transfer during reflow
- Inadequate component lead or termination solderability
- Poor pad solderability
- Insufficient flux activity
- Excessive pad to lead dimension ratio, enabling device to "float"

All of these factors, both individually and collectively, can lead to component misalignment, which is usually due to what is termed *preferential wetting*. This means, for example, that in the case of a leaded surface-mount device, one or more leads "wet" or are reflowed more rapidly than the others, causing the whole component to be pulled to the area where wetting first takes place. This can occur because of insufficient paste under one or more leads or poor solderability of the leads and/or pads. Misalignment resulting from the pick-and-place operation can be self-correcting due to the action of the molten solder when reflow takes place, if not too severe.

Dewetting

Dewetting is a phenomenon associated with poor solderability whereby the solder actually initially coats the surface but then is repelled to form droplets, much as water does on a freshly waxed automobile. Dewetting can be corrected either by improving the surface and/or solder powder solderability or by increasing the activity of the paste flux.

Discoloration

Discoloration here refers to that of epoxy/glass printed wiring board materials, which is especially prevalent in infrared reflow. This discoloration can be remedied by adjustments in heating profiles as well as component layout, as discussed in Chapter 6. The same chapter describes claims that the use of nitrogen greatly inhibits discoloration.

Dull Joints

Dull solder joints can result from contamination of the solder by metallic impurities as well as from the use of a non-eutectic alloy that has undergone vibration or other stresses while still in a plastic state following reflow.

Dullness also results from too slow a cooling rate following reflow, as frequently evidenced by assemblies subjected to vapor-phase soldering in a batch unit. In this case, the appearance of the joints is not indicative of any quality defects, although a shiny, lustrous surface is normally preferred.

Nonwetting

Nonwetting describes the partial or complete failure of molten solder to coat a metallic surface. While this can most frequently be attributed to extremely poor solderability, there have been instances of use of paste flux supplied with an insufficient activity level. Occasionally, due to a fault in the reflow equipment or incorrect temperature setting by the operator, the paste alloy simply has not melted in order to perform its wetting action.

Opens

This is a term used to describe a gap that can occur between a component lead and a board or substrate pad after reflow. Such opens are usually associated with "J"- or "knucklebone" -leaded devices.

There are two main reasons for this phenomenon. First, coplanarity of the leads might be so poor that one or more leads is not in contact with the solder paste when reflow takes place. Second, the leads might heat up much more rapidly than the pads during reflow because of the greater thermal mass of the printed wiring board. This is especially prevalent when vapor-phase reflow is employed.

In the first example, the solder cannot jump the distance between lead and pad. The only solution to this problem, apart from insisting on better coplanarity tolerances from the component suppliers, is to increase the thickness of solder paste deposit to compensate for the gap.

In the second example, the flux in the paste is attracted to the hottest area, in this case the lead. Assuming that the lead temperature has risen above the melting point of the paste alloy, the solder follows, leaving the pad barely wetted or "starved." Virtually all the solder has wicked up the leads, providing an insufficient amount to form a bond at the interface between the lead and pad.

Four potential solutions can solve this problem of opens. One, developed by Texas Instruments, involves the use of a so-called "ratioed alloy" solder paste. It contains separate constituents of pure tin and 10 percent tin/90 percent lead. The percentages of these are calculated to result in a 63 percent tin/37 percent lead alloy when they combine during reflow, at which time the melting point should be 183 degrees C (361 degrees F). The theory behind this development was that the delayed wetting of the tin-lead powder by the pure tin allows more time for the pads to heat up to a temperature similar to that of the leads so that much more of the molten solder remains at the lead/pad interface. This concept did, in fact, prove very successful.

The second solution is simpler. Also, it does not involve a change to solder constituents that are not the easiest to use in powder form, mainly because of inferior surface solderability. In this case, the RMA flux can be modified by removing one of the activators. The wetting action of the flux on the tin-lead coating of the lead sharply reduces, which inhibits wicking of both the flux and, therefore, the 63 percent tin/37percent lead solder.

Thirdly, the temperature differential between the leads and the pads can be minimized to some extent by careful preheat profiling.

If a vapor-phase reflow process has been used (and there is reason to believe that this might contribute to the problem), a fourth alternative is to convert to infrared or another soldering method that provides a less severe technique for heating the component leads.

Interestingly enough, the 62 percent tin/36 percent lead/2 percent silver alloy does not behave in the same way as the eutectic tin-lead, which might be due to its different flow characteristics and higher melting point.

Pinholes and Blowholes

Pinholes and blowholes differ from each other only in size, and both are normally caused by the outgassing of solvent or moisture that has penetrated the surface of the solder joint.

These contaminants can normally be eliminated by adequate preheating of the assembly prior to reflow in order to expel volatiles present in the solder paste and any moisture entrapped in the board material.

Solder-balling

As already indicated, there are numerous reasons for solder-balling. Since the inception of surface-mount technology, it has been the most highly-publicized problem connected with solder paste reflow.

Some of the principal causes of solder-balling are:

- Fines in the powder
- Oxidized powder
- Slump
- Solder spatter
- Inadequate flux activity

Fines are a persistent cause of solder balls. Figure 9-1 illustrates the effect of heat on a solder paste containing such powder particles, which are typically less than 20 microns (0.0008 inch) in diameter. The flux carries these away from the main deposit of paste onto nonwettable areas of the board or substrate, where upon reflow, surface tension causes them to form tiny spheres. Invariably, these fines are readily removed during the cleaning process with the flux to which they adhere. However as pad sizes become smaller in the future, their presence will be tolerated less than it is now, particularly in difficult-to-reach areas. This type of solder-balling can be reduced by careful temperature profiling to inhibit the spread of the flux.

Oxidized powder is a more serious cause of solder-balling, because it is normally responsible for the creation of large solder balls of up to 0.010 inch (0.254 mm) and more. These large balls are especially dangerous if they get stuck under components because they will generally not be dislodged by the cleaning process.

The surface oxides on such powder particles prevents them from coalescing on reflow with the main paste deposit. Hence, they might float on the flux into inaccessible areas or remain close to, or even on, the solder joint.

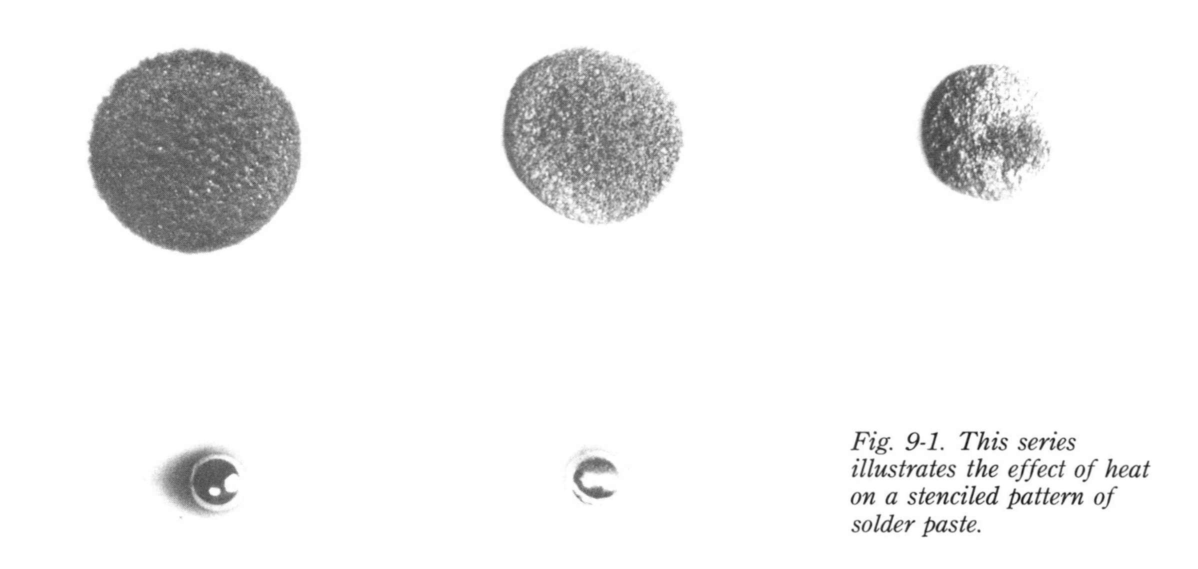

Fig. 9-1. This series illustrates the effect of heat on a stenciled pattern of solder paste.

Such powder can be made to reflow reasonably well by adding activator to the flux, but this is not always a wise practice because of chemical reactions the oxides can prompt, causing viscosity instability and other problems.

Slump is another reason for solder-balling. Paste that is deposited on nonwettable areas sufficiently distant from the metallized pads always forms itself into one or more solder balls. As explained in Chapter 9, slump can be fairly easily controlled by increasing viscosity to a level compatible with the method of deposition or by increasing metal content, and thus reducing the volume of flux.

Finally, the presence of moisture or solvent in the paste or on the board or substrate can lead to outgassing during reflow, which if explosive enough will spatter molten solder. This results in either solder balls or flakes of solder. Such an occurrence can be minimized or eliminated by modifying the preheat profile to ensure removal of the offending moisture or solvent.

Tombstoning

Also known as *drawbridging* or the *Manhattan effect*, tombstoning is when a chip component is standing up on one end during reflow.

The causes are generally the same as those of Component Movement/Misalignment, but in the case of the lightweight capacitors and resistors, the problem can be more difficult to resolve. In some instances, chip re-orientation might be required in relation to the direction of travel of the assembly through the reflow process to equalize the heat exposed to all leads.

Voids

Voids arise from the same causes as tombstoning, but differ from pinholes and blowholes in that they are confined inside the solder joint with no exit hole apparent following outgassing of solvent or water. What normally happens is that the solder solidifies before the vapors or steam can escape.

These hidden cavities are of obvious concern to any manufacturer, particularly if high-reliability equipment is involved. Voids significantly reduce a solder joint's ability to survive thermal cycling stresses. Voids frequently act as a starting point for cracks in the solder joint structure. There are techniques to discover voids, but x-ray equipment is expensive and costly to operate, and the infrared-laser system has not yet been widely accepted.

The conclusion is that most voids originate from solvents used in the paste flux. Work is being done to see if replacement of these solvents by even higher boiling varieties can reduce the extent of this threat to the integrity of the solder joint, if not entirely eliminate it. Voids can be so large as to reduce solder volume in a joint to an unacceptable level. It is theorized that by introducing a solvent with a boiling point well above the reflow temperature for the 63 percent tin/37 percent lead alloy, this will not vaporize to cause voids. Completed studies demonstrate that the use of such solvents significantly reduces the size of cavities within the solder joint, which presumably are now filled with unevaporated solvent rather than vapor.

Wicking

See the section entitled Opens.

10

Test Methods

The very first solder pastes were relatively crude mixtures of solder powder and a flux medium. The applications of these pastes were non-critical and there was little concern for these materials to meet exacting specifications. As their use became more widespread, more attention was paid to the quality of the raw materials. Likewise, test methods were developed to assure a quality product.

The definition of what an acceptable product is has also changed over the years, and more stringent requirements are being placed on the product as more demands are made of it. Many of the original tests that were devised to evaluate solder paste are still in existence today, and many new ones have been added to the list. The test methods in use today have been developed by both the manufacturers and users of solder paste. The test methods can broadly be thought of as evaluating some property of either the flux or the alloy (i.e. the solder powder). Many of the tests can be thought of as "functional tests" in that they directly indicate something about how the paste will perform in an actual application. For example, the ceramic or solder-ball test and solderability on copper are such tests.

The choice of what test to perform, if any, depends on many factors. The type of assembly being soldered is a major factor. The requirements for structural assemblies, consumer products, and high-reliability electronics all vary. A second factor is whether you are restricted by contract to conform to some standard specification. Such a specification can originate through the government (e.g. federal or military), industry (e.g. Bellcore and AT&T specifications), or private organizations (e.g. IPC, ASTM, and ISO). Requirements and methods vary with each specification.

SPECIFICATIONS

The following pages list the various specifications and point out the major differences between them. In some cases, a specification is still in a ''draft'' form; such cases give the date of the latest revision. The specifications listed are:

- Federal Specification QQ-S-571E
- IPC-SF-818
- IPC-SP-819
- AT&T Material Specification S9104
- Bellcore TA-TSY-000488
- IIW Sub-Commission Draft 1A (1987)
- ASTM B486-68 T

Each specification has a list of the required tests to be performed.

The remainder of the chapter discusses the details of the various tests (generic test methods). Note that different specifications might specify a halide test or an SIR measurement, and there might be some differences in the details of carrying out the test such as conditions of temperature and storage time, etc. In this respect, the IPC document and the proposed IIW specification provide for the user and vendor to mutually agree on the details of the test method.

Federal Specification QQ-S-571E
Solder, Tin Alloy: Tin-Lead Alloy; and Lead Alloy
Last Revision 1972

This specification gives requirements to classify a flux as R (rosin), RMA (rosin mildly activated), RA (rosin activated) or AC (inorganic acid activated). Specifications are provided for flux and alloy.

Tests for Paste

1. Resistivity of water extract
2. Chlorides and bromides (silver chromate test)
3. Solder powder mesh size
4. Viscosity
5. Flux percentage
6. Solder pool
7. Spread factor

8. Dryness
9. Effect on copper mirror

Alloy Composition

The alloy composition is determinable by any suitable method, including wet-chemical, spectrochemical techniques, or both. Allowable levels of impurities are given for various alloys for antimony, bismuth, silver, copper, iron, zinc aluminum, arsenic, and cadmium.

IPC-SF-818
General Requirements for Electronic Soldering Fluxes

IPC-SF-818 proposes a new system of nomenclature for flux classification using the letters L, M, and H to denote low, moderate and high activity. In association with the letters, numbers 1, 2, and 3 are used to denote the assembly class the flux belongs to. The classification 1, 2, or 3 depends on the results of SIR (surface insulation resistance) testing.

One of the major differences between this and the QQ-S-571E specification is that SIR is used as the crucial test of ionic contamination rather than water extract resistivity, test which is used in the QQ-S-571 specification. The IPC specification allows for non-rosin fluxes to be classified except for IA type. Another major difference is that either the flux or the flux residue after reflow can be used in certain tests. This is dependent on vendor and user agreement.

Flux Tests

1. Copper mirror
2. Halide (silver chromate test)
3. Halide content (by titration)
4. Corrosion test
5. Surface insulation resistance (SIR)

IPC-SP-819
General Requirements for Solder Paste

Alloy

Powder is to be separated from the flux and analyzed by accepted methods. Melting point or melting range is determined by an acceptable instrument to within 1 degree C.

Powder size is to be determined by any method that is agreed upon by used and vendor. Powder shape is specified to have a length to width ratio of 1.5:1 or less.

Tests for Paste

1. Metal content
2. Viscosity, Brookfield
3. Slump
4. Solder ball, visual standards supplied

5. Powder size
6. Tack
7. Wetting

AT&T Material Specification 59104

Flux classification is based on the IPC concept of low, moderate, and high activity.

Tests for Flux

1. Copper mirror
2. Halide content
3. Corrosion
4. pH
5. Flux content

The solder alloy must meet the requirements of Material Specification 58327, ASTM ASTM B32, QQ-S-571E, or IEC 68-2-20.

Tests for Paste

1. Metal content
2. Particle size
3. Shape (length to width ratio of 1.5:1 or less)
4. Viscosity, Brookfield
5. Spread/slump
6. Solder-ball
7. Tackiness
8. Wetting
9. Surface insulation resistance (SIR)

Bellcore TA-TSY-000488

Requires a sample of the flux from the supplier. The copper mirror and SIR tests use undiluted paste flux.

Flux Tests

1. pH
2. Halide (silver chromate)
3. Fluoride
4. Copper mirror
5. Surface insulation resistance (SIR)

IIW Sub-Commission 1A Ad Hoc Committee for Solder Pastes Fourth Working Draft Specification for Classification and Methods of Testing for Solder Pastes, 1987

This specification proposes a somewhat more complex flux classification based on flux type, flux basis, and flux activation. Fluxes would be classified with numbers only, for example 112 is resin type, rosin based, and halogen activated. The alloy is classified according to ISO (International Standards Organization).

Tests

1. Solder ball
2. Solder wetting
3. Viscosity (coaxial cylinder as described in DIN 53019). Possibility of measuring viscosity at shear rates of 0.1 to 50.0 sec^{-1}. Brookfield method is also specified. Dispensing under pressure. The amount of paste dispensed is an indirect measure of viscosity/rheology.
4. Tack (similar to IPC method)
5. Slump
6. Corrosion of flux residues
7. Surface insulation resistance (SIR)
8. Tack after reflow
9. Storage stability (Any separation must be readily dispersed, and viscosity must be within 20 percent of nominal after 3 months.)
10. Metal content

ASTM B486-68 T, Tentative Specification for Paste Solder

Solder alloy must conform to ASTM B 32 solder metal.

Paste

1. Stability (visual)
2. Metal content
3. Particle size of powder (ASTM B 214, sieve analysis)

SOLDER BALLS

The solder-ball test is carried out by stenciling a dot of solder paste onto a substrate and reflowing at the appropriate temperature. Typically, a dot 0.250 inches in diameter with a thickness of 10 mil is used. A small hand-held stencil is satisfactory for placing the deposits onto the substrate. Figure 10-1 shows a small stencil capable of placing one dot down at a time. The IPC-SP-810 standard provides a schematic of a stencil that allows the placement of three dots at once. This type of paste deposit is required in four different IPC tests including the solder-ball test.

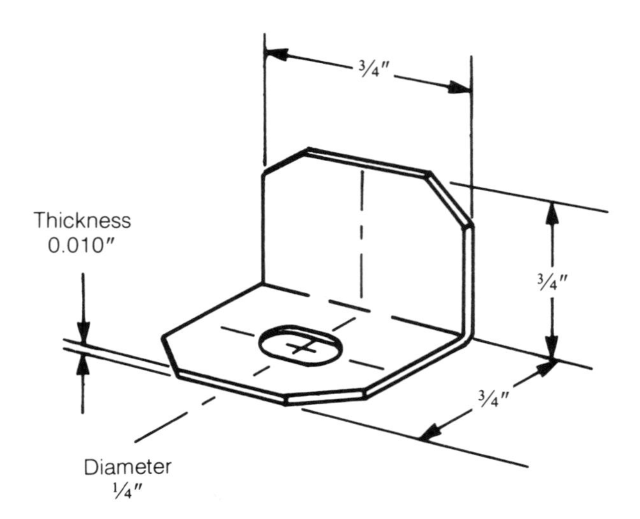

Fig. 10-1. Schematic of a stencil used to place a solder dot on a substrate.

A nonwetting substrate is necessary and a chip of 0.025-inch-thick alumina has been used for many years as the substrate for reflow. An alternate substrate (and less costly) is a glass microscope slide. The slide can be placed in an opaque and/or lighted background for observation. In the laboratory, reflow can be carried out on a solder bath or the surface of a hot plate. The hot plate's surface must be monitored with a surface thermometer so as to obtain the correct reflow temperature of 25 ± 3 degrees C above the liquidus of the solder paste alloy. A Brown reflow oven provides a convenient way to carry out the reflow, if available.

Another means of controlling the reflow temperature is provided by an electronically controlled hot plate such as the Fairweather PHP 66ST system as shown in FIG. 10-2. The surface can be controlled to ±1 degree C from ambient to 427 degrees C. A digital readout indicates the actual surface temperature. Reflow should occur within 20 seconds after coming in contact with the hot plate of solder bath. The substrate is then carefully removed from the surface and allowed to cool.

Upon solidification, a ball of solder surrounded by the nonvolatile flux components remains, as shown in FIG. 10-3A (*see color section*). The surrounding flux medium is observed for the presence of solder balls that have not coalesced into the body of the large ball. Ideally, a perfect solder ball test would show a single ball of solder surrounded by the remaining flux medium completely free of any solder balls. Depending on the solder powder quality, type of activator,

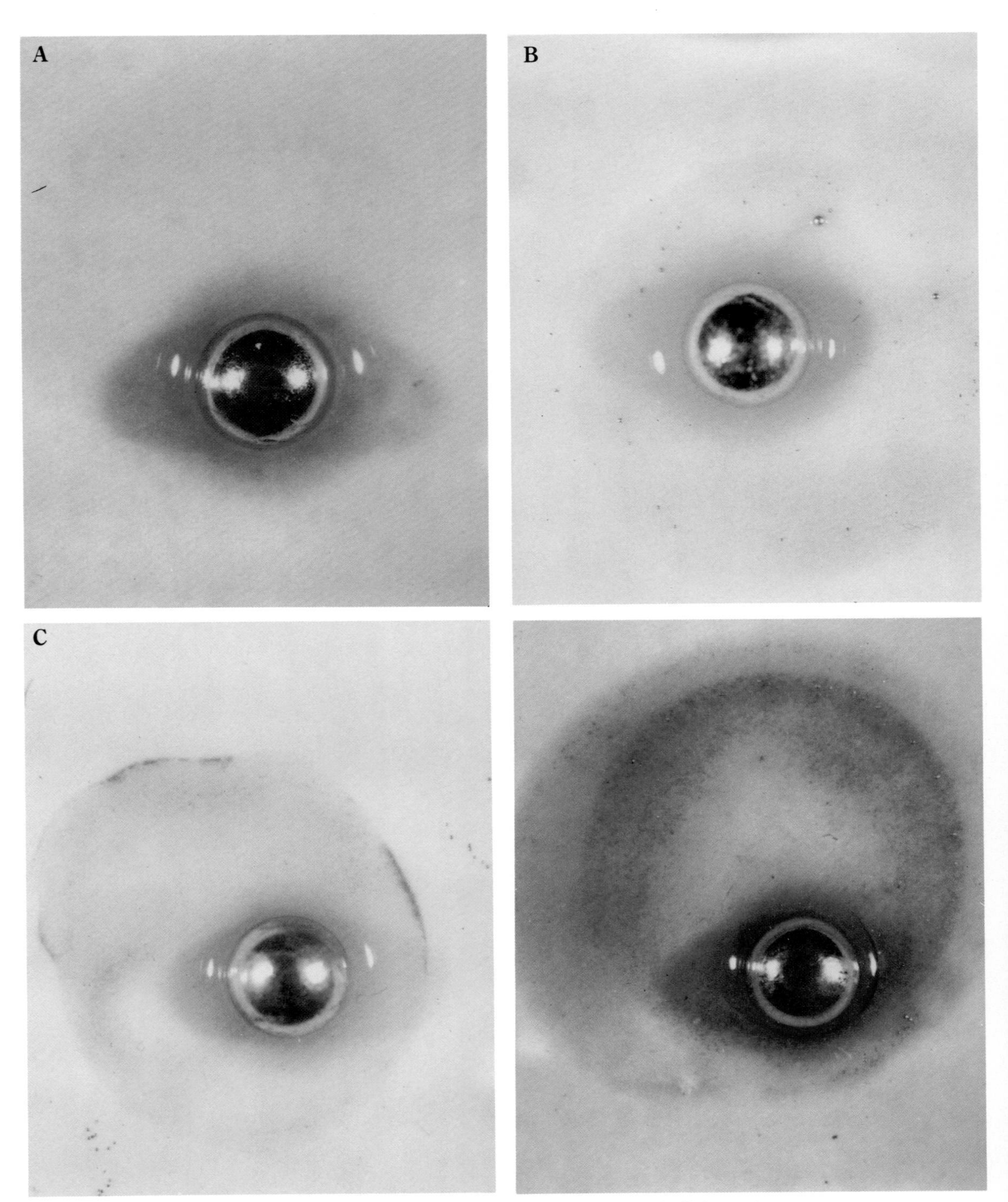

Fig. 10-3. Results of the solder-ball test showing varying degrees of solder ball formation.

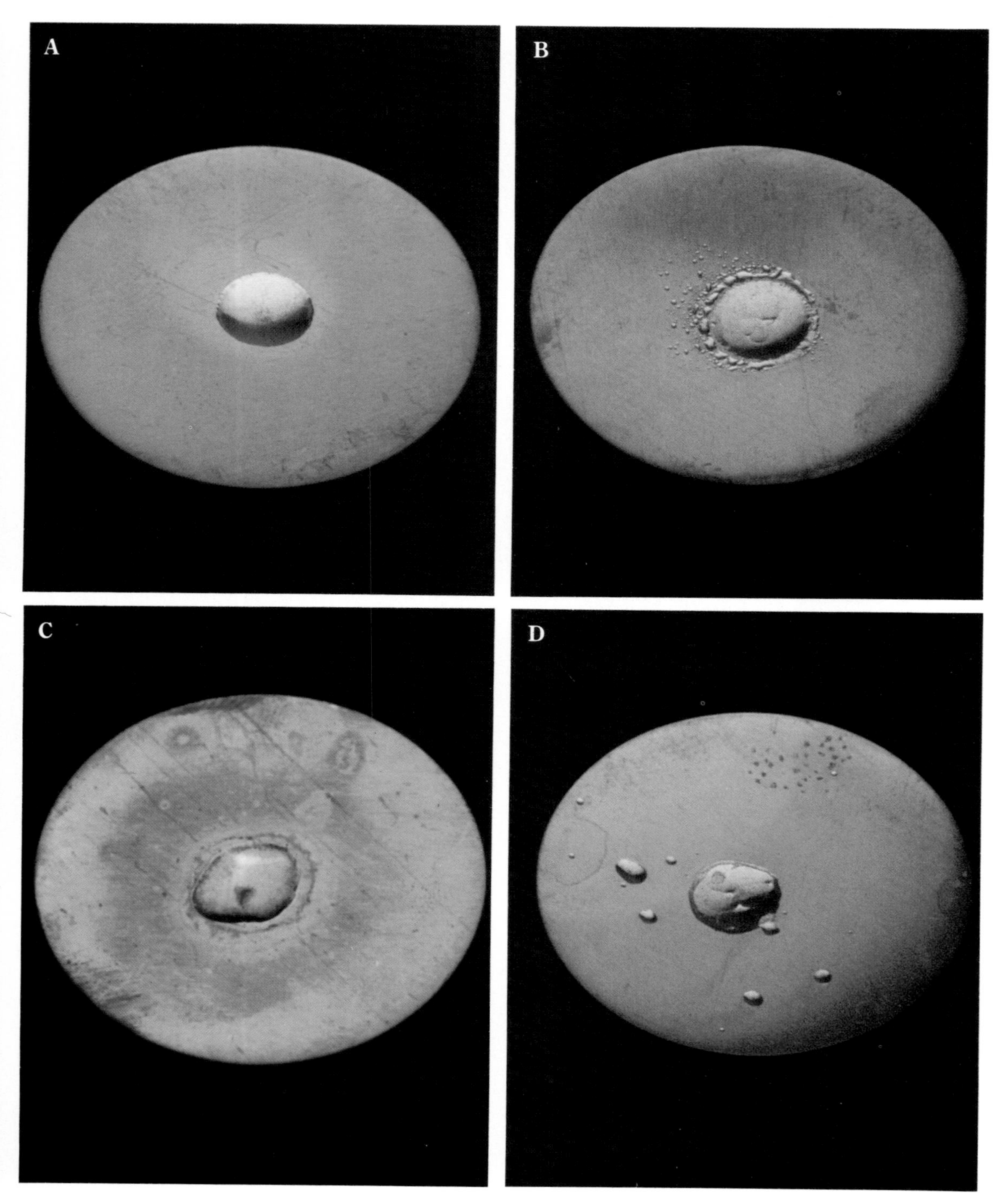

Fig. 10-4. Solderability on copper illustrating different types of wetting behavior.

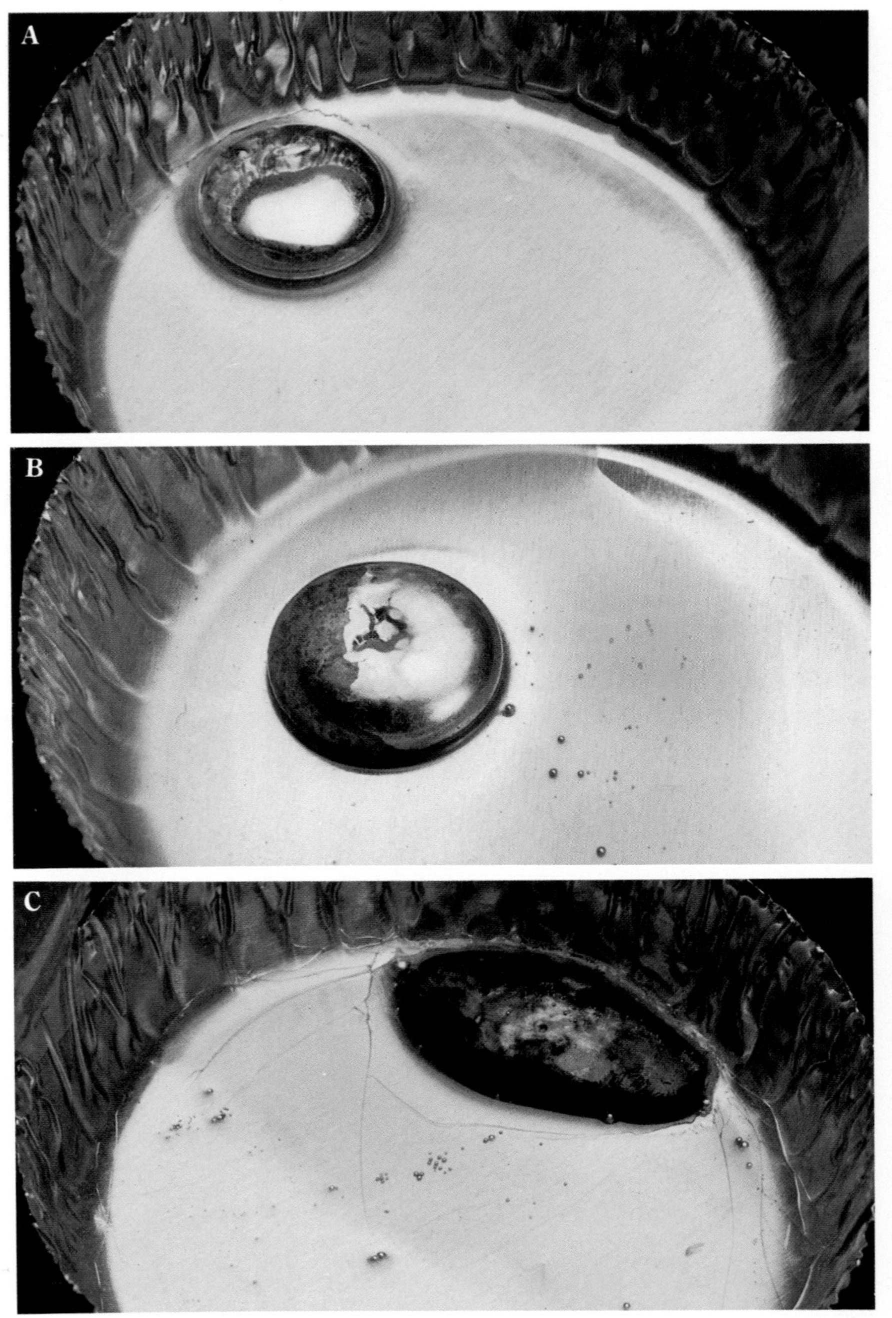

Fig. 10-15. Simple visual test for powder quality. (A) Example of excellent powder quality as exhibited by the total lack of any residues. (B) Some oxidation present as indicated by the dark residue. (C) Solder covered with a black residue due to highly oxidized powder.

Fig. 10-20. Examples of corrosion test failures as followed according to IPC-SF-818, test method 2.6.15.

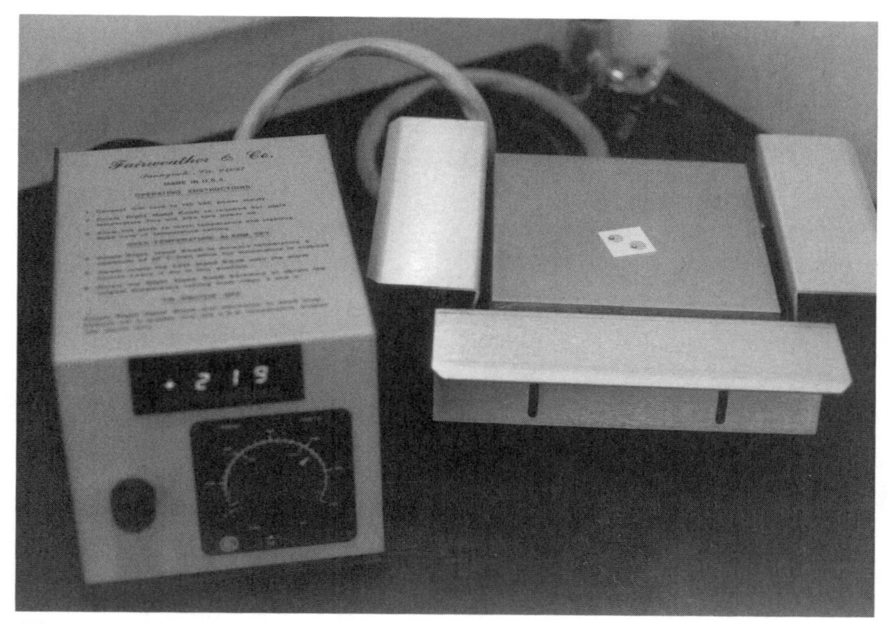

Fig. 10-2. Solder-ball test carried out on a hot plate.

resins in the flux, and viscosity and flow characteristics of the flux at high temperatures, there are various degrees of solder ball formation. A completely solder-ball-free flux residue is unrealistic, so as an aid in establishing pass/fail criteria, the IPC specification provides visual standards.

Figure 10-3 illustrates various degrees of solder ball formation. Figure 10-3A is an example of the ideal case where no solder balls are left in the surrounding flux residue. Figure 10-3B shows some isolated solder balls left in the flux residue and 10-3C shows uncoalesced solder balls on the periphery of the flux residue. This is usually due to the presence of fine particles.

The result obtained in FIG. 10-3B, which shows a relatively small number of isolated solder balls, is considered acceptable. When soldering metallic surfaces, the formation of solder balls usually decreases. However the test can signal a potential problem if an excess of solder balls are present. For example FIG. 10-3C shows an unacceptable amount of solder balls. Figure 10-3D is also an unacceptable result. This is typical of solder pastes that contain fines, especially below 20 microns in diameter. The residue appears as a gray mass consisting of a large mass of solder balls. Results such as these are typical of extremely poor powder quality (oxidized and/or containing a substantial amount of fines).

SOLDERABILITY

A solderability test gives some indication of the ability of the solder paste to wet a metallic surface. Note that the result of a solder ball test is not always sufficient by itself to predict the solderability properties of a paste. In many instances, a solder paste that is marginal or even failing with respect to the solder ball test could prove completely satisfactory for wetting a metallic substrate. A copper coupon (usually a circular disk of copper about 5 to 10 mil thick) or an epoxy/glass laminate (specified in IPC-SP-819) can be used for a solderability test. The most important factor to be controlled is the state of oxidation of the copper surface. This can be controlled by cleaning and then oxidizing under controlled conditions. Suggested methods are given in IPC and QQS-571E.

A dot of solder paste is placed on the copper coupon in a similar manner as described for the solder ball test. Reflow can be carried out on a solder bath or hot plate. After reflow, the coupon is removed and allowed to cool. The residual flux is cleaned off with an appropriate solvent (for common rosin-based fluxes, isopropyl alcohol should be sufficient) and the coupons are observed for wetting characteristics.

Figure 10-4 (*see color section*) shows various states of wetting. In FIG. 10-4A, wetting is uniform because all of the solder alloy has melted and observation of the outer edge of the reflowed solder shows no defects such as dewetting or nonwetting of the copper substrate. Figure 10-4B shows nonwetting while FIG. 10-4C shows dewetting, characterized by a small annular ring of solder that surrounds the main mass of the reflowed solder. Figure 10-4D shows the main mass of solder surrounded by relatively large "spots" of solder. On heating, the solder paste spattered, throwing molten solder to surrounding areas. This is typical of solder paste that has absorbed water. Such spattering might also occur if the heating cycle was too rapid for the particular solvent system. This might indicate that a pre-drying or cure is necessary.

FINENESS OF GRIND

Fineness of grind is a standard method used in the paint industry to assess the degree of pigment dispersion. A fineness of grind gauge provides a direct measure of the largest particles present in a dispersion. For solder paste, the upper limit of the particle size range of the solder powder can be crucial, because for screening or dispensing, the presence of oversize particles could result in screen or dispensing nozzle clogging.

A fineness of grind gauge determines whether oversize particulate matter is a solder powder particle, an agglomeration of particles, or particulate matter from the flux itself. Fineness of grind does not provide any information with regard to the overall particle size distribution in the powder. Fineness of grind is specified as a test method in IPC-SP-819 and refers to ASTM D 1210-79 [1] for the exact procedure. The following European specifications also provide a procedure for carrying out fineness of grind: ISO 1524, DIN 53-203, and DEF 1053-M82.

A fineness of grind gauge consists of a hardened steel block that is ground to provide one or two channels that are tapered to a specific depth. Many scales of measurement have been adopted in the paint industry to determine particle size. The most direct method of measurement, and the one best suited for solder paste, is a scale calibrated in microns. For example, the fineness of grind gauge type CMA 185 or equivalent provides a scale of 0 to 185 microns. To determine a fineness-of-grind reading, place about 5 grams of solder paste at the deep end of the channel, as shown in FIG. 10-5. Hold the scraping blade perpendicular to the grind gauge and draw it down the length of the channel so that the solder paste fills the channel. Typically, there is a series of long, isolated scratches near the top of the gauge and a point with a distinct speckled appearance. You might want to take a reading of one of the upper scratches and a reading at the point of a major break. This provides some information about the oversize particles and roughly indicates the upper cut-off of the particle

Fig. 10-5. The proper position for draw-down of solder paste using a fineness of grind gauge.

size distribution. For example, a reading of the fourth continuous scratch and the point of a major break could be specified as 80/70 for a 200 mesh powder and 55/40 for a 325 powder.

METAL CONTENT

The metal content of a solder paste is usually specified to be plus or minus 1 percent of the nominal value. Place a sample of the paste—10 to 50 grams—in a conveniently sized vessel—one that can withstand the reflow temperature of the solder—and weight it to the nearest 0.01 gram. Heat the paste until it melts and coalesces. After cooling, clean the metal thoroughly with an appropriate solvent and weigh again to the nearest 0.01 gram. The percent metal is simply.

$$\frac{\text{weight(metal)}}{\text{weight(paste)}} \times 100$$

Figure 10-6 shows a sample of solder paste before and after reflow.

VISCOSITY

The standard method of determining the viscosity of solder paste is based on the Brookfield viscometer. A solder paste is usually classified by the manufacturer using a single viscosity reading that is obtained under a given set of conditions. Since a particular viscosity is usually recommended by the manufacturer for a given type of paste (i.e. screening, stenciling, or dispensing), it is important that the viscosity be within the limits recommended by the manufacturer. Questions with regard to the significance and/or meaning of a relative single point measurement are addressed in Chapter 4. The concern here is mainly with the details of carrying out a viscosity measurement.

Fig. 10-6. Reflow of solder paste for determining metal content.

The Brookfield viscometer is specified in the QQ-S-571E and the IPC-SP-819 specification. Some European specifications also recommend that viscosity is measured with a Brookfield viscometer. Although the actual procedure for carrying out a viscosity measurement is straightforward, it probably gives more problems, confusion, and disagreement that the measurement of any other parameter. There are three main factors that can be considered crucial in determining viscosity:

1. Thixotropic nature of solder paste
2. Inherent stability of solder paste
3. Variables that effect the viscosity measurement

As discussed in Chapter 4, solder paste is thixotropic by nature. Its rheological behavior, and in particular its viscosity, is dependent on its prior shear history (the rate and time duration of the shear). Thus, any handling or mixing could cause changes in viscosity and yield point. This is one of the most common causes of discrepancies in viscosity readings taken by different people. Sample pre-treatment must be specified as part of the viscosity procedure.

Other factors can contribute to changes in viscosity. Depending on the particular paste formulation, viscosity could increase or decrease. This can be accounted for by various factors such as storage time and temperature. Factors such as evaporation of solvents, crystallization or precipitation of flux components, rebuilding of thixotropic agents, and reaction products of activators and powder can all change the original nominal viscosity. In some cases, excessively high temperatures have the effect of decreasing viscosity due to breakdown of rheological agents. In such cases, check a sample of the manufacturer to confirm that a change has taken place. If not, check to see that all parties involved are using the same procedure and conditions of measurement.

To measure solder paste viscosity, a Brookfield RVTD viscometer with a Helipath stand is recommended. T-bar spindles are used in conjunction with the Helipath stand. If you are following the IPC method, a chart recorder is also necessary. A sample of solder paste equivalent to a volume of 5 cm minimum diameter and 5 cm minimum depth is gently mixed (or treated as recommended by the manufacturer) to produce a homogeneous sample. The sample must be brought to a constant temperature of 35 ± 0.25 C. IPC-SP-819 calls for a minimum of two hours stabilization time; however, depending on the particular nature of the solder paste, the stabilization time could vary according to the manufacturer's recommendations. For low-viscosity solder pastes (less than 300,000 cps), the viscosity measurement is taken within 20 minutes after mixing so that powder settling is minimized. The particular T-bar required depends on the viscosity range. For viscosities between 300,000 and 1,400,000 cps a T-F spindle is used. For viscosities between 50,000 and 300,000 cps a T-C spindle is used. In all cases, a rotational speed of 5 rpm is used.

Place the rotating and descending T-bar in the solder paste. At this point, you must be careful because there is some arbitrariness as to when to take the viscosity reading. If the IPC method is not followed, use the procedure recommended by the paste manufacturer. If the IPC method is followed, set the Helipath stand so that it cycles up and down for 10 minutes (5 cycles). See IPC-SP-819 for exact settings of the Helipath stand. In this case, a chart recorder is necessary to record the cycling. If you are taking a single reading, record the viscometer digital readout and multiply by the appropriate factor.

For example, a reading of 42 (assuming a rotational speed 5 rpm and a T-F spindle) yields a viscosity of 42 × 2,000,000 × 0.01 = 840,000 cps. With the IPC method, the output of data is as shown in FIG. 10-7. Viscosity is calculated as the average of the peak and valley of the last two cycles:

$$\text{viscosity} = \frac{(A + B + C + D) \times K}{4}$$

where K is the combined spindle/speed factor for the particular T-bar used. In the example just cited, K = 2,000,000 × 0.01 = 20,000. If the average of the first two cycles is more than 10 percent higher than the last two cycles, the test is invalid and additional requilibrium time is required.

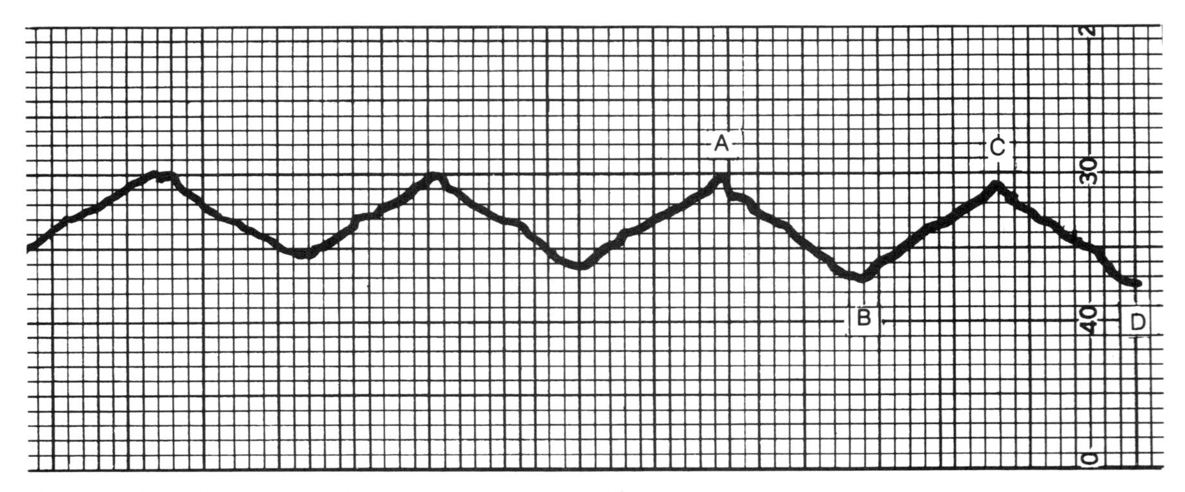

Fig. 10-7. Recorder output using the IPC method to determine viscosity.

Common Errors

To avoid some common errors that can affect the outcome of the viscosity reading, keep the following in mind.

- Make sure that the viscometer is accurate by periodic calibration with viscosity standards. Standards are available from the viscometer manufacturer.
- The viscometer should be level. Check and adjust by using the built-in level on the viscometer.
- Make sure the T-bar spindle is placed securely in the chuck. Otherwise, slipping will give erroneous results.
- Be sure that the solder paste has a uniform temperature and is at 25 ± 0.25 degrees C (77 ± 1 degree F).
- Use an accurate thermometer. An ASTM #17F thermometer is especially suited for the temperature range required in the above procedure.

Viscometer Accuracy

The accuracy of the viscometer is 1 percent of full scale. In the case of a TF spindle at 5 rpm, this means ± 20,000 cps. The actual percent error increases as lower scale readings are taken. For example:

2,000,000 cps ± 20,000 1% error
200,000 cps ± 20,000 10% error
100,000 cps ± 20,000 20% error

Conducting viscosity measurements is an active area of study by both the manufacturers and users of solder paste. Other types of viscometers and measurement at different shear rates are being evaluated for feasibility. Note that the IPC specification realizes this situation because it also states that other types of viscometers may be used if the results can be correlated and mutually agreed upon by the user and supplier.

TACK RETENTION

After a solder paste has been deposited on a substrate, components must then be placed into the solder paste that will eventually undergo reflow. The parts can be placed into the paste immediately after deposition or at some later time. Solder paste must have the property of being able to hold the parts in place until the time of reflow. Therefore, it must have some adhesive property, usually referred to as tack. In many cases, parts might not be placed onto the substrate until 4 to 8 hours after the paste has been applied. Thus, it is important to have a test method to determine how long the paste will remain tacky.

Note that tack is not a fundamental material constant of a substance. Tackiness depends on the exact experimental conditions including instrumental design, dimension, and rate. Thus, an exact definition of tack depends on the application you are interested in. Tack related to a printing ink might not be the same as for a solder paste. Tack is a measure of the stickiness or adhesive qualities of a substance and can be quantified under a given set of specific conditions. Furthermore, any method that quantifies tack must ultimately be related to the performance of that substance for some particular given process. In the case of solder paste, the tack force is ultimately related to the paste's ability to hold a particular device. Thus, tackiness of a solder paste is often dependent on the design and size of the particular electronic component and on the method and force used to place the part.

The importance of tack an as industrial parameter is by no means of recent concern. For example, Green undertook a detailed study of tack as related to printing inks.[2,3] Green describes the Tackmeter he that developed to study tack parameters of printing inks. Essentially, it is a mechanical finger that measures the force required to separate a plate from a given thickness of ink. Green proceeded with a detailed analysis and defined tack with equations relating it to viscosity. He found that tack or pull resistance is a function of the film thickness of the material and the area and speed of finger tip. He developed an equation that allows comparison of the tack of a substance to a standard. Voet gave arguments to suggest that viscosity is not the fundamental tack-determining property of highly viscous liquids, and that viscoelastic rupture can come into play.[4]

IPC Tack Test

The idea of tack as being a pulling force has been developed into a method for solder paste tack evaluation as described in the IPC paste specification IPC-SP-819. In this method, a probe of specified size is placed into a dot of solder paste with a specified loading force. The probe automatically retracts and the unit records the maximum pull force.

J. Chatillion and Son make the Digital Gram Gauge (DFG-2) and motorized stand LTCM-3) shown in FIG. 10-8. The test method consists of stenciling a dot of solder paste 0.245 mm (0.010 inch) thick and 6.35 mm (0.25 inch) in diameter on a glass microscope slide 76 mm by 25 mm (3 by 1 inch). Enough dots should be prepared so that there are six replicates per tack force measurement. Prepared samples should be stored at 25 ± 2 degrees C and 40 ± 10 percent relative humidity until tested. Storage in an enclosed container is not

Fig. 10-8. J. Chatillon and Son's Digital Gram Gauge (DFG-2) and motorized stand (model LTCM-3) as used in the IPC-SP-819 tack test.

recommended because open storage better simulates the actual conditions the paste will be in. A stainless steel test probe with a nominal diameter of 5.08 ± 0.127 mm (0.200 ± 0.005 inch) is attached to the force gauge. This is a special probe size and is available from Chatillion and Son.

Place the solder paste specimen slide under the test probe. The probe lowers into the solder paste at a rate of 2.54 ± 0.51 mm per minute (0.1 ± 0.02 inches per minute) and the test probe stops at a force reading of 300 ± 30 grams. Within 5 seconds the probe withdraws from the paste at a rate of 2.54 ± 0.51 mm per minute, and the unit records the maximum force to break contact. The loading of 300 grams ± 30 grams is not arbitrary—it is based on the work of Morris, who studied the typical placement forces using actual placement equipment. Measurements of the tack force are taken immediately after printing the paste and every 20 minutes for the first hour until a peak force is reached. Readings are taken every 3 hours until a decline is observed from which point tack force measurements are taken every 30 minutes. Tack force can be plotted as a function of time. The specification also suggests data be reported as

1. Time to reach 80 percent of the peak value
2. The peak tack force in grams with the expected variation
3. Time over which the peak value is maintained or for the tack force to decline to 80 percent of its peak value

At present, the IPC specification does not specify actual values of tack force as described above as being pass or fail. For a given solder paste, you must correlate the tack measurements with the assembly process.

A comparison of the tack properties of solder paste of various manufacturers was also undertaken by Morris.[5] Using the method of measurement previously described, eight different solder pastes from five different manufacturers were evaluated. A plot of the tack force versus the time is shown in FIG. 10-9 for four

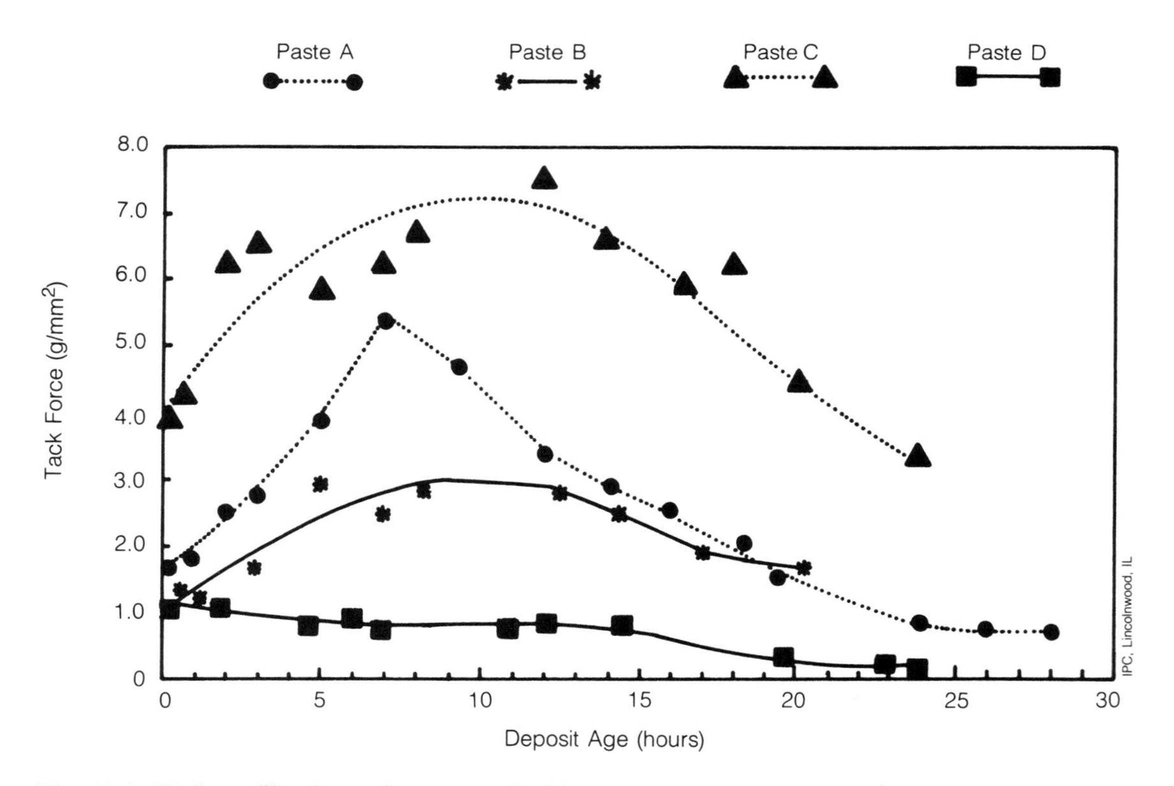

Fig. 10-9. Tack profiles for various types of solder paste.

of the samples. The tack profiles show a general trend where there is some initial tack force that increases over time until some maximum value is obtained after which the tack force decreases. The general shape of the curves reflects tack force as a function of viscosity.

Evaporation of solvent results in the paste having a higher viscosity. As evaporation continues, the paste becomes more of a solid than a liquid and its adhesive properties are lost, hence the tack force decreases. As this drying continues, the probe does not penetrate the surface and the paste is no longer suitable for part placement. Based on this, it was recommended for most automated lines that deposits should not be used after times that correspond to 80 percent of the minimum tack force. Apparently, this procedure is sensitive enough to distinguish different types of paste. The question to be answered next is: if definite values can be adopted as pass/fail criteria, which will relate directly to a manufacturing process?

Tacky Tester

A commercial instrument has been developed by Austin American Technology called the Tacky Tester (shown in FIG. 10-10). The principle of determining tack with this instrument is entirely different from the IPC method. The Tacky Tester simulates the shear forces that could be experienced by the paste due to the electronic component if the substrate were to undergo motion. The solder paste is stenciled onto a steel substrate. A stencil is attached to the top of the instrument and a squeegee is also supplied. The stencil supplied with the instrument is a typical footprint pattern (a concentric pattern of a PCC 68 and PCC 49). The stencil thickness and pattern can be customized. The steel substrate is placed in a grooved holder. A ceramic substrate is placed over the steel substrate but does not touch the solder paste. When activated, the steel plate moves up through an arc-like path. The ceramic then contacts the paste due to its own weight. Depending on the properties of the solder paste, the ceramic substrate might or might not slide off. If it does, the beam of a photoelectric cell is broken and the steel plate stops moving. You then measure the angle at which the steel plate stopped. The lower the angle, the less the solder paste can hold the component. A plot of angle versus time is shown in FIG. 10-11. Obtain a tack profile for a paste that has been judged as satisfactory in actual production use. This is then used as a standard to compare other batches or different paste samples. It is also possible to subject the solder paste to a heating cycle.

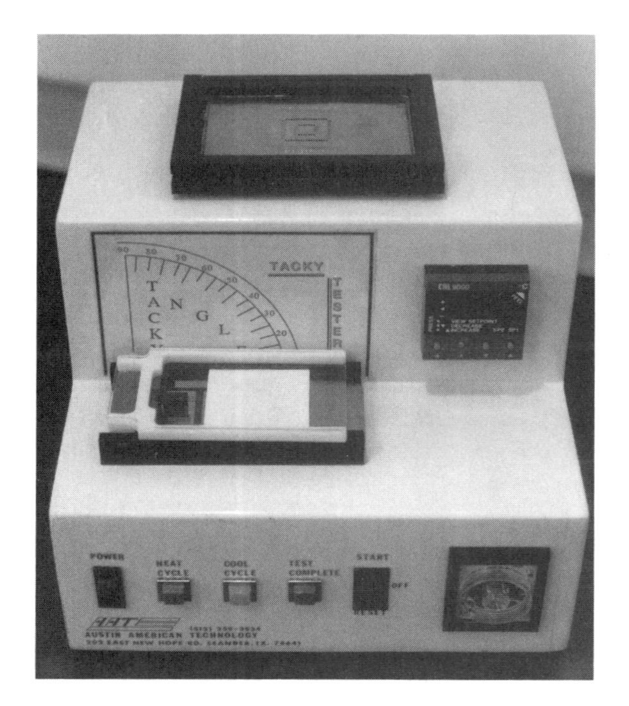

Fig. 10-10. Tacky Tester, manufactured by Austin American Technology, Austin, Texas.

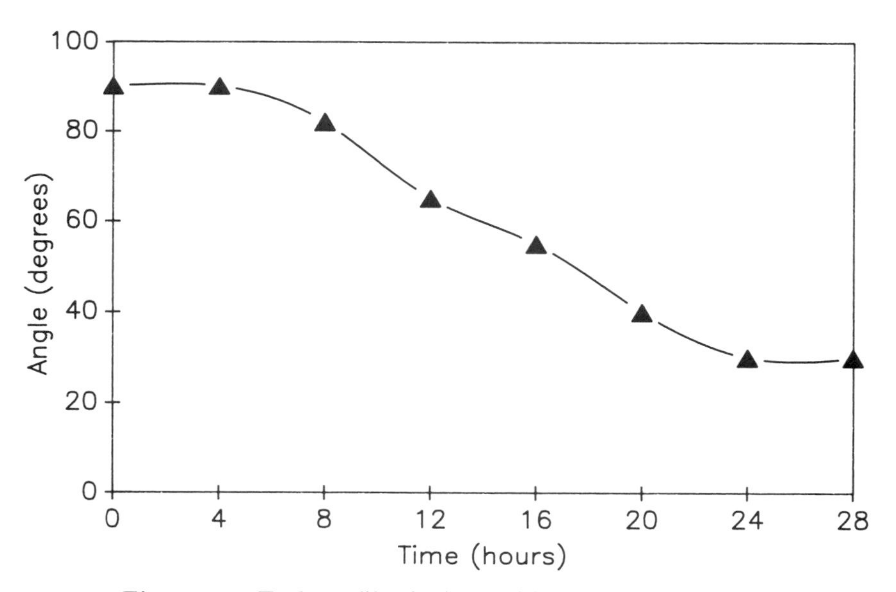

Fig. 10-11. Tack profile obtained with the Tacky Tester.

Solder Paste Quality Assurance Analyzer

Solder paste tack can also be measured with the Solder Paste Quality Assurance Analyzer, as shown in FIG. 10-12. This instrument is based on measuring tack as a pull or tensile force and is similar to the procedure used in the IPC method. The Solder Paste Quality Assurance Analyzer can be interfaced with a chart recorder or a microcomputer. The software package provides statistical process-control charts. Data can be displayed on the screen and stored if desired.

To carry out a measurement, a probe 5 mm in diameter lowers into a solder paste deposit at a rate of 0.2 mm per second. The total preload weight is 250 grams. Since the weight is provided by the loading mechanism, the probe applies the same force every time a measurement is made. After a 5-second rest period, the probe is automatically removed at a rate of 0.02 mm per second. The tensile force is measured as the probe is withdrawn as a function of the distance from its rest position. You can obtain a plot of the tensile force versus displacement. Thus, in doing a ''single'' tack measurement, you can get a characteristic profile of the paste's behavior for the complete withdrawal process. The area under the force- versus-displacement curve represents the tensile energy required to remove the probe.

Fig. 10-12. Solder Paste Quality Assurance Analyzer.

Figure 10-13 shows the force-versus-displacement curves for two different solder pastes. Similar measurements can be made for solder paste deposits that have been stored for various time periods so that tack behavior can be monitored over time. The maximum tack forces can also be plotted over time to obtain a tack profile similar to that plotted in FIG. 10-9.

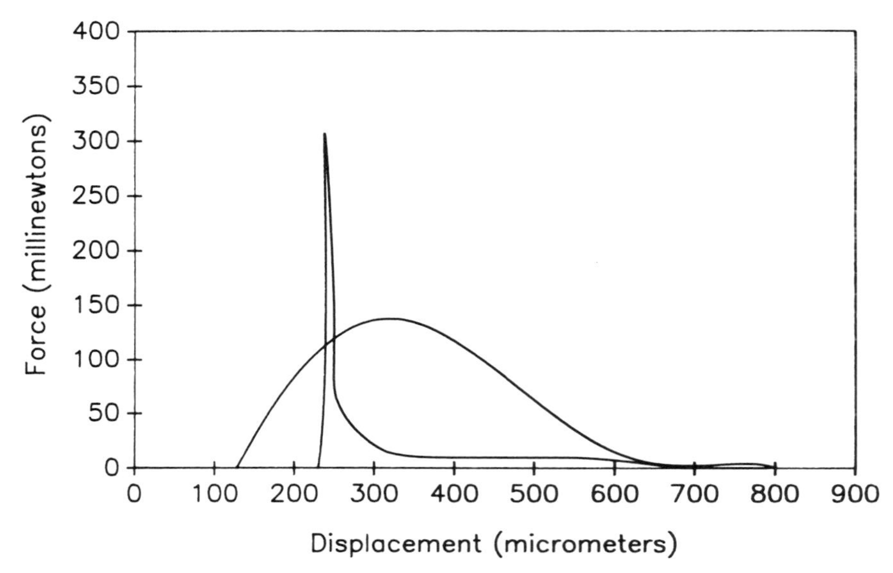

Fig. 10-13. Examples of force versus displacement curves for two different solder pastes using the Solder Paste Quality Assurance Analyzer.

Factors that Affect Tack

The tack life of a solder paste is a function of its formulation. Obviously the choice of solvent(s) plays a role since evaporation ultimately causes a dry surface. Activators would not be of major concern unless they are so reactive with moisture and/or the atmosphere that they effect tack by chemical reaction. Additives known as *tackifiers* can be used to prolong tack life. Finally, particle size and shape could be factors due to volume and surface area effects.

MacLeod and Hoover did a study to determine the effect of solvent, activator, tackifier, and viscosity on tack and printability.[6] Analysis of the data showed that viscosity had a significant effect on tack and surprisingly the tackifier additive did not. Also there was no significant correlation between tack and printability. Other studies have shown that a tackifier can extend the tack life of a paste.[7] These studies also indicate the expected decrease in tack life due to increased metal loading as demonstrated by the decrease in probe penetration for high metal (91 percent by weight) loadings.

SLUMP

The term *slump*, as applied to solder paste, refers to the spread or increase in substrate area covered by the paste due to a slow flowing motion under the action of gravity. To be precise, specification IPC-819 defines slump as "the distance the solder metal in solder paste spreads after printing during the drying and before the reflow process." The slump characteristics of a paste are especially important in dispensing, screening, and stenciling, where slump should be minimized. If after deposition, curing, or reflow that slump is great enough to cause contact between two paste deposits, unequal amounts of solder could result at different pad areas. Excessive slump can also cause the formation of solder balls after reflow. If the solder paste covers an area sufficiently greater than the nominal pad size, the reflowing paste might not be able to completely contract back to the pad area.

IPC-819 describes the measurement of slump using a test specimen consisting of three solder paste dots measuring 0.65 cm in diameter and 0.25 mm thick. One set of dots are stored for 50 to 70 minutes at 25 ± 5 degrees C and 50 ± 10 percent relative humidity. Another set is subjected to a temperature of 100 ± 5 degrees C for 10 ± 2 minutes. The samples must be placed in the oven within 5 to 10 minutes after printing. The spread or slump should not increase by more than 10 percent of the printed diameter.

With reference to screening and stenciling, slump or in particular the "instantaneous slump" has an effect on the overall print quality. In trying to assess print quality, Kevra and MacKay used a test patter for both screens and stencils.[8] The pattern consists of a elongated square detail aligned with and against the squeegee direction in two columns labeled A or B, and there are 10 sizes of detail in each column. The pattern also carries an array of larger squares ranging in size from 0.050 by 0.050 inch to 0.150 by 0.150 inch. Two footprint patterns are also included so that a performance comparison with a readily recognizable standard feature could be directly assessed. The "fine" footprint is for a 64 connection device with 0.015-by-0.100-inch pads and 0.015-inch gaps. The large pattern is for a 32 connection device with 0.030-by-0.100 pads and 0.030-inch gap. Finally, there are four dot arrays to indicate performance for leadless chip mounting. The array of six large dots is 0.030-by-0.060 inch with a 0.40-inch gap, and the array of 10 elongated pads is 0.015-by-0.030 inch with 0.020-inch gap. A square array of 0.015-inch pads with 0.032-inch gap is set above these while an "X" array of 0.007-by-0.007-inch pads with 0.016-inch gap is below them.

Assessment of screen printing quality is visual, based on criteria such as completed detail, no separation, no cross formation, no breaks in detail, and no lateral film separation, etc.

The larger land areas were examined for continuity of film and graded good, satisfactory, or bad. Both size of footprint were graded similarly. Patterns of dots were assessed collectively and graded L (large) if only the two central large arrays were fully printed, M (medium) if the top array also was fully printed, and S (small) if the 7-mil-square array was printed. To facilitate evaluation of the results, figures of merit were assigned to various patterns so that an overall or total figure of merit could be obtained. Use the above scheme to compare parameters of interest such as paste formulation, screen-versus-stencil thickness, etc.

To facilitate evaluating slump and print quality simply and quickly, Kevra and Mohoric have used a pattern as shown in FIG. 10-14.[9] The pattern consists of two pad sizes, typical of what might actually be found in surface mount

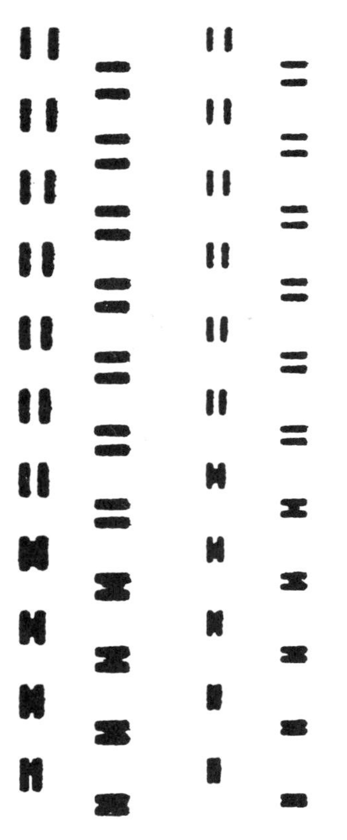

Fig. 10-14. Example of a solder paste exhibiting excessive slump.

assemblies. The nominal pad sizes chosen were 0.024-by-0.090 inch with 0.050-inch pitch and 0.016 by 0.070 inch with 0.032-inch pitch. For each pad size, a gradient was produced where the nominal pitch was increased and decreased by a constant factor. The following two sequences give the distances, in mils, between the pads.

Large pads	46	42	39	34	30	26	22	18	14	10
Small pads	31	28	25	22	19	16	13	10	7	4

Both pad sizes were aligned with and against the squeegee direction. A given solder paste printed with some given printing process allows immediate determination of slump. By visual inspection, record where the first pass occurs, i.e. the pad that shows no connection or running of the adjacent pads. Figure 10-14 shows an example of a paste graded as 5,5,6,6, (where the smallest spaced pads are designated as one.) If desired, calculate a slump factor as SF = d/2.

The previously described stencil provides a quick assessment of initial print quality by providing the slump factor, which represents the amount of slump for the large or small pads. The number of experiments executable is quite numerous and can lead to a 5- or 6-factor analysis of variance quite rapidly. However, emphasis is usually on a few specific variables to simplify data analysis. The variables are:

Solder paste formulation
Initial slump or print quality
Slump over time at ambient temperature
Slump over time at curing temperature
Stencil thickness
Direction of squeegee
Pad size
Printing conditions
(squeegee hardness, speed, and pressure)
Paste life
(how long after being on the screen printer failure occurs)

OXIDE CONTENT

The determination of the oxide content of a solder powder is of importance in that the powder quality relates to the ultimate quality of the solder paste. Various analysis methods exist such as SEM (scanning electron microscope) or chemical reduction. Little has been published in this area, especially with regard to the critical or allowable levels of oxide for given alloys.

A simple laboratory test can assess the relative state of oxidation of a solder powder by purely visual means. By following procedure to determine metal content (in the section under that name earlier in this chapter), you can observe the residue that surrounds the large mass of reflowed solder. A solder paste that contains a solder powder with a low level of oxidation will be surrounded by a clear flux residue, and the surface of the mass of reflowed solder will be free of any discoloration. If significant amounts of oxides are present, they will be visible in the surrounding flux, and after reflow, the surface of the reflowed powder might be partially covered with a dark film, indicating an oxidized powder. Figure 10-15 (*see color section*) shows various states of oxidized solder powders.

MELTING POINT

Cooling Curve

Solidus and liquidus temperatures can be determined from an analysis of a cooling curve. A cooling curve is a temperature time curve of the alloy during the process of solidification. Solder paste is placed in an inert crucible (fireclay or graphite) and then melted in an electric furnace. A thermocouple is inserted in the melt and the furnace contents are cooled. Temperature and time readings are taken at intervals of about every 30 seconds until solidification is complete. A recorder automatically obtains the cooling curve.

Figure 10-16 shows a typical cooling curve of a simple binary alloy. Point A represents the molten alloy. As the liquid starts to cool, the temperature

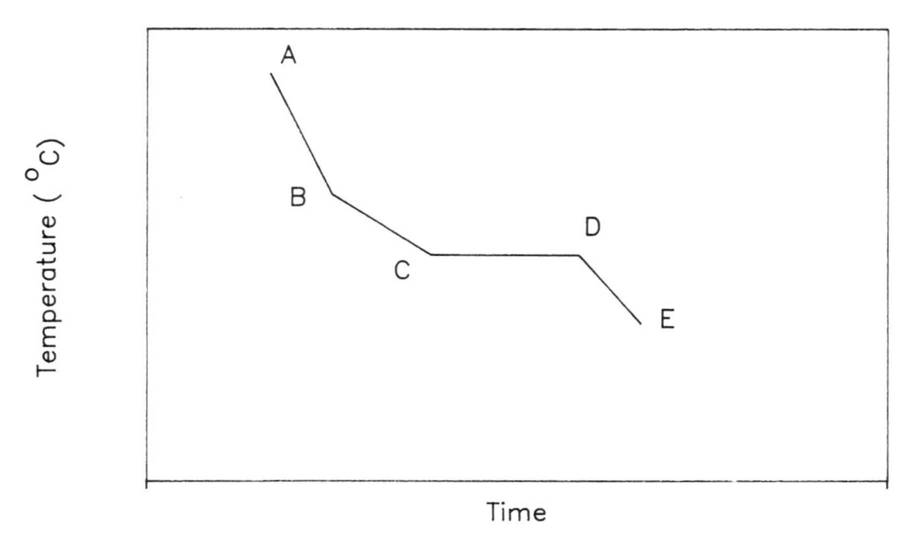

Fig. 10-16. Cooling curve for a 60 tin/40 lead solder alloy.

decreases until point B is reached. At B, solid material starts to form. As the composition of the liquid phase continues to change, the temperature decreases from B to C. At C, (the eutectic composition) the temperature remains constant. The segment DE represents any further cooling of the solid.

Thermal Analysis

Thermal analysis, in particular DTA (differential thermal analysis) and DSC (differential scanning calorimetry), can also be used to determine the thermal properties of a substance.[10, 11] The heat liberated or absorbed due to melting or to a chemical reaction or a transition in crystalline structure is measured as a function of temperature. DTA is a technique in which the temperature difference between a substance and a reference material is recorded as a function of time or temperature as the substances are heated or cooled at a controlled rate. A DTA curve consists of peaks indicating an exothermic or endothermic transition. The area under the curve is a measure of the heat evolved during the transition.

DSC differs from DTA in that a recording is made of the energy necessary to maintain a substance and a reference material at the same temperature. This rate of energy or power is plotted over time or temperature during which the sample and reference have been subjected to the same heating or cooling cycle. For most applications, both DTA and DSC are equally applicable. Care must be taken in the interpretation and comparison of thermograms, because a change in operating parameters such as heating rate, sample size, and atmosphere can affect the appearance and numerical values of the parameter being measured.

As an example of the application of thermal analysis to solder alloys, Kuck has constructed the tin-lead phase diagram using DTA and has also presented a method for detecting the composition of tin-lead alloys.[12] Figure 10-17 shows four distinct types of DTA curves that can be obtained for various compositions of tin and lead. In determining the composition, a knowledge of the liquidus temperature does not suffice. Since tin and lead have a eutectic composition, the liquidus temperature for alloy compositions in the range 185 to 232 degrees C is not unique. This can be seen from the phase diagram in FIG. 1-5. An alloy with a liquidus temperature between 185 and 232 degrees C could either be a *hyper*eutectic alloy (higher in tin than the eutectic) or a *hyper*eutectic composition (higher in lead than the eutectic). Also, the ratio of the heat absorbed from the melting of the secondary component to that of the eutectic was determined for known ratios of tin-lead alloy. Since this ratio is different for hypoeutectic and hypereutectic alloys, this provides a means of choosing the correct alloy composition for a given liquidus temperature.

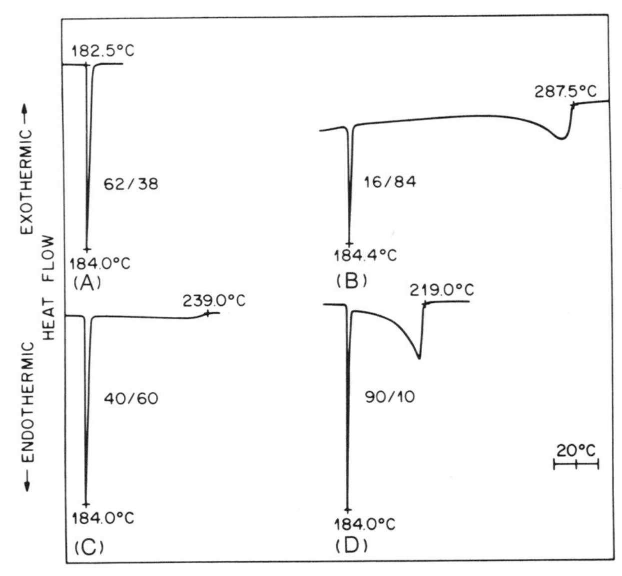

Fig. 10-17. Representative DTA races of (A) eutectic, (B) α-solid solution alloy, (C) hypoeutectic alloy, (D) hypereutectic alloy. (Elsevier Science Publishers, Physical Sciences and Engineering Division)

DETECTION OF HALIDES

As discussed in Chapter 3, halide-containing fluxes (in particular, free or ionizable halides) are quite effective as activating agents and are probably used more than any other type of chemical species. The presence of halides in a flux is of concern because of their potential corrosiveness. Thus, fluxes intended for high-reliability electronic applications usually have restrictions as to the amount of halide allowed. The specifications QS-571-E and IPC-SF-818 specify methods and criteria for classification of flux type with respect to halide content.

A simple test originally introduced in the American military specifications many years ago is based on a spot test using silver chromate reagent paper. This provides a qualitative or semi-quantitative determination of halide. One of the conditions of the QQS-571-E specification is that an RMA flux must show no evidence of halide using this test. Silver chromate paper is brick red in color. The ionic halides—iodide, bromide, and chloride—react with the silver chromate to form insoluble silver halides that are white to off-white in color, so the presence of these halides is indicated by a bleaching or whitening of the paper. The degree of discoloration is proportional to the amount of halide present. Some examples are shown in FIG. 10-18.

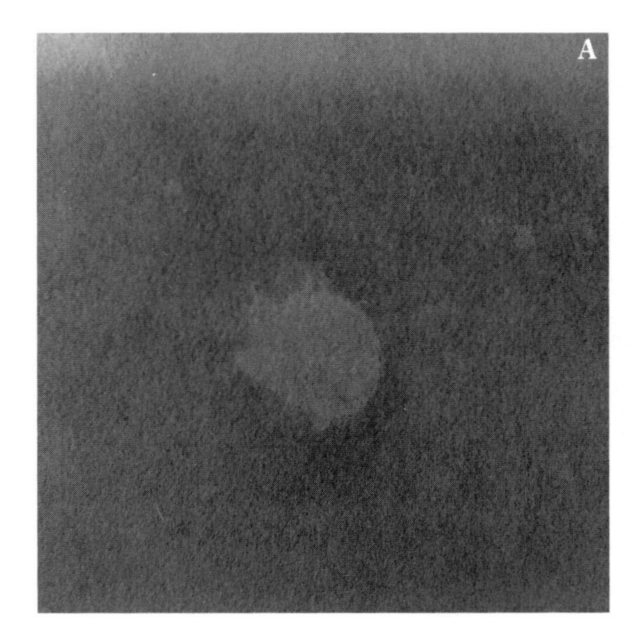

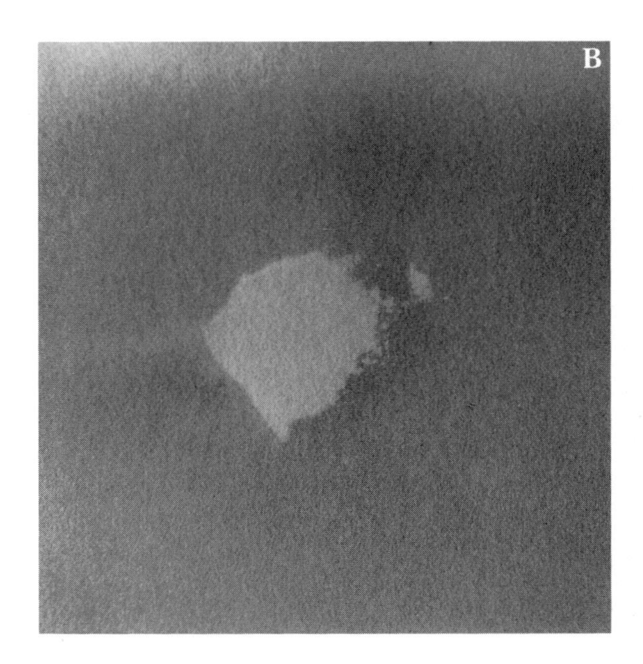

Fig. 10-18. Determination of the presence of halides using silver chromate paper. (A) Shows a slight discoloration, which indicates the presence of halide. (B) Indicates a relatively high concentration of halide as indicated by the distinct white discoloration.

When testing flux, especially RMA types, the amount of actual free halide is usually only a small percentage of the flux and the discoloration might be faint, especially for bromide-containing fluxes. The QQS-571 and IPC specification provide photos to use as pass/fail criteria. The fluoride ion does not cause discoloration of the chromate paper and is not detected.

In order to carry out the above spot test for a solder paste, the flux must be extracted from the paste first. An extraction is carried out by mixing and heating the solder paste with isopropyl alcohol (for rosin-based fluxes). The flux dissolves in the IPA and the solder powder can be separated from the solution by filtering the hot solution. The IPA is adjusted as needed so that the resulting solids content is approximately 35 percent by weight. Details of this procedure are given in QQS-571, paragraph 4.7.3.1.2. A drop of the solution is placed on the test paper for 15 seconds and then washed with IPA, dried, and observed for color change. A test is positive if there is any observable bleaching or whitening of the paper. Be careful to ensure that all flux residues are washed from the paper so as not to confuse a true bleaching with unwashed solids.

Flux classified as L3 in IPC-SF-818 must pass the silver chromate test. If the chromate test fails, a tritation (according to 2.3.3.5 of IPC-TM-650) determines the halide content and thus can further classify the flux as either L1, L2, M, or H. In the IPC specification, two methods are given for carrying out the silver chromate test. In the preferred method, a sample of actual solder paste flux from the manufacturer is tested by directly placing the paste flux on moistened chromate paper. If the flux is not available, an extraction procedure must be conducted so that the flux can be separated.

The above tests for halides are all done on the paste flux before reflow. Since it is only the ionic or ionizable halide ions that are detected by the above methods, a halide-containing activator might not liberate its halide in ionic form until reflow temperatures are reached. IPC-SF-818 allows for these tests to be carried out on the reflowed flux residues and is at the discretion of the user.

COPPER MIRROR

The copper mirror test, like the silver chromate test, is a method that has been used for many years and originated in the Bell/AT&T 1700 specification. The copper mirror test is a test for "corrosivity" and is one of the many criteria used to classify flux activity.

A copper mirror is a glass substrate with a vacuum-deposited film of pure copper. According to QQ-S-571, the thickness of the film should be such that it permits 10 ± 5 percent transmission of normal incident light with a wavelength of 5,000 angstroms. Details for the preparation of a copper mirror are given in the last revision of the QQ-S specification, i.e. QQ-S-571D paragraph 4.7.7.3. However, copper mirrors that meet the requirement of the military and IPC standards are available commercially.

In applying the test to solder paste, the flux must be separated from the metal powder (QQ-S-571E paragraph 4.7.3.1.2) to provide a solution of approximately 35 percent by weight of flux solids. Before placing the test solution on the mirror, inspect the surface for discoloration (oxidation). In the event of any oxidation, a rinse with a 50 gram per liter solution of EDTA (ethylenediamine-tetra-acetic acid) to remove surface oxides. Copper mirrors should be stored in a non-oxidizing environment such as a container that has been flushed with nitrogen.

A drop of the test solution is placed on the copper mirror. Care should be taken to ensure that the test solution is placed on the "copper" side of the copper mirror. Determine which side is which by placing a pencil point on the surface of the glass. If the reflection of the pencil point appears to touch the actual point of the pencil, that is the "copper" side (if the pencil point is placed on the reverse side, the points appear to be separated).

A drop of control flux (35 percent by weight of water white gum rosin in isopropyl alcohol) then goes alongside the test solution. If the control flux results in a failure, the test must be redone with a new copper mirror. According to IPC-S-818, the preferred method for solder paste is to place the actual solder paste flux on the copper mirror. If the paste flux is not available, then you might need to do an extraction.

The copper mirrors are then stored at 23 ± 2 degrees centigrade in 50 ± 5 percent relative humidity for 24 ± 1/2 hours. If a constant temperature/humidity environment is not available, a laboratory set-up can be used with aqueous solutions of certain salts, as given in ASTM E104.

At the end of the 24-hour storage period, the mirrors are removed from the humidity chamber and rinsed with isopropyl alcohol to remove all traces of the flux. The mirrors are then observed visually to determine if any copper has been removed. Placing the copper mirror on a piece of white paper may help in determining if any copper has been removed. A flux is said to have failed the copper mirror test if there is any complete removal of the copper film, indicated by visibility of the original glass surface. Figure 10-19 shows examples of partial removal (pass) and complete removal (fail). Interpretation of the results can be difficult and different observers come to different conclusions. As stated in QQ-S-571E, any superficial reaction or any partial reduction of the thickness of the copper film is not a cause for failure.

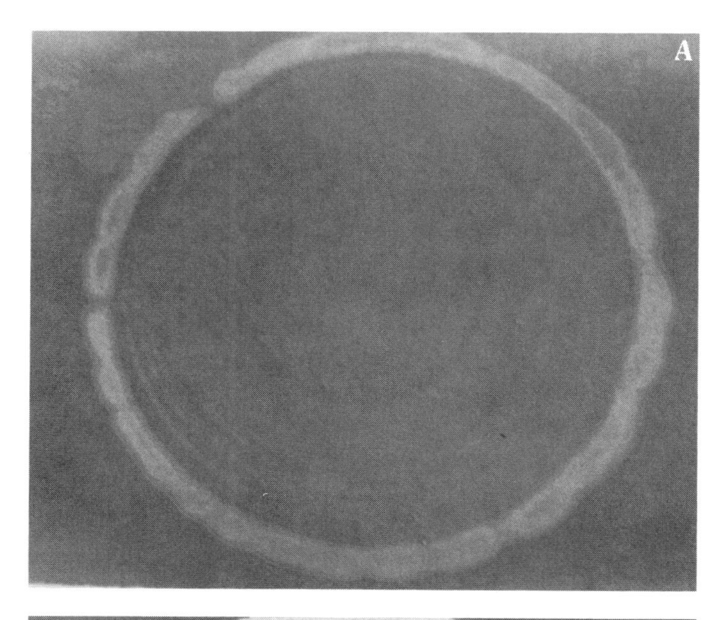

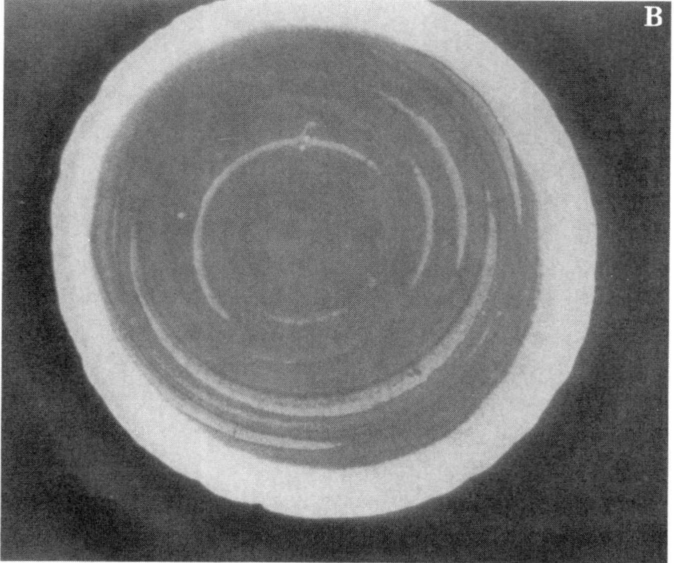

Fig. 10-19. Example of pass and fail using the copper morror test. In (A), there is only partial removal of the copper mirror indicating a "pass." In (B), there is complete removal of the copper indicating a "fail."

According to QQ-S-571E, a flux must pass the copper mirror test to be classified as either R or RMA while the IPC-SF-818 requires an "L" type flux to pass the copper mirror test; otherwise it must fall into the category M or H. The IPC specification allows, as an option, for the test to be carried out with reflowed solder paste flux residues. This may be used for acceptance or rejection but not for activity classification.

FLUX CORROSION TEST

The flux corrosion test is specified in IPC-SF-818 but not QQ-S-571E. This is a test designed to determine the effect of flux residues on copper under more extreme environmental conditions.

For this test (IPC-TM-2.6.15), obtain a sample of flux with all volatile solvents removed. This is actually a by-product of the procedure for the solids content of the flux described in IPC-TM-2.3.34. One must prepare a copper test panel, which is a 2-by-2-inch sheet of 0.020 ± 0.002-inch thick piece of 99 percent pure copper and contains a 1/8-inch deep circular depression in the center of the panel. The test panel must be pretreated immediately before performing the test. The pretreatment consists of degreasing and etching. When the panels are ready, about 1 gram of solder alloy (60/40) is placed in the depression of the panel along with the flux solids (at least 0.035 gram) and heated in a solder pot so that the solder specimen in the depression melts. The resulting test specimen is examined under 20X magnification and any discoloration or reaction products are noted, because this same test specimen will be evaluated after humidity exposure.

The test panel is stored at 40 ± 1 degree C and 93 ± 2 percent relative humidity. Type M and H fluxes store for 3 days and type L flux stores for 10 days. After the proper exposure time, the test panels are removed, examined at 20X magnification, and compared with the initial observation before storage. Then determine if any corrosion has taken place. According to IPC-TM-2.6.15, corrosion is described as a. excrescences at the interfaces of the flux residue and copper boundary or discontinuities in the residues, b. discrete white colored spots in the flux residues. Figure 10-20 (*see color section*) shows some test panels and various states of corrosion.

WATER EXTRACT RESISTIVITY

Water extract resistivity was developed as a test to assess the potential corrosiveness of a flux. The ionic content is determined by measuring the conductivity of an aqueous extract of the flux. In particular, those substances that form positively and negatively charged ions enable the conduction of an electric current. The more ionic material available, the greater the conductivity. This fact is the basis for the water extract resistivity test as given in QQ-S-571E. The ionic constituents are extracted from the flux by treatment with water, and the resistivity (reciprocal of the conductance) is measured. The units of resistivity are ohms-cm. The resistance, R (ohms), of a substance is directly proportional to its length, l (cm), and inversely proportional to its cross sectional area A, (cm^2), so that

$$R = \varrho \frac{l}{A}$$

where the proportionality constant ϱ (ohms-cm) is called the resistivity and is the resistance of a conductor 1 cm in length with a cross-sectional area of 1 cm^2. Its value depends on the nature of the conducting substance.

To conduct the test (QQ-S-571E, 4.7.3.2), carefully clean and thoroughly rinse a set of beakers with distilled or deionized water. The water itself must test not less than 500,000 ohms-cm. Extract a sample of flux according to QQ-S paragraph 4.7.3.1.2 to produce a solution that is approximately 35 percent flux solids dissolved in isopropyl alcohol. Place a specimen of 0.100 ± 0.005 ml of the flux solution in 50 ml of conductivity water (prepare five samples this way). Bring the contents of the beakers to a boil and boil for one minute followed by cooling in a water bath to 23 ± 2 degrees C. Then measure the resistivity of the samples. According to QQ-S-571E, to qualify as an R or RMA type flux, the mean of the extract resistivities must be at least 100,000 ohms-cm. To qualify as type RA, the mean must be at least 45,000 ohm-cm.

SURFACE INSULATION RESISTANCE (SIR)

For assemblies in which the substrate is a fabricated board, the dielectric (insulation) properties are of importance in determining the ultimate reliability of the product. Circuit boards are usually laminates of synthetic resins and filler materials that impart electrical insulation between the conductors. For example, FR-4 type boards are glass-filled epoxy laminates. The board is designed to conduct electric signals between particular components through specific conductive paths. These conductive paths must be properly insulated from each other. *Insulation* refers to the inherent resistance of the material, which in general is termed insulation resistance. Be careful not to confuse volume resistance, surface resistance, volume resistivity, and surface resistivity.[13] In the evaluation of the effect of flux on insulation resistance, concern is specifically with surface resistance. This is the resistance in ohms over a surface that is between pair of conductors of some particular geometry. A typical pattern used is the *comb pattern* shown in FIG. 10-21. This is reproduced from IPC-B-25.

During its lifetime, the circuit board is subjected to various treatments and environments including fabrication, handling, processing (soldering), and cleaning. Each of these stages represents a potential source of dielectric breakdown.

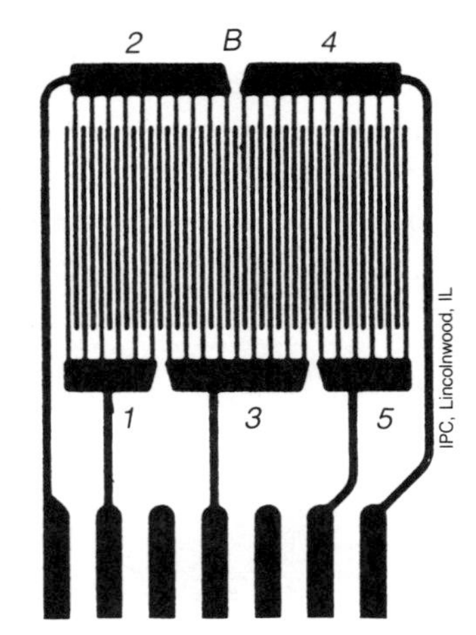

Fig. 10-21. Comb test pattern used for the surface insulation resistance test as specified in IPC-SF-818, method no. 2.6.3.3.

The effect of the flux on electrical insulation is of importance in producing assemblies of high reliability. Surface insulation resistance (SIR) is a more effective test than water extract resistivity because it evaluates the effect of the flux on the circuit board as opposed to water extract resistivity, for example, which is simply an indication of the ionic content of the flux. A flux might contain non-ionic substances that can degrade surface insulation resistance.[14]

The measurement of surface insulation resistance is not a simple routine test. It requires careful temperature and humidity control. The test itself could take up to 7 days. IPC-SF-818 uses SIR as one of the requirements to classify a flux as being either L, M, or H for a given type of assembly. Details of the test procedure age given in IPC-TM-2.6.3.3.

The following outlines the main features of the method. It is recommended that comb pattern B or E of IPC-B-25 be used. Obtain the appropriate type of boards with the comb pattern. A test chamber capable of at least 85 ± 2 degrees C and 90 percent relative humidity is required. Also required are a power supply that produces a standing bias potential of 100 volts dc with a tolerance of ± 10 percent and a resistance meter that can measure a resistance of 10^{12} ohms or greater with a test voltage of 500 volts.

Coat the comb pattern with a thin layer of flux. For paste flux, use a 35 percent solution of flux solids in isopropanol. Expose the flux-coated comb pattern to solder at 500 ± 10 degrees F for 5 ± 1 seconds. Depending on the flux classification, you might have to evaluate both cleaned and uncleaned boards. Paragraph 5.3.1.4 of IPC-SF-818 gives a step-by-step cleaning procedure.

Take an initial resistance measurement at ambient temperature. Then place the test specimens in the humidity chamber in a vertical position under a protective condensation drip shield. Attach the connectors at the appropriate test points and apply a voltage of 45 to 50 volts. When preparing to take resistance measurements inside the chamber at 24, 96, and 168 hours, first disconnect the 50-volt dc polarized voltage. Measure resistance at 100 volts reverse polarity. Classification according to IPC-SF-818 requires a minimum resistance of 100 megohms after 7 days. Further subclassification depends on whether the boards are cleaned or not cleaned as specified in TABLE 2-2.

Appendix A

Solder Paste Selection Guide

Paste Characteristics	Application Requirements					
	Dispensing	Screen Printing	Stencil Printing	Long Tack ($>$ 8 hours)	Low Slump	Fine-Pitch Leads ($\leq$ 0.025″/0.635 mm)
Metal Content (by weight)						
85	•	•		•		
86	•	•		•		
87	•	•		•		
88	•	•	•	•		
89		•	•	•	•	•
90		•	•	*	•	•
91			•	*	•	•
Powder Particle Size						
−200	•	•	•	*	*	
−325	•	•	•	•	•	•
−400	•	•	•	•	•	•
Viscosity (centipoises)						
300,000 – 350,000	•					
350,000 – 450,000	•					
450,000 – 600,000		•		•		
600,000 – 700,000		•		•		
700,000 – 900,000			•	*	•	*
900,000 – 1,200,000			•	*	•	•
1,200,000 – 1,500,000			•	*	•	•

* Special modification of standard formulations might be required.

Guide refers to printed wiring boards in surface-mount technology.

Appendix B

Solder Paste Troubleshooting Guide

PROBLEM	SOLUTIONS
Deposition	
Bridging	Increase paste metal percentage Increase paste viscosity Reduce paste powder particle size Reduce ambient temperature Reduce deposited paste thickness Improve accuracy of paste deposition Adjust paste deposition parameters Reduce component placement pressure Adjust preheat/reflow profiles
Clogging (Dispensing)	Reduce paste powder particle size Reduce paste viscosity Reduce paste metal content Shorten needle length Increase needle orifice diameter Change syringe/cartridge design Increase air pressure

Crusting	Prevent exposure of paste to moisture Reduce paste flux activity level Reduce lead content of paste alloy
Excessive Paste	Reduce deposited paste thickness Improve accuracy of paste deposition Adjust paste deposition parameters
Insufficient Paste; (Printing)	Increase deposited paste thickness Improve accuracy of paste deposition Adjust paste deposition parameters
Insufficient Tack Retention	Remove causes of paste flux solvent loss (heat, air movement, etc.) Reduce paste metal content Reduce paste viscosity Reduce paste powder particle size Vary paste particle size distribution
Slumping	Increase paste metal percentage Increase paste viscosity Reduce paste powder particle size Reduce ambient temperature Reduce deposited paste thickness Reduce component placement pressure
Smearing	Increase paste metal percentage Increase paste viscosity Reduce ambient temperature Adjust paste deposition parameters

Reflow

Blowholes	Adjust preheat profile Increase paste viscosity Increase paste metal content

Component Movement and Misalignment	Improve accuracy of paste deposition Improve accuracy of component placement Adjust preheat and reflow parameters Improve component/board solderability Increase paste flux activity Improve component/pad design dimensional "ratios"
Dewetting	Improve component/board solderability Increase paste flux activity
Dull Joints	Prevent vibration of assembly during and immediately after reflow Accelerate cool-down rate after reflow
Nonwetting	Increase reflow temperature Improve component/board solderability Increase flux activity
Opens	Improve component lead coplanarity Increase deposited paste thickness Adjust preheat profile Reduce paste flux activity level Reduce board pad size Change reflow method Change alloy composition
Pinholes	Adjust preheat profile Increase paste viscosity Increase paste metal content
Solder-balling	Adjust preheat profile Reduce fines in paste powder Increase paste viscosity Increase paste metal content Increase paste flux activity

	Improve accuracy of paste deposition Improve accuracy of component placement Adjust preheat and reflow parameters Improve board/component solderability Increase flux activity Improve component/pad dimensional ratios Change component chip orientation
Voids	Adjust preheat profile Increase paste viscosity Increase paste metal content
Wicking	Adjust preheat profile Reduce paste flux activity level Change reflow method Change paste alloy composition

Notes

Chapter 1. POWDER PROCESSING AND CLASSIFICATION

1. J.S. Hirschhorn, *Introduction to Powder Metallurgy*, (New York: American Powder Metallurgy Institute, 1969).

2. J. Langan, "Improvements in the Physical and Mechanical Properties of Solder Connections by the Use of Ternary Alloys," IPC Fall Meeting, San Francisco, September 1976.

3. B.D. Dunn, "The Properties of Near-Eutectic Tin/Lead Solder Alloys Tested Between +70° and −60°C and the Use of Such Alloys in Spacecraft Electronics," European Space Agency, Neuilly, France, 1975.

4. J. Langan and L. Souzis, "The Functional Alloy Approach to Soldering," Fifth International American Welding Society Soldering Conference, St. Louis, Missouri, 1976.

5. D.R. Olsen and K.G. Spanjer, "Improved Cost Effectiveness and Product Reliability through Solder Alloy Development," *Solid State Technology,* September 1981.

6. R.N. Wild, "Some Fatigue Properties of Solder and Solder Joints," Internepcon, Brighton, England, October 1975.

7. M.M. Karnowsky and A. Rosenzweig, "Trans. TMS-AIME," 242, 2257 (1968).

8. M.M. Karnowsky and F.G. Yost, "Trans. TMS-AIME," Aug. 1976.

9. F.G. Yost, "Aspects of Lead-Indium Solder Technology," Proceedings International Microelectronics Symposium, 1976.

10. R.A. Bulwith and C.A. MacKay, "Silver Scavenging Inhibition of Some Silver Loaded Solders," *Welding Journal Research Supplement* 64 (1985).

11. R. Kurz and E. Kleiner, "Uber das Anlaufverhalten von flussigen Zinn-Bei-Loten," *Zeitschrift fur Werkstofftechnik*, nr. 8, 1971.

12. W.R. Lewis, *Notes on Soldering* (Greenford, England: Tin Research Institute, 1948).

Chapter 2. FLUX CLASSIFICATION

1. H.H. Manko, *Solders and Soldering* (McGraw-Hill, 1979).

2. R.J. Klein Wassink, *Soldering in Electronics* (Ayr, Scotland: Electrochemical Publications Limited, 1984).

3. R. Woodgate, *Handbook of Machine Soldering* (New York: John Wiley and Sons, 1983).

4. R.J. Klein Wassink, "Wetting of Solid-Metal Surfaces by Molten Metals," *Journal of the Institute of Metals* 95 (1967): 38.

5. A. Bondi, "The Spreading of Liquid Metals on Solid Surfaces," *Chem. Rev.* 52 (1953): 417.

6. C.A. MacKay, "Comparison of Solderability Values as Measured by Different Solderability Test Methods," Printed Circuit World Convention, London, June 1978.

7. M.L. Ackroyd, C.A. MacKay, and J. Thwaites, "Effect of Certain Impurity Elements on the Wetting Properties 60% tin-40% lead Solders," *Metals Technology*, February 1975.

8. F. Grunwald and J. Lowell, "Aqueous Cleaning of Reflowed Surface Mount Assemblies," Nepcon Proceedings (Nepcon West), 1989.

Chapter 3. FLUX CHEMISTRY

1. N.M. Joye and R.V. Lawrence, "Resin Acid Composition of Pine Oleoresins," *Journal of Chemical and Engineering Data* 12 (1967): 279.

2. E. Knecht and E. Hibbert, *Jour. Soc. of Dyers and Colorists* 35 (1919): 148.

3. F.P. Veitch and W.F. Sterling, *Jour. Ind. Eng. Chem.* 15 (1923): 576.

4. G. Du Pont and J. Levy, *Bull. Inst. du Pin.*, 2d ser., No. 8, (1930): 79.

5. S.P. Mitra, *Proc. Natl. Acad. Sci. India* 20A (1951): 140.

6. J. Minn, "Determination of Oxidative Stability of Rosin Products by High-Pressure Differential Scanning Calorimetry," *Therm. Acta.* 91 (1985): 87.

7. D.G. Lovering, "Rosin Acids React to Form Tan Residues," *Electronic Packaging and Production* (February 1985): 233.

8. W.L. Archer and T.D. Cabelka, "Behavior of Rosin Fluxes and Solder Paste During Soldering Operations," *ISHM Proceedings* 353 (1986).

9. A. Latin, "The Influence of Fluxes on the Spreading Power of Tin Solders on Copper," *Trans. Faraday Soc.* 34 (1938): 1384.

10. T. Oyama, "Study of the Structure of the Alloy Layer of Hot Dip Tinplate by the use of the Radio-active Isotope of Tin (Sn^{13})," English Summary in *Tin and Its Uses* 65 (1964): 4.

11. W.R. Lewis, "The Action of Fluxes that Assist Tinning and Soldering," *Tin and Its Uses*, no. 72 (1966): 3.

12. I. Okamoto, A. Omori, and H. Kihadra, "Studies on Flux Action of Soldering (Report II) -Amine hydrochloride-," *Transactions of JWRI 2* (1973): 113.

13. L. Weinberger and D. Audette, "Energies of Activation and Rates of Solder Wetting with Activated Fluxes," *Insulation/Circuits* 26 (1980): 12.

14. J.F. Shipley, "Influence of Flux, Substrate and Solder Composition of Solder Wetting," 4th AWS International Soldering Conference, Cleveland, Ohio, April 1975.

15. I. Onishi, I. Okamoto, and A. Omori, "Studies of Flux Action of Soldering (Report I)," *Transactions of JWRI 1* (1972): 23.

16. H. Manko, "Color, Corrosion and Fluxes," *Electronic Packaging and Production*, February 1969.

17. F.C. Disque, "The Use of Rosin and Activated Fluxes," ASTM Proceedings, 1956.

Chapter 4. Rheological Considerations

1. R.H. Sabersky, A.J. Acosta, and E.G. Hauptman, *Fluid Flow* (New York: Macmillan Publishing Co., 1971).

2. E.U. Condon and H. Odishaw, eds., *Handbook of Physics* (New York: McGraw-Hill, 1958).

3. M. Stippes, G. Wemper, M. Stern, and R. Beckett, *An Introduction to the Mechanics of Deformable Bodies* (Columbus, Ohio: Charles E. Merrill Publishing Co., 1961).

4. Markus Reiner, *Lectures on Theoretical Rheology*, (Amsterdam: North-Holland Publishing Co., 1960).

5. H.A. Barnes and K. Walters, *Rheol. Acta.* 24 (1985): 323–326.

6. D. C-H. Cheng, *Rheol. Acta.* 25 (1986): 542–554.

7. F.R. Eirich, ed., *Rheology: Theory and Applications*, vol. 3 (New York: Academic Press, Inc., 1960).

8. R.E. Trease and R.L. Dietz, *Solid State Technology*, 1 (1972): 39.

9. D.E. Riemer, "Ink Hydrodynamics of Screen Printing," Proceedings, ISHM Symposium, 1985.

10. D.E. Riemer, "The Function and Performance of the Stainless Steel-Print Ink Transfer Process," Proceedings, ISHM Symposium, 1986.

11. D.E. Riemer, "The Shear and Flow Experience of Ink During Screen Printing," Proceedings, ISHM Symposium, 1985.

12. D.E. Riemer, "More Comments on 'The Function and Performance of the Stainless Steel Screen During the Screen-Print Ink Transfer Process,'" *International Journal for Hybrid Microelectronics* vol. 11, no. 1 (1988).

13. H. Rangchi, B. Huner, and P.K. Ajmera, "A Model for Deposition of Thick Films by the Screen Printing Technique," Proceedings, ISHM Symposium, 1986.

14. B. Huner, "Comments on 'The Function and Performance of Ink During Screen Printing' (ISHM vol. 10/2, 1-8, 1987)," *International Journal for Hybrid Microelectronics*, vol. 10, no. 4, 4th quarter (1987).

15. B. Huner, P. Ajmera, and H. Rangchi, "On the Release of the Printing Screen from the Substrate in the Breakaway Region," ISHM Symposium, 1987.

16. R.M. Stanton, "Rheological Aspects of Thick Film Technology Tack and Paste Transfer during Screen Printing," Proceedings, ISHM Symposium, 1976.

17. N.J.A. Sloane, "The Packing of Spheres," *Scientific American* 250 (January 1984): 116.

18. K. Gotoh and J.L. Finney, "Statistical Geometrical Approach to Random Packing Density of Equal Spheres," *Nature* 252 (November 1974).

19. T.C. Patton, *Paint Flow and Pigment Dispersion*, (New York: John Wiley & Sons, Inc., 1979).

20. N. Ouchiyama and T. Tanaka, *Ind. Eng. Chem. Fundam.* 23 (1984): 490.

21. R.K. Gupta and S.G. Seshasri, "Maximum Loading Levels in Filled Liquid Systems," *J. Rheol.* 30 (1986): 503.

22. R. Furth, ed. *Investigations on the Theory of Brownian Movement by Albert Einstein*, (New York: Dover Publications, Inc., 1956).

23. M. Mooney, "The Viscosity of a Concentrated Suspension of Spherical Particles," *Colloid Sci.* 6 (1951): 162.

24. C.R. Wildmuth and M.C. Williams, *Rheol. Acta.* 23 (1984): 627.

25. C.R. Wildmuth and M.C. Williams, *Rheol. Acta.* 24 (1985): 75.

26. D. Quemada, "Comments of the Paper 'Viscosity of Suspensions Modeled with a Shear-Dependent Maximum Packing Fraction' by C.R. Wildmuth and M.C. Williams," *Rheol. Acta.* 25 (1986): 647.

27. K.H. Sweeny and R.D. Geckler, "The Rheology of Suspensions," *J. Appl. Physics* 25 (1954): 1135.

28. J. Mewis, "Flow and Microstructures of Concentrated Suspensions," I. Chem. E. Symposium, series no. 91.

29. K. Weissenberg, "A Continuum Theory of Rheological Phenomena," *Nature* 159 (1947): 310.

30. H.H. Hull, "The Normal Forces and Their Significance," *Trans. of the Soc. of Rheol. V* 115 (1961).

31. J.D. Ferry, *Viscoelastic Properties of Polymers, Third Edition* (New York: John Wiley and Sons, 1980).

32. J.J. Aklonis and W.J. MacKnight, *Introduction to Polymer Viscoelasticity, Second Edition* (New York: John Wiley and Sons, 1983).

33. K.F. Hsu, "The Viscoelastic Properties of Thick Film Pastes (Inks)," Proceedings, ISHM Symposium, 1985.

Chapter 5. Methods of Deposition

1. Carl Missele, "Screen Printing Primer—Part 3," *Hybrid Circuit Technology*, May 1985.

2. A.W. Dobie, "Manufacturing Considerations for Metal Stencils and Emulsion Screens Designed for Solder Paste Deposition," Expo SMT '88 Proceedings, 1988.

3. D. Utz, "Better Solder Printing with Stencils," *Circuits Manufacturing*, October 1985.

4. John D. Borneman and Royce L. Rennaker, "Paste Printing from a Pro," *Circuits Manufacturing*, February 1987.

5. S. Ruback, "Surface Mount Technology: Part II—Deposition of Solder," *S.I.T.E.*, October 1987.

Chapter 6. Reflow Procedures

1. T.Y. Chu, J.C. Mollendorf and R.C. Pfahl, Jr., "Soldering Using Condensation Heat Transfer," Nepcon/West, February 1974.

2. M.J. Ruckriegel, "The Benefits of Dual Vapor In-Line Vapor Phase Soldering," *Hybrid Circuit Technology*, November 1985.

3. N.R. Cox, "Near IR Reflow Soldering of Surface Mounted Devices" *Surface Mount Technology*, October 1986.

4. N.R. Cox, "Lamp IR Soldering," *Circuits Manufacturing*, May 1988.

5. A.N. Arslancan and D.K. Flattery, "Considering IR Reflow Soldering," *Circuits Manufacturing*, November 1987.

6. S.J. Dow, "Use of Radiant Infrared in Soldering Surface Mounted Devices to Printed Circuit Boards," *Solid State Technology*, November 1984.

7. C. Lea, *A Scientific Guide to Surface Mount Technology* (Ayr, Scotland: Electromechanical Publications Limited).

Chapter 7. Residue Removal

1. H.H. Manko, *Soldering Handbook for Printed Circuits and Surface Mounting*, (New York: Van Nostrand Reinhold Company, 1986).

2. C. Lea, "A Scientific Guide to Surface Mount Technology," (Ayr, Scotland: Electrochemical Publications Limited, 1988).

3. C.K. Ellenberger, Safety, Performance and Cost—Factors in Evaluating Defluxing Solvents for Military PWA's.

4. A.E. Wang and K.C. Seghal, "Effects of Wetting and Capillary Action on the Cleaning of SMA's, *Printed Circuit Assembly*, August 1988.

5. W.G. Kenyon, "New Ways to Select and Use Defluxing Solvents," Nepcon West/East Proceedings, 1979.

6. J. Brous and A.F. Schneider, "Cleaning Surface-Mounted Assemblies with Azeotropic Solvent Mixtures," *Electri-onics*, April 1984.

7. J.E. Hale and W.R. Steinacker, "Complete Cleaning of Surface Mounted Assemblies," Nepcon/West Proceedings, February 1985.

8. W.G. Kenyon, "SMT Forum: Cleaning Surface Mount Assemblies," *Surface Mount Technology*, December 1988.

9. J.K. Bonner, "Cleaning Surface Mount Assemblies: The Challenge of Finding a Substitute for CFC-113," Nepcon West Technical Proceedings, March 1989.

10. F. Grunwald and J. Lowell, "Aqueous Cleaning of Reflowed Surface Mount Assemblies," Nepcon West Technical Proceedings, March 1989.

11. R. Rich, "New Cleaning Method Sends Circuit Boards Spinning," *Circuits Manufacturing*, November 1988.

12. H.H. Manko, "How Clean is 'Clean' in Electronics?" *Quality Assurance,* June 1966.

13. T.F. Egan, *Plating*, April 1973.

14. J. Brous, "Evaluation of Post-Solder Flux Removal," *Welding Journal Research Supplement*, December 1975.

Chapter 10. TEST METHODS

1. ASTM Standard Test Method D 1210, "Test for Fineness of Dispersion of Pigment-Vehicle Systems," 1979.

2. H. Green, "Industrial Rheology and Rheological Structures," (New York: John Wiley & Sons, Inc. 1949).

3. H. Green, *Ind. Eng. Chem.* 14 (1942): 576.

4. A Voet and C.E. Geffken, "The Nature of Tack," *Ind. Eng. Chem.* 43 (1951): 1614.

5. J. Morris, "Solder Paste Tackiness Measurement," *IPC Technical Review*, September/October 1987.

6. N. MacLeod and L.P. Hoover, "Factors Affecting the Tackiness and Printability of Resin Based Solder Paste," 10th Annual Soldering/Manufacturing Seminar, Naval Weapons Center, China Lake, California, 1986.

7. J. Kevra and D. Mohoric, ''Solder Paste: Tack and Cure,'' Proceedings, American Society of Metals, April 1989.

8. J. Kevra and C.A. MacKay, ''Comparing Screen Printing Qualities of Solder Creams,'' Proceedings of the Technical Conference, 3rd Annual International Electronics Packaging Conference, 1983 p. 325.

9. J. Kevra and D. Mohoric, ''Solder Paste Slump Evaluation,'' *Printed Circuit Assembly*, September 1988.

10. T. Daniels, *Thermal Analysis*, (New York: Halsted Press, a division of John Wiley & Sons, 1973).

11. Wesley Wm. Wendlandt, *Thermal Analysis, Third Edition,* (New York: John Wiley & Sons, 1986).

12. V.J. Kuck, ''Determination of the Liquidus Temperature and Composition of Tin/Lead Solders using Differential Thermal Analysis,'' *Thermo. Acta.* 99 (1986): 233.

13. C.J. Tautscher, ''Cleanliness and Electrical Insulation Resistance Testing of Printed Wiring,'' distributed by Western Marketing, Washington.

14. J. Brous, ''Water-Soluble Flux and its Effect on PC Board Insulation Resistance, *Electronic Packaging and Production* (July 1981): 79.

Index

T

Help Us Help You

So that we can better provide you with the practical information you need, please take a moment to complete and return this card.

1. I am interested in books on the following subjects:

- ☐ architecture & design
- ☐ automotive
- ☐ aviation
- ☐ business & finance
- ☐ computer, mini & mainframe
- ☐ computer, micros
- ☐ electronics
- ☐ engineering
- ☐ hobbies & crafts
- ☐ how-to, do-it-yourself
- ☐ military history
- ☐ nautical
- ☐ other ______________________

2. I own/use a computer:

- ☐ Apple/Macintosh ______________________
- ☐ Commodore ______________________
- ☐ IBM ______________________
- ☐ Other ______________________

3. This card came from TAB book (no. or title):

____

4. I purchase books from/by:

- ☐ general bookstores
- ☐ technical bookstores
- ☐ college bookstores
- ☐ mail
- ☐ telephone
- ☐ electronic mail
- ☐ hobby stores
- ☐ art materials stores

Comments ______________________

____

____

Name ______________________

Address ______________________

City ______________________

State/Zip ______________________

TAB BOOKS Inc.